Adhesive Bonding in Five Steps

This book is dedicated to Prof. Dr. Walter Brockmann, a great scientist, teacher, and promoter of adhesive-bonding technology, who unfortunately passed away far too early in June 2011 at the age of 72.

In various positions, Prof. Brockmann significantly advanced the corresponding research and application engineering activities in Germany and Europe over decades. Before his death, he had headed the Materials and Surface Engineering Group (AWOK) at the University of Kaiserslautern, Germany, since 1990. Prior to that, he worked for more than 20 years as a scientist and department head at the Fraunhofer Institute for Applied Materials Research (IFAM) in Bremen, Germany.

As a co-founder of the "European Adhesion Conference" (EURADH), he has significantly promoted European and international cooperation in the field of adhesive-bonding technology and with his significant involvement in the initiation of the "World Congress on Adhesion and Related Phenomena" (WCARP) Prof. Brockmann strongly supported the globalization of research in adhesive-bonding technology.

His open, uncomplicated, and friendly manner, coupled with his boundless knowledge and experience, has enabled Prof. Brockmann to convey the many aspects and advantages of adhesive-bonding technology to me and many others, including R&D managers, scientists, students, and users in industry. In my case, he triggered a great enthusiasm for this technology that remains unbroken to this day.

Adhesive Bonding in Five Steps

Achieving Safe and High-Quality Bonds

Jürgen Klingen

WILEY-VCH

Author

Dr. Jürgen Klingen
41366 Schwalmtal
Germany

Cover Design: SCHULZ Grafik-Design
Cover Image: Atlas Copco

Library of Congress Card No.: applied for

British Library Cataloguing-in-Publication Data A catalogue record for this book is available from the British Library.

Bibliographic information published by the Deutsche Nationalbibliothek
The Deutsche Nationalbibliothek lists this publication in the Deutsche Nationalbibliografie; detailed bibliographic data are available on the Internet at <http://dnb.d-nb.de>.

Print ISBN: 978-3-527-34914-2
ePDF ISBN: 978-3-527-83224-8
ePub ISBN: 978-3-527-83225-5
oBook ISBN: 978-3-527-83226-2

Typesetting Straive, Chennai, India
Printing and Binding CPI Group (UK) Ltd, Croydon, CR0 4YY

Printed on acid-free paper

C113795_110122

Contents

Preface *xv*
Author Biography *xvii*

1 Introduction *1*
1.1 The *Art* of Adhesive Bonding *1*
1.2 Adhesives *1*
1.3 Adhesive Bonds *2*
1.3.1 Metal Bonds *3*
1.3.2 Plastic Bonds *4*
1.3.3 Glass Bonds *4*
1.3.4 Wood Bonds *5*
1.4 Adhesive Bonding in Industry and Craft *6*
1.5 An Example for Adhesive Bonding in Nature *7*

2 History of Adhesive-Bonding Technology *9*
2.1 First Adhesives *9*
2.2 Adhesive-Bonding Technology: 1845–1930 *10*
2.3 Adhesive-Bonding Technology: 1930–1960 *14*
2.4 Adhesive-Bonding Technology from 1960 Until Today *18*
References *20*

3 Wetting, Adhesion, and Cohesion *21*
3.1 Introduction *21*
3.2 Wetting of Surfaces *21*
3.2.1 Young's Equation and Surface Tension *21*
3.2.2 Surface Tension of Liquids *23*
3.2.3 Surface Energy of Solids *23*
3.2.4 Determination of the Critical Surface Tension *24*
3.2.5 Polar and Dispersive Interactions *24*
3.3 Adhesion to Surfaces *26*
3.3.1 Intermolecular Interactions *26*

3.3.2 Summary Adhesion *28*
3.4 Cohesion of Adhesive Layers *29*
3.4.1 Summary Wetting, Adhesion, and Cohesion *29*
References *30*

4 Adhesive Bonding in Industry and Craft *31*
4.1 Introduction *31*
4.1.1 Role of Adhesive Bonding in Production Today *31*
4.1.2 Limitations of Adhesive Bonding *32*
4.2 The Performance of Standard Adhesives Today *32*
4.2.1 Vespa is Held by a 1 cm^2 Adhesive Layer *32*
4.3 Advantages and Disadvantages of Adhesive Bonding *33*
4.4 Structural and Elastic Bonding *36*
4.4.1 Structural Bonding *38*
4.4.1.1 Limits of Structural Bonding *39*
4.4.2 Elastic Bonding *39*
4.4.2.1 Elastic Bonding for Heavier Loads *39*
4.5 Selected Adhesive Systems used in Industry and Craft *40*
4.5.1 "High-Speed" Cyanoacrylate Adhesives for Small Bonding Areas *41*
4.5.2 "Easy-to-Use" Hotmelt Adhesives for Rapid Further Processing *41*
4.5.3 "All-Purpose" Adhesives for Numerous Applications *42*
4.5.4 The "All-Rounders" for Structural Applications *43*
4.5.5 "User-Friendly" Conformable High-Performance PSA Foam Tapes *44*
4.5.6 PSA Tapes for "Quick Connect and Disconnect" *46*
4.5.6.1 High-Strength Fasteners *46*
4.5.6.2 Hook-and-Loop Fasteners *46*
4.5.7 "The Strong PSA Tape" for All Cases *47*
References *48*

5 The 5 Steps for Achieving Safe and High-Quality Bonds (Overview) *49*
5.1 Introduction *49*
5.2 Requirements for the *Special Process* *49*
5.3 Quality Standards *50*
5.4 The 5-Step Project Management System *51*
5.4.1 Introduction *51*
5.4.1.1 Process Map *52*
5.4.2 Planning, Concept, and Feasibility (Part 1) *52*
5.4.3 Development and Launch (Part 2) *53*
5.5 Gate Reviews *53*
5.5.1 The Gatekeeper *54*
5.5.2 Color Coding *54*
5.6 Quality Requirements According to DIN 2304 *55*
5.6.1 Part 1 – Classification of Adhesive Components into Safety Classes *55*
5.6.2 Part 2 – Assignment of the Responsible Supervisors (SIC) *57*

5.6.3 Part 3 – Proof of Applied Load < Load-Bearing Capacity of the Adhesive Layer *57*
5.6.3.1 Measurements *57*
5.6.3.2 Component Testing *58*
5.6.3.3 Documented Experience *58*
5.6.3.4 Combinations of 1–3 *58*
References *58*

6 Planning (Step 1) – Preparing the Plan and the Project Contract *59*
6.1 Introduction *59*
6.2 The Project Plan and Contract *59*
6.3 Roles of Project-Relevant People *60*
6.3.1 Project Sponsor (Client) *60*
6.3.2 Project Manager (Planner) *61*
6.3.3 Project Champion (Advocate) *61*
6.3.4 Supervisor In Charge (SIC) *61*
6.3.5 Operating Staff *62*
Reference *62*

7 Concept (Step 2a) – Substrates, their Surfaces, and their Treatment *63*
7.1 Introduction to Chapters 7 and 8 *63*
7.2 Substrates and their Surfaces *64*
7.2.1 Introduction *64*
7.2.1.1 Metals *64*
7.2.1.2 Plastic Materials *65*
7.2.1.3 Glasses *66*
7.2.1.4 Wood *66*
7.2.2 Ferrous Materials *67*
7.2.2.1 Introduction *67*
7.2.2.2 Mechanical Properties of Ferrous Materials *68*
7.2.2.3 Classification of Steel *69*
7.2.2.4 Surface Improvement of Steel *71*
7.2.3 Aluminum and their Alloys *74*
7.2.3.1 Introduction *74*
7.2.3.2 Properties of Aluminum *74*
7.2.3.3 Applications of Aluminum *75*
7.2.3.4 Surface Improvement of Aluminum *75*
7.2.4 Plastic Materials *76*
7.2.4.1 Introduction *76*
7.2.4.2 Properties *76*
7.2.4.3 Chemistry of Plastic Materials *78*
7.2.4.4 Thermomechanical Properties of Plastic Materials *80*
7.2.4.5 Viscoelastic Properties of Plastic Materials *82*

7.2.4.6 Plastic Materials Used in Industry and Craft *85*
7.2.5 Glass *93*
7.2.5.1 Introduction *93*
7.2.5.2 Classification of Glass *94*
7.2.5.3 Manufacture of Glass *94*
7.2.6 Wood *95*
7.2.6.1 Introduction *95*
7.2.6.2 Properties of Wood *95*
7.2.6.3 Wood Surfaces *96*
7.3 Surface Cleaning and Treatment *96*
7.3.1 Introduction *96*
7.3.1.1 Importance of Surface Treatment *97*
7.3.2 Surface Characteristics *100*
7.3.2.1 Metals *100*
7.3.2.2 Plastics *102*
7.3.2.3 Glasses *105*
7.3.3 Surface Cleaning *105*
7.3.3.1 Cleaning with Organic Solvents *105*
7.3.3.2 Cleaning with Aqueous Systems *108*
7.3.3.3 Cleaning of Metal Surfaces *108*
7.3.3.4 Cleaning of Plastic Surfaces *109*
7.3.3.5 Cleaning and Degreasing Processes *109*
7.3.4 Surface Treatment *110*
7.3.4.1 Mechanical Surface Treatment *111*
7.3.4.2 Chemical Surface Treatment *115*
7.3.4.3 Physical Surface Treatment *118*
7.3.4.4 Summary: Physical Surface Treatment Methods *125*
References *127*

8 Concept (Step 2b) – Adhesives, their Selection, and Definition of Concepts *129*
8.1 Introduction *129*
8.2 Adhesives for Industry and Craft *129*
8.2.1 Introduction *129*
8.2.1.1 Adhesive Classification by Curing Mechanism *130*
8.2.1.2 Adhesive Classification by Number of Components *132*
8.2.2 1K Adhesive Systems *132*
8.2.2.1 Cyanoacrylate Adhesives *133*
8.2.2.2 1K PU Adhesives *137*
8.2.2.3 1K Silicone Adhesives (RTV-1) *143*
8.2.2.4 Hotmelt Adhesives *146*
8.2.2.5 Photoinitiated Adhesives *155*
8.2.2.6 Anaerobic Adhesives *161*
8.2.2.7 1K Epoxy Adhesive *164*
8.2.2.8 Solvent-Borne and Dispersion Adhesives *169*

8.2.2.9 Pressure-Sensitive Adhesives (PSAs) *174*
8.2.3 2K Adhesive Systems *189*
8.2.3.1 Introduction *189*
8.2.3.2 Curing and Processing of 2K Adhesives *190*
8.2.3.3 Applications of 2K Adhesives *190*
8.2.4 2K Epoxy Adhesives *190*
8.2.4.1 Overview *190*
8.2.4.2 Chemistry and Composition *190*
8.2.4.3 Curing/Processing *191*
8.2.4.4 Properties *192*
8.2.4.5 Applications *192*
8.2.5 2K Acrylate Adhesives *193*
8.2.5.1 Overview *193*
8.2.5.2 Chemistry/Composition *193*
8.2.5.3 Processing *193*
8.2.5.4 Properties *194*
8.2.5.5 Applications *194*
8.2.6 2K PU Adhesives *195*
8.2.6.1 Overview *195*
8.2.6.2 Curing/Processing *195*
8.2.6.3 Properties/Applications *196*
8.3 Adhesive Bonding on Metal, Plastic, Glass, and Wood Surfaces *196*
8.3.1 Adhesive Bonding on Metals *196*
8.3.1.1 Adhesive Bonding on Steel Surfaces *197*
8.3.1.2 Bonding on Steel Sheet *197*
8.3.1.3 Bonding on Oily Steel Sheet *198*
8.3.1.4 Bonding on Galvanized Steel *199*
8.3.1.5 Surface Treatment of Galvanized Steel *200*
8.3.1.6 Bonding on Stainless Steel *200*
8.3.1.7 Surface Treatment of Stainless Steel *201*
8.3.1.8 Bonding on Aluminum Surfaces *202*
8.3.1.9 Surface Treatment of Aluminum *202*
8.3.1.10 Surface Treatment for Long-term Aging-resistant Bonds *203*
8.3.2 Adhesive Bonding on Plastic Materials *204*
8.3.2.1 Role of Diffusion and Solubility Characteristics *205*
8.3.2.2 Bonding on Recycled Plastic *205*
8.3.2.3 Long-term Resistance of Plastic Bonds *206*
8.3.3 Adhesive Bonding on Glass *207*
8.3.3.1 Selection of Adhesive for Glass Bonding *207*
8.3.3.2 Glass Bonds with Long-term Resistance *208*
8.3.3.3 Surface Treatment of Glass *208*
8.3.3.4 Adhesives for Glass Bonding *210*
8.3.4 Adhesive Bonding on Wood *210*
8.3.4.1 Requirements for Wood Bonding *211*
8.3.4.2 PVAC Dispersions (White Glue) *212*

8.3.4.3 Hotmelt Adhesives *212*
8.3.4.4 Formaldehyde Resin Dispersions *213*
8.3.4.5 Post-curing PVAC Dispersions *213*
8.3.4.6 Moisture-curing 1K PU Adhesives *213*
8.3.4.7 Bonding Wood to Other Materials *213*
8.4 Loads on Adhesive Bonds *214*
8.4.1 Introduction *214*
8.4.2 Long-term Behavior of Adhesive Bonds *214*
8.4.3 Mechanical Loads *215*
8.4.3.1 External Loads *215*
8.4.3.2 Internal Loads *215*
8.4.4 Environmental Loads *219*
8.4.4.1 Temperature Changes *219*
8.4.4.2 Water Diffusion *220*
8.4.4.3 Corrosive Media Impact *220*
8.5 Adhesive Specification *222*
8.5.1 Introduction *222*
8.5.2 Definition of Substrates *223*
8.5.3 Determination of Loads on the Joint *223*
8.5.4 Definition of Adhesive Requirements *224*
8.5.5 Definition of Surface Treatment Method *225*
8.5.6 Definition of Manufacturing Process *226*
8.5.7 Adhesive Specification *228*
8.5.8 Some Practical Guidelines for Adhesive Selection *229*
8.6 Definition of Concepts *230*
References *231*

9 Feasibility (Step 3) – Demonstrating Feasibility and Validation of the Preferred Concept *235*
9.1 Introduction *235*
9.2 Production of Bonded Joints in the Laboratory *236*
9.2.1 Preparations Before Starting Work in the Laboratory *236*
9.2.1.1 Provision of Personal Protective Equipment *237*
9.2.1.2 Preparation of the Workplace *239*
9.2.1.3 Preparation of the Adhesive *241*
9.2.2 Execution of the Work for the Production of a Bonded Joint in the Laboratory *241*
9.2.2.1 Surface Treatment *241*
9.2.2.2 Bonded Joints with 1K Adhesives *244*
9.2.2.3 Bonded Joints with 2K Adhesives *256*
9.3 Test Methods for Adhesive Bonds *263*
9.3.1 Introduction *263*
9.3.1.1 Technological and Physical Tests *263*
9.3.2 Technological Test Methods for Adhesive Bonds *264*
9.3.2.1 Shear Test *265*

9.3.2.2 Tensile Test *267*
9.3.2.3 Peel Test *268*
9.3.2.4 Cleavage Test *270*
9.3.2.5 3-Point Bending Test *270*
9.3.2.6 Wedge Test *270*
9.3.3 Aging Conditions for Adhesive Bonds *272*
9.3.3.1 Immersion Test *272*
9.3.3.2 Climate Test *272*
9.3.3.3 Climate Change (VW PV 1200) *272*
9.3.3.4 Salt-Spray Test *273*
9.3.3.5 Climatic Corrosion Test (KKT Test) *273*
9.3.3.6 Cataplasma Test *274*
9.3.3.7 Outdoor Weathering Test *274*
9.3.3.8 Fracture Surface Evaluation *274*
9.3.4 Physical Test Methods *275*
9.3.4.1 Differential Scanning Calorimetry (DSC) *276*
9.3.4.2 Dynamic Mechanical Thermal Analysis (DMTA) *276*
9.3.4.3 Thermo Gravimetric Analysis (TGA) *279*
9.3.4.4 Fourier Transform Infrared Spectroscopy (FTIR) *280*
9.4 Properties of Adhesive Bonds *280*
9.4.1 Introduction *280*
9.4.1.1 Structural Adhesive Bonds *280*
9.4.1.2 Elastic Adhesive Bonds *281*
9.4.2 Thermomechanical Properties of Adhesives *281*
9.4.2.1 General Thermomechanical Properties of Adhesives *281*
9.4.2.2 Thermomechanical Properties of PSAs *284*
9.4.3 Shear Properties *286*
9.4.3.1 Shear Modulus *287*
9.4.3.2 Shear Stress–Strain Behavior *287*
9.4.4 Creep and Flow *289*
9.4.4.1 Creep *289*
9.4.4.2 Flow *289*
9.4.4.3 Influence of Creep on the Bonded Joint *289*
9.4.4.4 Phases and Types of Creep Behavior in Adhesive Joints *290*
9.4.4.5 Creep Curves *292*
9.4.4.6 Stress Relaxation Curve *294*
9.5 Designing Adhesive Bonds *295*
9.5.1 Introduction *295*
9.5.2 Provision of Sufficient Bonding Area *297*
9.5.2.1 Introduction *297*
9.5.2.2 Butt Joints *297*
9.5.2.3 Single Lap Joints *298*
9.5.3 Avoiding Stress Peaks *304*
9.5.3.1 Peel Loads *305*
9.5.3.2 Bending Loads *306*

9.5.3.3 Cleavage Loads *306*
9.5.3.4 Plastic Deformation of the Substrates *307*
9.5.4 Dimensioning of Bonded Joints *307*
9.5.4.1 Reduction Factors *309*
9.5.4.2 Dimensioning of Elastic Adhesive Bonds *310*
9.5.4.3 Dimensioning of Adhesive Bonds with Plastics *312*
9.5.4.4 Concluding Remarks on the Structural Design and Dimensioning of Bonded Joints *313*
9.5.5 Examples for Dimensioning of Adhesive Bonds *313*
9.5.5.1 Example 1 – Joining Two Stainless-Steel Plates (Structural Bonding) *313*
9.5.5.2 Example 2 – Bonding of a GFRP-Element for Interior Design (Elastic Bonding) *315*
9.5.5.3 Example 3 – Bonding of a Glass Pane with an Aluminum Frame (Elastic Bonding) *317*
References *319*

10 Development (Step 4) – Establishing a Robust Manufacturing Process Ready for Production *321*
10.1 Introduction *321*
10.1.1 Status of the Project *321*
10.1.1.1 Objectives of the Development Step *321*
10.1.1.2 Requirements for the Successful Completion of the Development Step *323*
10.2 Adhesive-Bonding Manufacturing Process (Overview) *323*
10.2.1 Process Objectives for the Production of a Bonded Joint *325*
10.3 Processes for Adhesion Build-up *326*
10.3.1 Adhesive Preparation *326*
10.3.1.1 Storage/Tempering *326*
10.3.1.2 Homogenization *327*
10.3.1.3 Setting the Viscosity *327*
10.3.1.4 Newtonian and Non-Newtonian Fluids *328*
10.3.1.5 Viscosity of Adhesives *329*
10.3.1.6 Thixotropy of Adhesives *330*
10.3.1.7 Setting the Room Humidity *330*
10.3.2 Adhesive Delivery and Application *331*
10.3.2.1 Adhesive Delivery to the Surface *331*
10.3.2.2 Adhesive Application to the Substrate *336*
10.3.2.3 Bead Application *338*
10.3.2.4 Spot Application *338*
10.3.2.5 Screen-Printing Application *338*
10.3.2.6 PSA Tape Lamination *339*
10.3.2.7 Brushing, Troweling, and Squeegeeing *339*
10.3.2.8 Spray Application *339*
10.3.2.9 Roller Application *340*

10.3.2.10 PSA Film Lamination *340*
10.4 Processes for Cohesion Build-up *340*
10.4.1 Adhesive Component Mixing *341*
10.4.1.1 Mixing in Batch *341*
10.4.1.2 Static Mixing *342*
10.4.1.3 Quality of the Mixture *342*
10.4.1.4 Process Technology of Static Mixing *344*
10.4.1.5 Dynamic Mixers *345*
10.4.2 Joining and Fixation of the Bond *346*
10.4.2.1 Joining *346*
10.4.2.2 Fixation *346*
10.4.3 Curing of the Adhesive Layer *346*
10.4.3.1 Physically Curing Systems *347*
10.4.3.2 Chemically Curing Systems *347*
Reference *348*

11 Start of Production (Step 5) *349*
11.1 Work Done So Far *349*

12 Contemporary Adhesive-Bonding Applications *351*
12.1 Introduction *351*
12.1.1 Adhesive Market Development *351*
12.1.2 Customer Requirements for Adhesive Bonding *352*
12.2 Adhesive Bonding in Lightweight Construction *352*
12.2.1 Introduction *352*
12.2.2 Lightweight Requirements in Transportation *352*
12.2.3 Lightweight Bonding with Modern "Crash-Resistant" 2K Adhesives *353*
12.2.4 Elastic Bonding in Truck Construction *354*
12.2.5 Printable Adhesives *355*
12.3 The Role of Adhesives in Modern Facade Construction *355*
12.3.1 Structural Glazing with Silicone Adhesives *356*
12.3.2 Structural Glazing with PSA Tapes *356*
12.4 Adhesive Bonding on Low-Energy Plastics *357*
12.4.1 Introduction *357*
12.4.2 R&D for Bonding to Low-Energy Surfaces *358*
12.4.3 Self-Adhesive Transfer Tapes *358*
12.4.4 Structural Adhesives for Low-Energy Surfaces *359*
12.5 Adhesive Bonding with Structural 2K Adhesives *360*
12.5.1 Guinness World Record *360*
12.5.2 Structural Bonding in Sports *360*
12.5.3 Structural Bonding in Art *361*
12.5.4 Bonding with 2K Acrylates in Advertising *362*
12.5.5 Highly Transparent Bonding *363*
12.6 Inductively Curing Adhesives *363*

12.6.1 Advantages of Inductively Curing Adhesives *363*
12.6.2 User Benefits *364*
12.6.3 Potential for Innovative Bonding Solutions *365*
12.7 Adhesive Bonding with High-Performance Acrylic Foam Tapes *366*
12.7.1 Characteristics and Properties *366*
12.7.2 User Benefits *366*
12.7.3 Acrylic Foam Tapes for Powder-Coating Processes *366*
12.7.4 Flame-Retardant Acrylic Foam Tapes *367*
12.7.5 Visually Demanding Applications *367*
12.7.6 Bonding of Touchscreens *368*
12.8 Closing Wounds with Hotmelt Adhesives *369*
12.8.1 User Benefits *369*
References *369*

General References *371*
Index *373*

Preface

There is no doubt that adhesive-bonding technology is one of the key technologies for manufacturing products in the twenty-first century. The reasons for this are the need to simplify production processes, implement modern and innovative design concepts, reduce costs in the manufacture of products, and make greater use of lightweight construction methods to save energy.

However, the bonding process is a "special process" that cannot be inspected without destroying the products it creates, the bonded components. In addition, its design is decisively, indeed almost exclusively, responsible for the quality of the bonded joint. The reason is that in adhesive bonding, unlike alternative joining processes such as welding, screwing, and riveting, the joining agents used here – the adhesives – are not present in their final form, but are physically and/or chemically modified by the user during the manufacture of the components. To guarantee the required product quality, the process used for this purpose must be developed in a detailed, careful, and resilient manner, from the initial assessment of the materials involved to the successful completion of the bonded joints.

This book is intended for engineers, chemists, scientists, technicians, foremen, and students who are charged with the development of such a process or who, for other reasons, would like to inform themselves about the necessary prerequisites and measures for the production of an optimally bonded component. It describes in detail the steps required to set up an adhesive-bonding process to produce a high-quality component. This is done with the help of a specially developed 5-step project management system tailored to adhesive-bonding technology, which accompanies the development team from the initial idea for the adhesive-bonding process to its successful introduction into production. The requirements of DIN 2304 – a standard developed specifically for adhesive-bonding technology – are observed and ensure the establishment of suitable organizational structures in the manufacturing plant, the design of the environment required for adhesive-bonding technology, a high-quality production facility, and execution in line with quality standards. The tools and quality techniques required for planning and executing the 5-phase management process are provided by the Six Sigma methodology.

When working through the reading, the reader is taken by the hand, so to speak, and guided step by step through this process. Right at the beginning of the book – in the concept stage – the necessary basic knowledge of adhesive-bonding technology, technical information about the substrates used, the methods for treating their surfaces, and knowledge about the properties and behavior of adhesives are provided in detail. This enables the reader to outline bonding concepts as a basis for the further development steps.

In the subsequent feasibility demonstration step, practical work begins in the laboratory, initially focusing on the production and testing of laboratory samples of all outlined concepts. The aim here is to identify the most suitable candidate and then validate it after preparation and intensive testing of the corresponding practical components. At the end of the feasibility step, a concept validated for the stresses occurring in practical use is available, which is described in detail by the substrates to be used, the required surface treatment, the most suitable adhesive, and the necessary manufacturing steps for producing the bonded component.

The goal of the development step following the feasibility stage is to establish a robust process for manufacturing the already-validated bonded component. To this end, once the manufacturing process suitable for production scale has been outlined, its suitability in principle must be demonstrated by appropriate pilot runs using statistical methods. Subsequently, after the standards for production and quality control have been developed and the personnel performing the work on the production machines have been trained and instructed, production of the bonded component can be started.

This book is intended to help strengthen the already-outstanding status of adhesive-bonding technology as a modern joining method. To this end, it enables users in industry and the trades to plan the process of bonding components systematically and thus to make it highly efficient. Thus, the application of the proposed management system enables the reproducible production of safe and high-quality bonded joints.

I would like to thank Wiley-VCH for their willingness to publish the book and for their great help in realizing it. I would also like to thank Prof. Dr. Paul Ludwig Geiß, head of the Materials and Surface Engineering Group in the Department of Mechanical Engineering (AWOK-Arbeitsgruppe Werkstoff- und Oberflächentechnik) at the University of Kaiserslautern, for the interesting and stimulating discussions on the conception of the book. I hope that it will be of great help to many users in industry and trade in the development of appropriate bonding processes and thus make a significant contribution to the further positive development of bonding technology in the twenty-first century.

Schwalmtal, Germany
January 2022

Dr. Jürgen Klingen

Author Biography

Dr. Jürgen Klingen studied chemistry at the University of Duisburg, Germany, and received his doctorate from Prof. Robert Gillard at the University of Wales (Department of Applied Chemistry) in the field of crosslinking of polyisoprene-based adhesives.

For more than 35 years, he held various positions in research, development, and application engineering for the company 3M Deutschland GmbH, 1 year in corporate research in St. Paul, Minnesota, USA, where he worked on new high-performance adhesive systems for the European market. He received the 3M Corporate Circle of Technical Excellence Award for his research in this area. From 1996 to 2017, Dr. Klingen headed 3M's European Corporate Materials and Process Laboratory in Neuss, Germany, where he was responsible for technology development of new adhesives, tapes, films, coatings, and polymer processing for Europe.

Since the end of 2017, he has been working as a consultant for the development of bonding processes in industry and trade. Dr. Klingen is co-author of two technical books and sole author of one technical book as well as holder of several patents in the field of adhesive-bonding technology.

1 Introduction

1.1 The *Art* of Adhesive Bonding

In a broader sense, the word *art* means any developed action based on knowledge, training, perception, imagination, and intuition as on the initiative to perform it. This description also applies to adhesive bonding, since the development of a safe and high-quality bond also requires similar attributes such as appropriate knowledge, creativity, experience, and innovative strength. It is therefore permissible and appropriate to apply the term *art of adhesive bonding* to the creation of a safe and high-quality adhesive bond (Figure 1.1).

The classic joining technologies such as screwing, riveting, and welding are used today in numerous applications in industry and trade. However, there are some side effects, such as weakening of the materials involved, uneven stress distribution, and a high probability of corrosion, which the user has to accept. In contrast, adhesive-bonding technology, which can be used to join almost all different engineering materials, offers considerable advantages. Thus, in the early phase of component development, the designer enjoys the design freedom desired through the use of adhesive-bonding technology. And later, after the development of the bonding system has been completed, engineers in the manufacturing plant can easily implement it in existing production processes for individual and series production. The use of adhesives to join materials is characterized by the fact that identical or different substrates are joined over a large area by an organic material (the adhesive), and the resulting system (the bonded joint) is capable of transferring the acting forces from one substrate to the other. A special feature here is that the bond cannot be detached without destroying it.

1.2 Adhesives

Adhesives are nonmetallic organic materials with sufficient internal strength (cohesion) that are capable of bonding materials through intermolecular interactions occurring at substrate surfaces (adhesion) and transferring forces from one material to another.

Adhesive Bonding in Five Steps: Achieving Safe and High-Quality Bonds, First Edition. Jürgen Klingen.

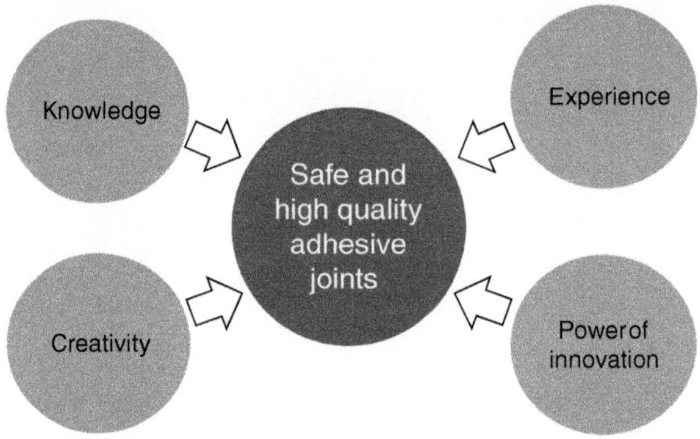

Figure 1.1 The creation of safe and high-quality adhesive joints through knowledge, creativity, experience, and innovation.

Two basic requirements must be met for a functioning adhesive, which are accomplished by appropriate adjustment of the chemical composition and physical properties:

- *good adhesion* – provided by sufficient molecular interactions with the material surfaces.
  During the bonding process, the adhesive must behave like a liquid, with a relatively low viscosity and the ability to wet the surface of the substrate to establish intermolecular interactions. This allows the molecules of the adhesive to approach the nanometer-scale molecular regions of the substrates.
- *good cohesion* – provided by sufficient molecular interactions within the cured adhesive layer.
  In application, the cured adhesive layer must behave like a strong solid with low-molecular flexibility. This is necessary for the transfer of tensile, shear, and peel forces from one substrate to another and to resist environmental influences. Therefore, for good cohesion, the adhesive chemistry must be adjusted to allow molecular interactions within the adhesive layer.

1.3 Adhesive Bonds

An adhesive bond is a two-dimensional connection of similar or dissimilar materials with the help of an organic material that adheres well to the surface of the two substrates to be joined. After the bonding has been prepared and the bonded component is in use, the task of the adhesive bond is to transfer forces from one substrate to the other.

In industry and craft usually, the following materials are used for the creation of an adhesive bond:

- metals,
- plastic materials,
- glasses, and
- wood.

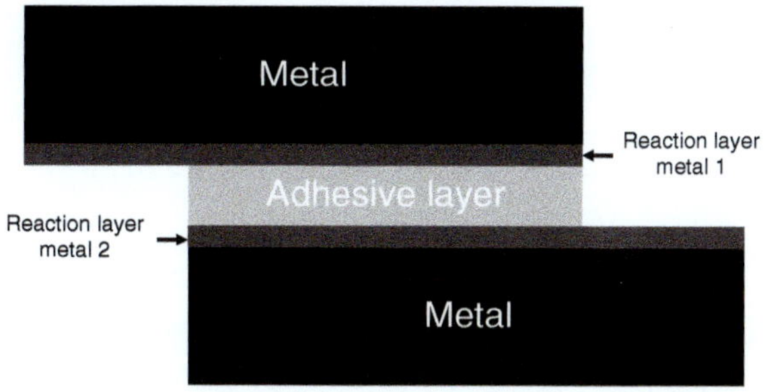

Figure 1.2 Structure of an adhesive bond with metal substrates, in which the adhesive is applied directly to the reaction layer.

1.3.1 Metal Bonds

When bonding to metal, it should be noted that contact is not usually made with the bare metal. Provided that the substrates are cleaned and, if necessary, surface-treated, a – so-called – reaction layer remains, which has been created by the reactions of the metal surface with the environment. This layer usually consists of reaction products of the materials with gases in the air such as oxygen, carbon dioxide, or even water. If the adhesion of this layer to the base material of the substrate and additionally its cohesion is sufficient, the adhesive can be applied directly to this layer (Figure 1.2).

Compared to plastics and glasses, the surface structure of metals is relatively complex and consists of:

- base metal material,
- interface layer (2000–5000 nm) with modified physical and/or mechanical properties, caused by the forming process during the production of the material,
- reaction layer (1–50 nm) formed by chemical changes at the interface and representing the actual adhesion layer,
- adsorption layer (0.1–1 nm), with adhering *non-material* molecules such as water and gases, and
- impurity layer, with solid materials such as dust or liquid contaminants such as oils, greases, or moisture.

In any case, for good adhesion, both the adsorption layer and the contamination layer must be removed by suitable cleaning and/or surface treatment methods before the adhesive is applied to the material surface (Figure 1.3). As already discussed, bonding is then usually carried out on the reaction layer. Examples for this case are aluminum and copper. In a normal atmosphere, aluminum always has an up to 50 nm thick porous oxide layer (reaction layer) that forms spontaneously in air. This surface can be bonded directly if no extreme high demands are made on long-term resistance. If this is the case, as often in the aerospace industry, however, this oxide layer must be removed from aluminum components. In the case of copper, a resistant brown copper oxide layer forms after a few hours in air and adheres well to the substrate. After months, this layer transforms by reaction with the CO_2 in the air to green basic copper(II) carbonate, the so-called patina, which additionally contains sulfate and/or chloride anions. Adhesive bonds of good quality can be

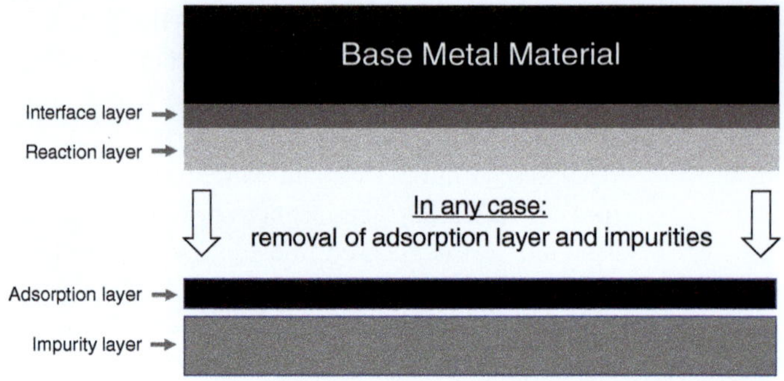

Figure 1.3 Detailed structure of a metal surface and removal of the adsorption layer/impurities to expose the reaction layer before bonding.

produced on both the brown copper oxide and the green copper carbonate layers. Therefore, the reaction layer does not have to be removed here either.

These are cases where, after cleaning the surfaces, adhesives can be applied directly to the reaction layer. However, this is not always the case. If the cohesion of the reaction layer is low and/or its adhesion to the base material is insufficient, it is necessary to remove the reaction layer from the surface as well. This is the case with low-alloyed or unalloyed steels, because the reaction layer (rust) does not adhere sufficiently to the base material. Another reason for such a measure may be the poor influence of the reaction layer on the aging resistance of the bond, as in the case of galvanized (zinc-coated) steel. Also, in this case, the reaction layer should also be removed from the surface before bonding. Zinc oxide initially forms on galvanized steel surfaces, which then reacts further with carbon dioxide in the air to form resistant zinc hydroxide/carbonate reaction layers. These layers bond very well with the base material even under the influence of temperature, but their hydrophilic (water-attracting) character leads to a negative influence on the quality of the bond, which is why they should be removed.

1.3.2 Plastic Bonds

As with metals, plastics are often not bonded to the pure material. In the case of softened plastics, for example, plasticizers can diffuse to the surface after some time, leading to a drop in strength and thus to failure of the bond. This effect occurs, for example, with polyvinylchloride (PVC) or ethylene-propylene-diene rubber (EPDM) materials and cannot be completely avoided by surface treatment. Therefore, when selecting materials for bonding, care should be taken to ensure that the parts to be bonded do not contain such substances or that these are crosslinked and thus chemically anchored in the polymer.

1.3.3 Glass Bonds

When bonding glass, the adhesive is usually not applied to the pure glass surface either, but finds a layer covered with OH groups generated by hydrolysis. These are created by so-called *glass corrosion*, in which the metal oxides contained in the glass are dissolved in water adsorbed on the surface. This increases the OH ion concentration on the glass surface and the pH value rises. Thus, glass corrosion has a considerable negative influence on the formation of adhesive forces on the glass surface and, especially in the case of glasses with a high alkali content, the surface energy is significantly reduced. For this reason, the surfaces of glass should be treated with suitable adhesion promoters, especially if the glass composites are required to have high resistance to aging.

1.3.4 Wood Bonds

Due to its pronounced pore structure, which offers the possibility of mechanical anchoring of the adhesive film in the structure, wood is a grateful substrate for bonding. In addition, the chemical structures of the main constituents of the wood matrix (cellulose and lignin), which to a large extent contain hydroxyl and other polar functional groups, result in numerous opportunities for the formation of hydrogen bonds and even covalent chemical bonds in the adhesive layer. However, the behavior of the adhesive at the boundary layer is strongly influenced by the wood density, which depends on the wood species, but within the wood species also on whether it is early or late wood. Since wood in practice always has a residual moisture content of 6–15%, the adhesive used must have a certain compatibility with water and water vapor, both during application and in continuous use.

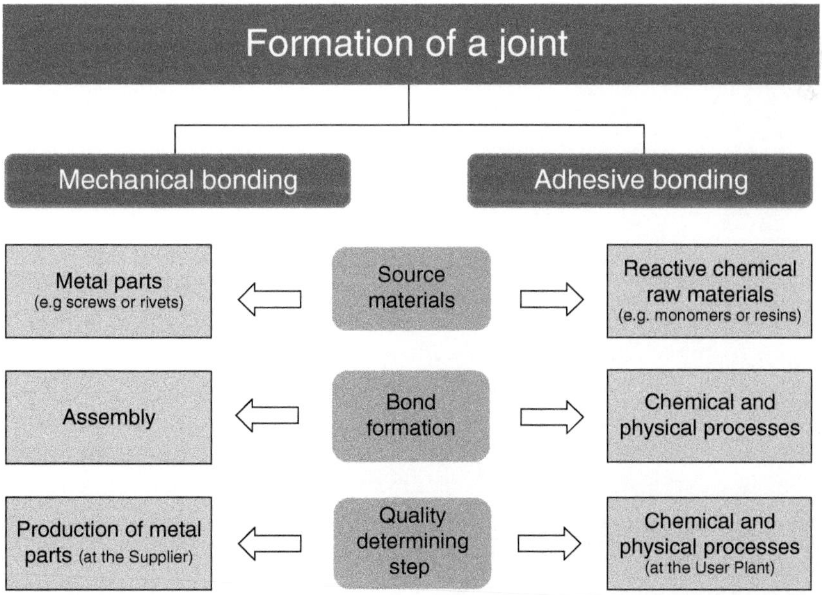

Figure 1.4 Formation of a joint by using the classical methods and adhesive bonding.

1.4 Adhesive Bonding in Industry and Craft

The production of high-quality bonded components in industry and crafts makes high demands, which applies to both the manufacturing process and the environment in which the parts are produced. A key difference from traditional joining techniques such as screwing or riveting, where the joining process has no or little influence on the quality of the joint, is that in adhesive bonding adhesion and cohesion are built up in the manufacturing process. Thus, the process parameters during the manufacture of bonded components are of extreme importance for the quality of the final bond (Figure 1.4).

The mechanism for adhesion formation on material surfaces and the chemical and physical processes that occur when loads are applied to the bond are not fully understood and are still the subject of current and future research. Therefore, the exact calculation of adhesion forces at the beginning and after loading is not possible. This is often seen as a disadvantage of the adhesive-bonding technique. However, it is very well possible to manufacture bonded components with good reliability if appropriate planning is carried out, taking into account existing experience, and extensive proof of the robustness of the bonded joint is provided by investigations on laboratory samples and prototypes.

Figure 1.5 Mussels adhesive bonded to a wooden pile in the North Sea near Domburg, the Netherlands. Source: Dr. Jürgen Klingen, Aboso-Consulting.

1.5 An Example for Adhesive Bonding in Nature

There are also excellent bonded joints in nature that have been optimized by evolution over long periods of time. One example is the bonding of mussels to different surfaces by means of an adhesive made of proteins, whereby this adhesive is produced by the mussels themselves with the help of their specialized glands. This enables the mussels to adhere to materials such as wood, metal, glass, and coral, and to withstand strong impacts from salt water (Figure 1.5). Due to the unique combination of properties, this adhesive is of interest to adhesive researchers and is a candidate for medical adhesive applications, particularly in surgery and regenerative medicine. For example, such biocompatible adhesives could enable the rapid treatment of complicated bone fractures instead of fixing them with screws, nails, or plates. Also, this adhesive developed by nature might be able to quickly close skin wounds and other injuries.

2
History of Adhesive-Bonding Technology

2.1 First Adhesives

"From birch tar for shafting weapons to animal glue for holding together the first books."

A corresponding *Guinness World Record* from 15 July 2019, shows very well how powerful modern adhesive systems are today [1]. An adhesive bond with a diameter of just 7 cm – the diameter of a standard soft drink can – held a 17.5-ton truck on a crane one meter above the ground for one hour. This corresponds to the current performance level of adhesives, which was achieved after numerous research and development efforts over many decades. However, after the first attempts in human history to create an adhesive bond between two material surfaces, the holding power of the bond was still very far from this level of performance. The first adhesive in human history can be considered to be birch tar, which was used in the Stone Age primarily for bonding tools and weapons. The black, tar-like distillate was obtained from birch bark by carbonization in the absence of air. The oldest evidence of this are two 200 000-year-old stone artifacts with traces of birch tar found in Campitello (Italy). Also, the *famous man* from the Tisenjoch – also called *Ötzi* – who died about 5300 years ago and was found not so long ago as a glacier mummy in South Tyrol (Italy), carried arrows made of flint and branches of the wool snowball. The materials were joined with a plant-fiber-reinforced birch tar glue (Figure 2.1).

As early as about 4000 years before Christ, the Sumerians deliberately developed and made an adhesive from animal hides (glutin glue) for building their houses and temples, which they called *segin*. To seal boats and build houses, they also used natural asphalt, which was available in their homeland, southern Mesopotamia, the region around present-day Iraq. Beginning around 1500 before Christ, the Egyptians discovered a concoction of sinew, gristle, and other animal waste as a suitable adhesive for carpentry. This art of *glue cooking* was further developed by the ancient Greeks and Romans, and the profession of glue cook – in Greek *kellopsos* – was introduced early in Greece. The Romans extended the range of glues used at that time to include fish glue produced by boiling fish waste, which did not appear in our latitudes until the sixth century Anno Domini. For special applications, glues were made from the swim bladders of fish. In the early Middle Ages, the development

Adhesive Bonding in Five Steps: Achieving Safe and High-Quality Bonds, First Edition. Jürgen Klingen.

Figure 2.1 Birch tar produced in a one-pot process. Source: Jorre, https://en.wikipedia.org/wiki/File:Birkenpech.JPG. Licensed Under CC BY-SA 3.0.

of adhesives in Central Europe stagnated, so that there were hardly any interesting innovations in early adhesive technology until the fifteenth century. It was not until the invention of letterpress printing by Johannes Gutenberg of Mainz in Germany, which took place around 1450, that a new activity began, as special adhesives were needed to finish the books in the bookbindery. In addition, the renaissance of veneering technology in the sixteenth and seventeenth centuries created a great need for suitable adhesives. Thus, the first glue factory was built in Holland in 1690. However, the first patent for a fish glue for joinery was first granted more than 60 years later, in 1754 in England. Although the demand for wood glues increased during this period, there were initially no significant new developments in the field of gluing technology.

2.2 Adhesive-Bonding Technology: 1845–1930

"From the first self-adhesive band-aid to cover wounds to the invention of masking tape that made two-tone painted cars possible."

1845 – The first self-adhesive plaster was patented by Horace Day and William Shecut from New York, USA. The related US Patent 3965 describes a new and improved process for making a plaster based on natural rubber and a cotton fabric that had special properties in covering wounds. Its formula was based on rubber, pine gum, Peru balsam, litharge (lead oxide), and turpentine solvent. By perforating the fabric, the new flexible material was able to drain sweat and wound fluid. After the patent was granted, the rights were sold to Thomas Allcock, who marketed the new self-adhesive plasters under the name *Allcock's Porous Plaster*.

At the end of the nineteenth century, the demand for goods of all kinds increased and, with the associated mass production, the packaging problems also grew. New adhesives had to be developed, as the existing products no longer met quality requirements.

1880 – Otto Ring, a Berlin merchant, developed the first ready-to-use glutin-based adhesive with the brand name *Syndetikon*. This adhesive quickly became the first *all-purpose adhesive* and developed into a worldwide success. Advertising slogans translated into several languages described a product with seemingly magical power. The somewhat foul-smelling adhesive was produced in the "Factory for Chemical-Technical Specialties," which Otto Ring had founded in Berlin-Schönefeld in 1878. Ring's first product, a stain remover, flopped. But two years later, the entrepreneur achieved a breakthrough with his fish glue, thanks in part to a consistently implemented advertising strategy. The exact recipe is no longer known, but the 1893 manual for druggists mentions that *Syndetikon* was produced *by boiling out fish parts of all kinds, entrails, and swim bladders.* Otto Ring died in 1937, and shortly after his death, market leadership of *Syndetikon* in Germany was over. General-purpose resin-based adhesives with improved performance came onto the market, which did not become brittle and stained over time, and had another major advantage: they simply did not smell of fish.

1882 – A German patent for a self-adhesive plaster that could deliver active ingredients to the skin was granted to pharmacist Paul Carl Beiersdorf (Figure 2.2). This was also the birth of the Beiersdorf company in Hamburg-Altona, Germany. The invention was the result of experiments to develop medicinal plasters using gauze

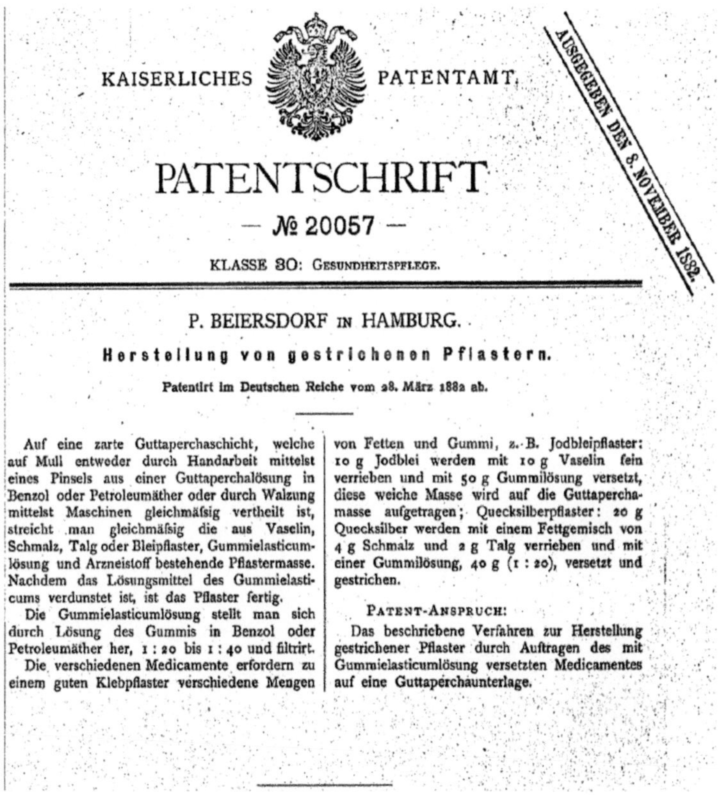

KAISERLICHES PATENTAMT.

AUSGEGEBEN DEN 8. NOVEMBER 1882.

PATENTSCHRIFT

— № 20057 —

KLASSE 30: GESUNDHEITSPFLEGE.

P. BEIERSDORF IN HAMBURG.

Herstellung von gestrichenen Pflastern.

Patentirt im Deutschen Reiche vom 28. März 1882 ab.

Auf eine zarte Guttaperchaschicht, welche auf Mull entweder durch Handarbeit mittelst eines Pinsels aus einer Guttaperchalösung in Benzol oder Petroleumäther oder durch Walzung mittelst Maschinen gleichmäfsig vertheilt ist, streicht man gleichmäfsig die aus Vaselin, Schmalz, Talg oder Bleipflaster, Gummielasticumlösung und Arzneistoff bestehende Pflastermasse. Nachdem das Lösungsmittel des Gummielasticums verdunstet ist, ist das Pflaster fertig.

Die Gummielasticumlösung stellt man sich durch Lösung des Gummis in Benzol oder Petroleumäther her, 1 : 20 bis 1 : 40 und filtrirt.

Die verschiedenen Medicamente erfordern zu einem guten Klebpflaster verschiedene Mengen von Fetten und Gummi, z. B. Jodbleipflaster: 10 g Jodblei werden mit 10 g Vaselin fein verrieben und mit 50 g Gummilösung versetzt, diese weiche Masse wird auf die Guttaperchamasse aufgetragen; Quecksilberpflaster: 20 g Quecksilber werden mit einem Fettgemisch von 4 g Schmalz und 2 g Talg verrieben und mit einer Gummilösung, 40 g (1 : 20), versetzt und gestrichen.

PATENT-ANSPRUCH:

Das beschriebene Verfahren zur Herstellung gestrichener Pflaster durch Auftragen des mit Gummielasticumlösung versetzten Medicamentes auf eine Guttaperchaunterlage.

Figure 2.2 Patent for a *self-adhesive plaster* that releases active ingredients, granted to Paul Carl Beiersdorf in 1882. Source: Dr. Jürgen Klingen, Aboso-Consulting.

soaked in hot ointments. Years after, Oscar Troplowitz, who had taken over the company from Beiersdorf, recognized the potential of Beiersdorf's patent for drug-free plasters to cover wounds.

Although the first prototypes irritated the skin, Troplowitz brought the product to market. However, the new application focus was not on skin, but on patching damaged bicycle inner tubes for repair. Many years later, these efforts led to the invention of a transparent rubber adhesive film, which has been marketed by Beiersdorf as *tesafilm*® since 1936.

1889 – Ferdinand Sichel, a master tapestry maker from Hannover, Germany, invented the first ready-to-use plant-based adhesive, replacing the less user-friendly glue based on animal waste. The result of his efforts was a durable adhesive made from starch that could be easily mixed with water and worked for a long time. Another advantage was that this new product could be packaged ready for sale. The idea for the new development project came during Sichel's training, when he learned about the difficult production and handling of glutinous adhesives made from animal waste. These products had to be recooked for each use and then quickly consumed. In addition, during prolonged storage, the acidic decomposition products of the glue discolored the glued wallpaper unsightly.

1901 – *Leukoplast*, a viscose-fiber-based fabric strip coated with a zinc-oxide-rubber resin compound, was launched and became one of Beiersdorf's most successful products (Figure 2.3). Before that could happen, Paul Carl Beiersdorf had founded a company to manufacture and further develop the first products based on his 1882 invention. By 1892, he was already able to offer plasters for more than a hundred different pharmaceutical applications, but had no idea at the time that he had laid the foundation for an entire new industry. It was Beiersdorf's successor, the pharmacist Oscar Troplowitz, who realized that there was even greater potential in these self-adhesive medical products. To further exploit Beiersdorf's technology, he therefore hired the chemist Isaac Lifschütz and commissioned him to develop new and more efficient adhesives for further applications. The reason for this was

Figure 2.3 Roll of *Leukoplast* from the 1930th found in a Red Cross first-aid bag. Source: ©Superbass/CC-BY-SA-4.0 via Wikimedia Commons.

also that at that time doctors had long been looking for a way to treat not only rashes but also wounds quickly and effectively. Lifschütz reacted to this and, to make the wound dressing antiseptic for wound healing as well, mixed zinc oxide into the rubber. He called this new dressing *Leukoplast* (from the Greek term *leukós* for white and the Latin word *emplastrum* for dressing). The new product was a breakthrough in wound treatment and not only helped doctors to treat wounds effectively but also made it possible to fix gauze bandages. Beiersdorf did not live to see this invention: he died in 1896. *Leukoplast* quickly found widespread use outside medicine. Thus, in addition to wound treatment, it was also used in the home for many applications and recommended for many handicrafts.

1922 – This year, wound care was revolutionized by the invention of *Hansaplast* (Figure 2.4). Already at the beginning of the twentieeth century, the development and introduction of *Leukoplast* by Beiersdorf had represented a significant advance in wound care. However, it took another 21 years before the invention of *Hansaplast*, in which a gauze bandage was simply added to the *Leukoplast* dressing. This created the plaster as we know it today; it was self-adhesive yet easy to remove. This new product revolutionized the medical tape business because it was ideal for treating open wounds. After decades of further development of backing materials and adhesive systems, there are now a large number of specialized types on the market.

1925 – Richard Drew of the 3M Company in St. Paul, USA, rather accidentally invented masking tape (often called crepe tape), which is used today for a variety of applications (Figure 2.5). While introducing a waterproof sandpaper to an

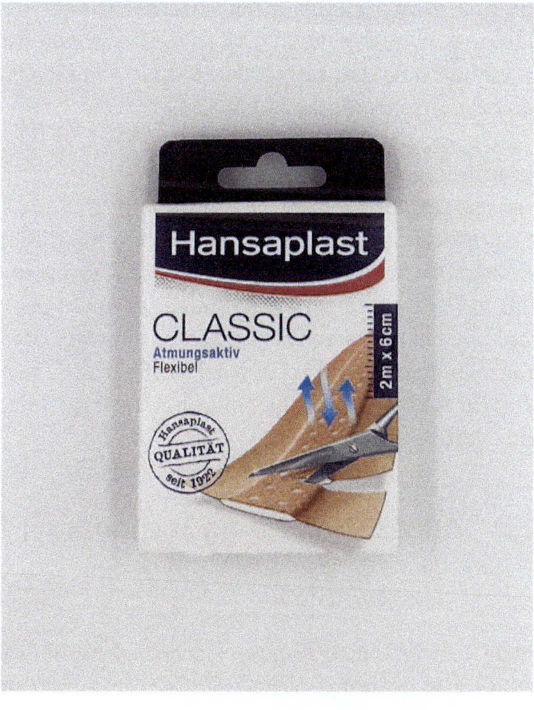

Figure 2.4 Package of *Hansaplast CLASSIC* as sold today. Source: Dr. Jürgen Klingen, Aboso-Consulting.

Figure 2.5 The first application of *masking tape* at a automobile manufacturer in Detroit. Source: 3M.

automotive paint shop, Drew learned of problems with covering partially finished components in two-tone automotive painting. In his lab, he began experimenting with adhesives based on vegetable oils, resins, and natural rubber. After two years of development, Drew successfully completed his work in 1925 and presented the first masking tape, a five-centimeter-wide strip of paper with a thin layer of adhesive, to automakers in Detroit. Customers from the automotive industry were so enthusiastic that they ordered three truckloads of the new product immediately after the first sampling.

1930 – Richard Drew continued to work in the lab to find more applications for his technology and, five years after inventing the masking tape [2], launched the first transparent tape (Figure 2.6). Originally, the new product was intended for bakers, meat processors, and grocers who were looking for a closure for their cellophane packaging. On 8 September 1930, the first samples of the new products were shipped and used by customers with a success not expected. As a result, the new product sold like hot cakes and was used for a variety of repairs as well as many office and household applications. During World War II, the demand for this *Scotch tape* was so great that the company could not keep up with the supply and apologized for it in advertisements.

2.3 Adhesive-Bonding Technology: 1930–1960

"From the first synthetic all-purpose adhesive to the silver bullet super glue."

1932 – Pharmacist August Fischer of Bühl, Germany, discovered that a 40% solution of polyvinyl acetate (PVAC) in a mixture of acetone and methyl acetate

Figure 2.6 Metal container with *Scotch® Cellophane Tape* from the 1950s. Source: Dr. Jürgen Klingen, Aboso-Consulting.

Figure 2.7 *UHU® – the all-purpose adhesive* as it is sold today. Source: Dr. Jürgen Klingen, Aboso-Consulting.

could produce very good adhesive bonds. Thus, the world's first ready-to-use, clear synthetic resin adhesive was born. Fischer had acquired the small chemical factory in 1905 and was initially involved in the production of printing inks, stamp pads and varnishes. His new adhesive, called *UHU®– the all-purpose adhesive*, turned out to be a decisive innovation and was able to form good bonds with most materials known at the time, even with the first plastics such as Bakelite (Figure 2.7).

1936 **–** The success story of *tesafilm* begins at Beiersdorf. The triumphant success of this transparent adhesive tape, which is based on acetate film, is closely linked to the industrial salesman Hugo Kirchberg (1908–1999). He applied for a job at Beiersdorf in 1934 and was entrusted with the task of developing and systematically driving forward the self-adhesive tape business at Beiersdorf. He firmly believed in the success of such products, especially for use in the office sector. Based on the knowledge and experience available at Beiersdorf after years of work in the field, a clear adhesive tape was developed by him and as this product was difficult to handle, he simultaneously designed and patented what was then called *a container for rolls of adhesive tape loaded with dry adhesive*. This provided the first dispenser for clear office tape, which can be found in modified form on almost every desk today. In 1936, *tesa*® was introduced as a brand for self-adhesive products, and *tesafilm* (so called since 1941) became an extremely successful product for household, office, and numerous other applications (Figure 2.8).

1942 – The invention of polyvinyl formal modified phenolic resin adhesives by Norman Adrian De Bruyne is considered a breakthrough in structural bonding in aircraft construction [3]. This adhesive – called *Redux* – contributed significantly to the improvement of aircraft structures. Primarily because of its excellent aging resistance, this class of adhesive is still highly valued today in the structural bonding of aluminum. The name *Redux* is derived from *Research at Duxford* and refers to the place where these exceptional products were developed (Aero Research Limited in Duxford, UK).

1943 – Based on a brilliant idea by Vesta Stoud of Dixon, Illinois, USA, the *Duck* or *Duct Tape* was invented. At that time, the U.S. Army used paper tapes to seal boxes containing cartridges. However, these had to be available very quickly during a mission so that soldiers could load their rifles quickly in combat. Paper tapes were not ideal for these missions, however, because they tended to tear off, making it difficult to remove them quickly enough, which in turn put soldiers in difficult situations. An inspector at a rifle cartridge factory, namely Vesta Stoudt, discovered the problem and suggested an alternative solution: A stronger, PE-coated fabric tape that could be easily torn off by hand and was more resistant to water and accidental

Figure 2.8 The first *tesafilm*® product of 1936. Source: Tesa.

damage. She made this suggestion to her supervisor, but did not find the support she wanted. Therefore, she decided to write President Franklin D. Roosevelt directly and describe the problem and her solution to him. The President was persuaded by her strong arguments and so the first *Duct Tape* was created, which became the basis for today's *Duct Tapes* having cult status. Thus, there are numerous applications for this tape in many industrial and craft sectors, as well as in almost every household, where it is used as an all-purpose weapon (Figure 2.9). But these products are indispensable not only here, but also in space travel. When an oxygen tank in the Apollo 13 space shuttle exploded in April 1970, *Duct Tape* acted as a lifesaver for the three astronauts by enabling the necessary repair of a CO_2 filter, thus ensuring the astronauts' return to Earth.

1946 – Pierre Castan's research on epoxy resins, conducted in the early 1940s, formed the basis for the development of a new type of adhesive by Swiss researcher Eduard Preiswerk of Ciba in Basel, Switzerland [4]. This new epoxy adhesive – called *Araldit 194* – was capable of bonding various materials, especially metals, resulting in high-strength composites. The adhesive was launched by Ciba in 1946 and successfully used in various industries, including aerospace. To this day, epoxy adhesives in both their two component (2K adhesives) and one component (1K adhesives) versions are preferred products for structural bonding.

1953 – The invention of anaerobic adhesives, which cure in the absence of oxygen, was made by Vernon Krieble in Hartford, USA, and represents the foundation of the Loctite company. At the first official public appearance of this company in July 1956, the statement that *the age-old problem of loose nuts and bolts in machinery and equipment can be solved* opened the introduction of anaerobic adhesives and the

Figure 2.9 A roll of *Duct Tape* as it is widely used today made of a PE-coated fabric backing with a natural-rubber-based PSA. Source: Dr. Jürgen Klingen, Aboso-Consulting.

great success of this technology. Today, anaerobic adhesives are firmly established in adhesive-bonding technology and are used to fasten and secure bolts and to bond hubs in industry and trade.

1958 – Under the product name *Eastman 910* the first cyanoacrylate adhesive, also called *Super Glue,* was introduced. This unique adhesive was invented during World War II by the American chemist Harry Coover, who was working at Eastman Kodak in New York on the development of optical prisms for weapons systems. The extreme tackiness of the cyanoacrylates he used – the basic substance for the Super Glue – prevented it from being used for its intended application. However, after further development activities, a variety of industrial applications were soon found for this technology. Cyanoacrylate adhesives are now firmly established not only in industry and the trades. Special types are used in medicine for wound closure in accidents or seamless surgical procedures, as well as to assist in the visualization of fingerprints in forensic technology.

2.4 Adhesive-Bonding Technology from 1960 Until Today

"From Post-it® to modern 2K epoxy adhesives for lightweight construction."

1980 – The invention of *Post-it®* sticky notes by 3M came about rather by accident. Back in the late 1960s, 3M scientist Spence Silver was working at 3M Company's central laboratory in St. Paul, USA, to develop a new process for manufacturing adhesives. The result of his efforts was unexpected and initially rather disappointing, as he obtained materials with relatively low adhesive strength. However, the adhesive-coated items could be easily removed from many surfaces without leaving any residue. No practical application was initially seen, and – in keeping with the 3M philosophy – Spence presented the material internally to other scientists. So, Art Fry was brought in, who, as a member of a church choir, was constantly annoyed that his bookmarks had fallen out of their notebooks at church. He took the new adhesive, applied it to small pieces of paper, and tested his invention in church the next Sunday. Actually, his bookmarks were reliable, but still came off easily without destroying the sheets of paper. That's how Post-its were invented and came onto the market in the early 1980s in the USA and also in Germany. Today, they are a standard product in every office and are also used in almost every household to convey brief information (Figure 2.10).

1980 – As a result of research and development efforts in the late 1970s, a new type of adhesive tape technology was developed by 3M. As a result of these efforts, the first so-called *acrylic foam tapes* (*VHB™ tapes*) (VHB, very high bond) were introduced to the market to meet the demand for double-sided adhesive tapes for bonding applications with high requirements. Due to their viscoelastic properties, these new self-adhesive tapes are highly conformable to surfaces, exhibit excellent stress distribution in the adhesive bond, and have the advantage over liquid adhesives of easier processing. The tapes are marketed under the brand name *VHB* and are now used successfully in many areas of industry as well as in the trades (Figure 2.11).

Figure 2.10 A selection of *Post-it®* products as they are sold today. Source: Dr. Jürgen Klingen, Aboso-Consulting.

Figure 2.11 A roll of *acrylic foam tape VHB™* as it is sold today. Source: Dr. Jürgen Klingen, Aboso-Consulting.

Application examples include the bonding of high-rise building facades, seals in car doors, roof structures for truck trailers, and the production of solar modules.

1990 – The first curable self-adhesive tapes were developed by 3M in Germany and the first product was launched in 1997 under the name *Structural Bonding Tape* [5]. After simple application to the substrate, these pressure sensitive adhesive (PSA) tapes cure to structural strength at about 140 °C for about 20 minutes. They thus combine the advantages of self-adhesive tapes (immediate adhesive build-up and ease of application) with those of liquid structural adhesives (high strength). The technology is now widely used, for example, in bonding mirror frames to windshields. Here, the heat required to produce the windshield is also used to cure the adhesive tape.

1998 – New high-temperature-curing 1K epoxy adhesives for bonding metal parts in automotive construction are launched by Dow in Europe. These crash-resistant adhesives are designed to structurally bond oiled sheet metal directly without cleaning the surface and are also often used in the so-called hybrid bonding processes, which, e.g. combine spot welding with adhesive bonding. Significant and successful applications have been established in the automotive industry in body-in-white and in the assembly of parts such as doors, trunk lids, and hoods, where high adhesive performance and crash resistance are required in the case of car accidents.

2016 – The development of new types of room-temperature-curing 2K epoxy adhesives was completed at 3M in Germany and corresponding products were launched on the market [6]. The new technology is able to solve the warpage and tolerance problems in automotive production that are based on the different coefficients of thermal expansion of the materials used. Unlike hot-curing 1K adhesives, whose strength build-up does not begin until a late stage of automotive production at around 180 °C in the e-coat oven, the new 2K products start to cure at room temperature immediately after the joining process. The new systems are developed in such a way that rapid induction curing of less than one minute at approximate 130 °C, which takes place immediately after joining of the substrates, already achieves sufficient handling strength for subsequent further processing of the component. This is a decisive advantage over 1K epoxy adhesives, since the much earlier strength buildup means that substrate displacements and residual stresses in the metallic materials used are avoided. The result is flat components with the required low tolerances, leading to higher process quality with less scrap and thus lower manufacturing costs.

References

1 Press Release DELO (2019). DELO Sets World Record for Strongest Adhesive. www.delo-adhesives.com (accessed 11 June 2021).

2 Drew, R.G. (1930). U.S. Patent 1 760 820.

3 Bishopp, J.A. (1990). The history of redux and the redux bonding. *International Journal of Adhesion and Adhesives* 17 (49): 287–301.

4 Preiswerk, E. (1965). Die Bindefunktion der Äthoxylin-(Epoxy-) Harze. *Technica* 4: 247 and 5, 355.

5 Bennett, G., Geiß, P.L., Klingen, J., and Neeb, T. (1996). Structural bonding tape: an innovation in adhesive bonding. *Adhesive Age* 39 (10): 19–23.

6 Klingen, J. and Hasenberg, D. (2010). Punktuell Kleben mittels Induktion. *Adhäsion Kleben und Dichten* 54: 31–33.

3

Wetting, Adhesion, and Cohesion

3.1 Introduction

Good wetting of the substrate by the adhesive, its good adhesion to the material surface, and good internal strength (cohesion) of the cured adhesive layer are the basic requirements for a good bond. Adhesion is strongly dependent on wetting, because if it is not sufficient, the necessary molecular interactions cannot be established, and adhesion is then poor or non-existent. The third requirement, good cohesion, must ensure that the strength of the adhesive layer formed after curing is sufficient for the intended application. Thus, the primary objective in the development of an adhesive bond is to optimize these three influencing variables in such a way that the forces can be sufficiently transmitted even under the loads that actually occur and the bonded component does not fail in actual use.

3.2 Wetting of Surfaces

For the described interactions to take place on the materials to be joined, the adhesive molecules must be brought very close to the corresponding groups on the surface of the joined part. This is only possible if sufficient wetting of the joined part surface by the adhesive is ensured.

Wetting refers to the behavior of liquids when they come into contact with material surfaces. Depending on the type of liquid and solid, wetting is more or less pronounced, where the topology of the surface also plays an important role. A drop of liquid applied to a horizontal, flat surface forms a contact angle θ between the edge of the drop and the solid, which can theoretically range from 0°, complete wetting to 180°, complete non-wetting (Figure 3.1).

3.2.1 Young's Equation and Surface Tension

According to Young's equation [1], which has been known for more than 200 years, the following basic relationship exists between the contact angle θ, the surface energy of the solid σ_s, the surface tension of the liquid σ_l, and the interfacial tension σ_sl (Figure 3.2):

$$\sigma_s = \sigma_\mathrm{sl} + \sigma_\mathrm{l} \cdot \cos\theta$$

Adhesive Bonding in Five Steps: Achieving Safe and High-Quality Bonds, First Edition. Jürgen Klingen.

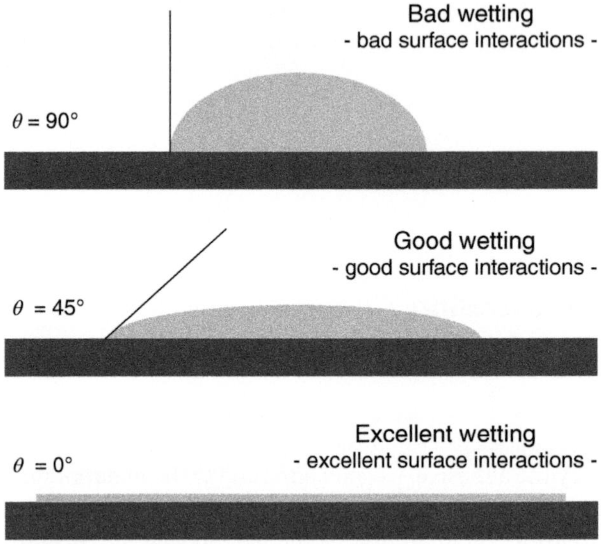

Figure 3.1 Bad, good, and excellent wetting of a liquid on a solid surface.

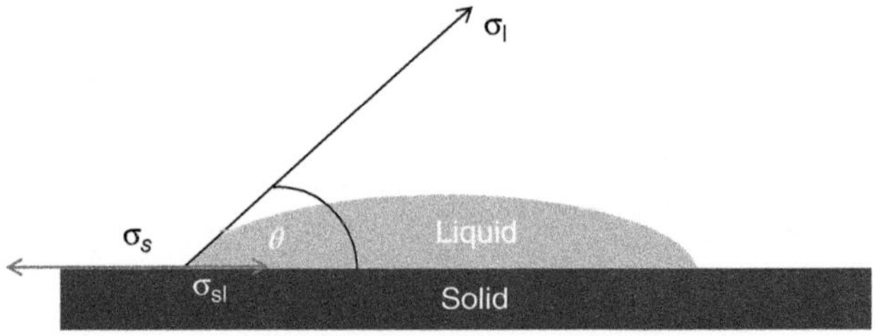

Figure 3.2 Schematic representation of the parameters for the Young equation.

The interfacial tension σ_{sl} depends on the interactions between the liquid and the solid. If there were no attractive interactions between the solid and the liquid, the liquid would try to minimize its interface with the solid and thus avoid wetting. In this case, the interfacial tension between solids and liquids σ_{sl} is maximum. It decreases with increasing interaction between solids and liquids and reaches 0 when complete wetting of the surface has occurred.

Depending on the size of the angle θ, the following surface tensions of the substrate surface are given:

- if the angle is large and thus $\cos\theta$ is small, the surface tension of the substrate is low,
- if the angle is small and thus $\cos\theta$ is large, the surface tension of the substrate is high, and
- when the angle is 0° and thus $\cos\theta$ is 1, $\sigma_{sl} = 0$ and there is complete spreading on the substrate surface, resulting in $\sigma_s = \sigma_l$.

The spreading of a liquid on a solid given by cos θ depends largely on the surface tension of the liquid, the surface energy of the solid, and the specific interfacial tension. A technically usable adhesion between the wetting liquid and the solid is only possible if the angle is 90° or smaller and the criterion of complete wetting of the surface is fulfilled if the surface energy of the solid σ_s is equal to or greater than the surface tension of the liquid σ_l.

3.2.2 Surface Tension of Liquids

The surface tension (with the unit being $J/m^2 = N/m$) of a liquid is a measure of the work that must be done to increase the surface area of a liquid by one unit. This physical property enables non-wetted objects to float on water surfaces. However, this is only possible as long as their weight is not great enough to overcome the surface tension of the liquid. This effect is demonstrated, for example, in Figure 3.3, where a paper clip made of a silicone-coated iron wire floats on a water surface, which is only slightly deformed. Water walkers take advantage of this effect, which allows them to walk on a water surface. For example, the surface tension of water at 20 °C is about 73 mN/m.

3.2.3 Surface Energy of Solids

The surface energy of a solid is a measure of the work required to break the chemical bonds per unit and create a new surface. The unit of surface energy is – as in the case of surface tension of liquids – $J/m^2 = N/m$, and it is always positive because energy is always required to break bonds. Since the thermodynamically stable state of a system is the one with the lowest (free) energy, every liquid-solid system strives to avoid or minimize surfaces with high surface energy (Figure 3.4). For this purpose, solid materials with high surface energy (e.g. steel) can be easily wetted by liquid

Figure 3.3 Surface tension allows a paper clip made of a silicone-coated iron wire to float on a water surface. Source: Dr. Jürgen Klingen, Aboso-Consulting.

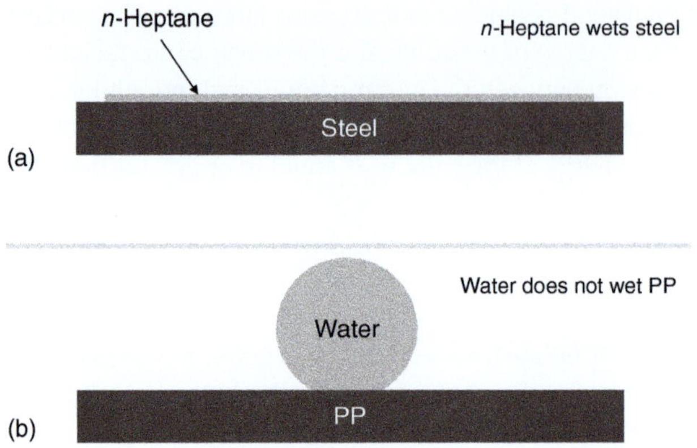

Figure 3.4 Good wetting of the low-energy liquid *n*-heptane on the high-energy solid steel (a) and poor wetting of the high-energy liquid water on the low-energy solid PP (b).

materials with low surface energy (e.g. *n*-heptane). However, to minimize the energy of the system, liquids with high energy (e.g. water) are not allowed to wet surfaces with low energy, e.g. made of polyethylene (PE) or polypropylene (PP).

The surface tension of liquids is easily accessible by appropriate measurements, but the surface energy of solids can hardly be measured directly. This is due to the fact that it is not possible to generate a new surface without spending energy on other processes, such as deformation of the body.

3.2.4 Determination of the Critical Surface Tension

The critical surface tension and thus the wettability of a solid is often determined by contact angle measurements using the Zismann method (Figure 3.5). Here, the critical surface tension σ_{crit} of a liquid is sought, which allows complete wetting on the corresponding solid. To determine it, the cos θ values of different contact angles are plotted against the surface tension of the corresponding liquids. The value of the surface tension extrapolated to $\cos \theta = 1$ (contact angle $= 0°$) is called the critical surface tension and reflects the surface energy of the solid that was evaluated.

This value is often referred to as the surface free energy (SFE) of the solid, and in practice is often determined using a series of test inks with known surface tensions. In determining the SFE, the analyst looks for the one that just completely wets the solid.

3.2.5 Polar and Dispersive Interactions

As discussed, the wetting of the surface by the adhesive depends on the surface tension of the adhesive and the surface energy of the substrate, whereby a distinction is made between disperse and polar interactions:

- *dispersive interactions* result from temporal fluctuations in the charge distribution of atoms/molecules (London interactions), and
- *polar interactions* are caused by permanent and induced dipoles as well as hydrogen bonds.

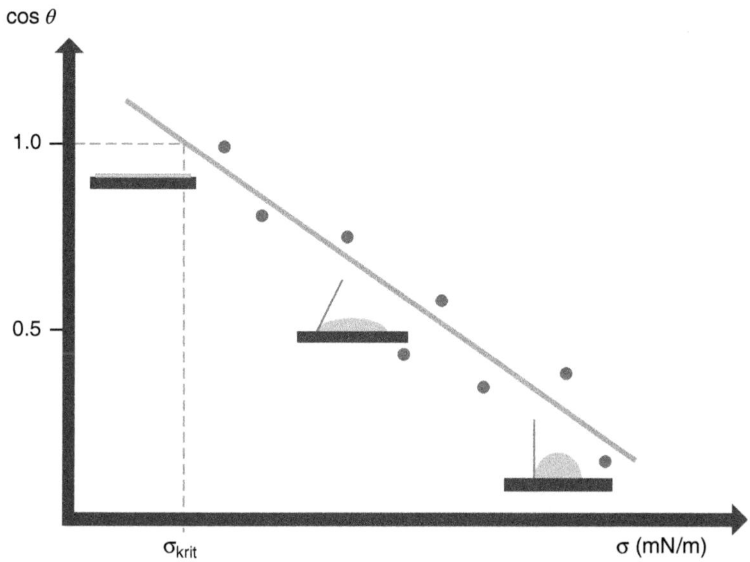

Figure 3.5 Determination of the critical surface tension σ_{crit} of a solid using the Zismann method.

Accordingly, the surface energy or surface tension is composed additively of a disperse and a polar component. Since London interactions occur in all atoms and molecules, there is no material whose surface energy has only a polar component. On the other hand, there are materials composed exclusively of hydrocarbon chains, e.g. PE and PP, which have no polar groups in the polymer chain and thus have only a dispersive component in their surface energy. Their surface energy is therefore purely dispersive in nature. On the other hand have polar materials a high proportion of polar interactions with about 50% for glass and 20% for polyamide 6 (PA 6), while the proportion is zero for PE and PP [2].

The solid with the lowest-known surface energy of about 18 mN/m is polytetrafluoroethylene (PTFE), but the surface tensions of commonly used organic materials are between 30 and 50 mN/m (compared to water with a surface tension of about 73 mN/m). Glasses and ceramic materials show much higher surface tensions of about 300–1500 mN/m, and metals are at the top with values of around 1000 and 5000 mN/m. This explains the excellent ability of glass, ceramics, and metals to allow very good and often excellent adhesion. Non-polar polymers – such as PE and PP – which are frequently used in industrial applications, have a relatively low surface energy of just over 30 mN/m. Therefore, without a surface treatment to increase the surface energy, these plastics will not be wetted by adhesives, which does not make for good adhesive bonds. Polar polymers such as polyamide (PA), polyethylene terephthalate (PET), and epoxy and phenolic resins with surface energies in the range of about 40–50 mN/m generally bond well without surface treatment.

3.3 Adhesion to Surfaces

Adhesion is responsible, for example, for the fact that mortar and bricks reliably hold together a building from Roman times to this day. A car tire only works if the rubber adheres very well to the organic or steel fabric, and even the corrosion protection of our cars, which is provided by paints, would not be possible without good adhesion of the paints to the metal surface. Adhesion is one of the standard terms used in adhesive-bonding technology and refers to the two-dimensional bonding of similar or dissimilar substances based on molecular interactions between the adhesive and the material surface. Cohesion, on the other hand, reflects the internal strength of the adhesive layer, which results from internal molecular interactions. Thus, cohesion, in combination with adhesion, largely determines the strength of an adhesive bond. Adhesion is one of the most important material phenomena in nature and technology, and the question is how it comes about. Often, the interlocking of the adhesive with the surface of the substrate is thought to cause adhesion, which can play a role in bonding rough or porous surfaces such as paper, wood, and ceramics. However, this theory fails on smooth surfaces.

3.3.1 Intermolecular Interactions

It is generally accepted that adhesion is generated to a very large extent by intermolecular interactions at the surface of materials. These interactions are caused by permanent or induced dipoles, hydrogen bonds, and dispersion forces (London forces), whereby the partially charged positions in the polymer chains come into contact with the corresponding positions on the surface of the substrate. In contrast, molecular interactions within the polymer matrix of the adhesive are responsible for the build-up of cohesion (Figure 3.6).

These molecular interactions can be described as follows:

- *Permanent dipoles* – These are present in molecules where electrons are more strongly attracted to one atom than the other, giving them a permanent electric dipole moment. This causes other molecules in the vicinity to align their permanent dipole moments so that the positively charged side always faces the negative side of the neighboring molecule and vice versa (Figure 3.7a).
- *Induced dipoles* – As the name indicates, these species are not present in the system from the beginning, but are induced when the substrate and adhesive come together. For example, the good bonding between polar adhesives and metals can be explained by the creation of induced dipoles. Although purely metallic surfaces are nonpolar, the induction of dipoles in the metal builds up charges that lead to the attraction of adhesive molecules and thus to good adhesion (Figure 3.7b).
- *Hydrogen bonds* – They are formed when a hydrogen atom is bonded to particularly electronegative atoms. In this case, a very polar hydrogen-containing group is formed, and a strong partial positive charge appears on the hydrogen atom. Such groups are capable of attracting oxygen- and nitrogen-containing molecules with free electron pairs. When they meet, the hydrogen atom takes on a bridging

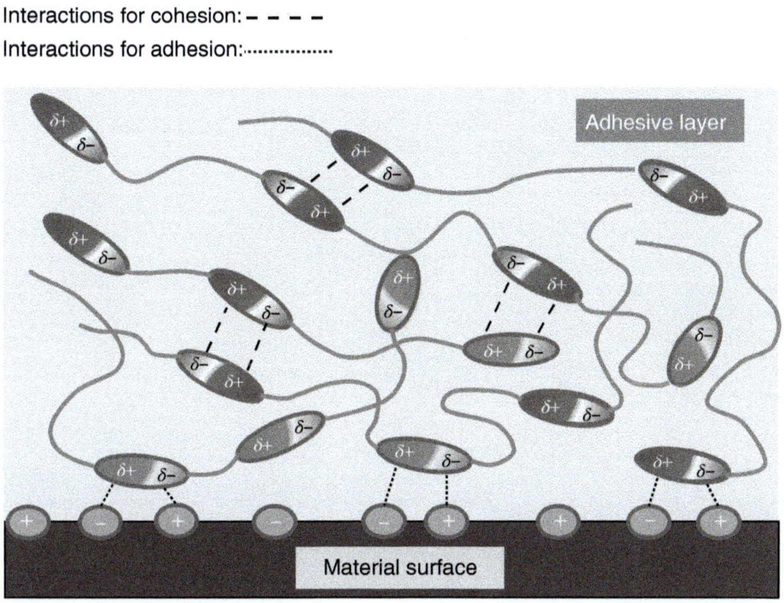

Figure 3.6 Molecular interactions of the adhesive polymer chains for the creation of adhesion and cohesion (schematic and not to scale).

function and is located between its original bonding partner and the new partner, which has a free electron pair. The much stronger interaction here compared to the dipole–dipole interaction can be explained by the small size of the hydrogen atom, which causes an excellent convergence of the dipoles (Figure 3.8).

- *Dispersion forces* (*London forces*) – These intermolecular forces are significantly smaller than the others and arise from spontaneous fluctuations in electron density in molecules. This creates a dipole moment that is capable of inducing another dipole moment in an adjacent group. It is important for adhesive bonding that dispersion forces also act between materials with completely nonpolar molecules. This allows bonding of, e.g. low-energy surface materials like PE and PP.

The molecular interactions described, which are responsible for adhesion (and also cohesion), only occur when the binding partners are very close to each other. They generally have lower binding energies than purely chemical bonds and are effective between both similar and dissimilar materials.

As a rule, the interactions necessary for the formation of adhesion take place in the range from 0.3 to 0.5 nm and have the following binding energies in kJ/mol:

- permanent dipoles: <20,
- induced dipoles: <2,
- hydrogen bonds: <50, and for
- dispersion forces (London forces): 0.1–40.

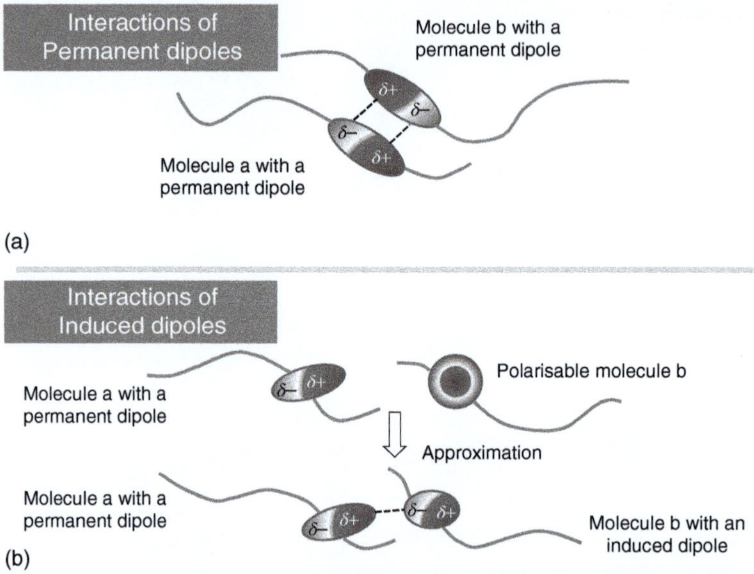

Figure 3.7 Interactions of permanent (a) and induced dipoles (b).

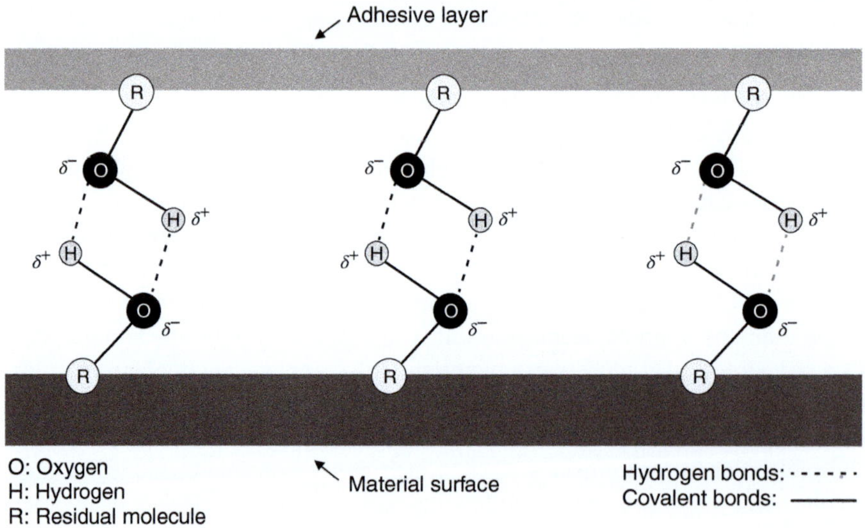

Figure 3.8 Formation of hydrogen bonds between the adhesive and the material surface to establish adhesion (schematic and not to scale).

3.3.2 Summary Adhesion

Several intermolecular forces are involved in the formation of adhesion between the molecules of the adhesive and the surface of the substrate to be bonded. Due to their higher binding energies, the interactions attributable to dipole–dipole and hydrogen bonds are likely to account for the largest proportion compared to dispersion forces, with the particular nature and strength of the bond depending on the ambient temperature and the distances between the molecules involved. To enable the necessary interactions for the formation of good adhesive properties, many molecules of the adhesive suitable for interaction must be brought close to the corresponding groups of the substrate surface. To ensure this, good wetting of the surface is required. Achieving this through the selection of appropriate adhesives, joining parameters, and surface treatments is one of the most important tasks in designing good bonded joints.

3.4 Cohesion of Adhesive Layers

In addition to adhesion, the cohesion of an adhesive layer – also called internal strength – is of great importance for the quality of an adhesive bond. It is determined by attractive forces between atoms and molecules that take place within the adhesive layer (Figure 3.6). The cohesion of adhesives is significantly influenced by the type of polymer, temperature, molecular weight, and degree of crystallization as well as crosslinking of the adhesive polymer. Good cohesive strength of the adhesive layer is indicated by high tensile strength and low elongation, as well as low creep or flow under mechanical stress. Thermoplastics and thermosets exhibit large differences in cohesion, with the relatively loose structure of the linear molecules of thermoplastics, possibly provided with side chains, resulting in relatively low internal strength. Thermosets are cross-linked by covalent bonds and therefore exhibit high cohesive strengths, which also explains the difference in mechanical properties of these polymer systems at high temperatures. The presence of hydrogen bonds in the functional groups of acrylates and epoxy-based adhesive polymers also has a very positive effect on cohesion and therefore plays a major role, especially for high-performance adhesives. Since these hydrogen bonds are not present in non-polar polymers, much higher-molecular weights are required to achieve sufficient strengths, which are then generated in such polymers by a large number of relatively weak intermolecular interactions. The cohesion of adhesives is also strongly influenced by temperature. Rising temperatures increase the mobility of the polymer molecules in the adhesive and thus reduce the internal strength, while falling temperatures lead to a solidification of the adhesive layer and thus to higher cohesion. For sufficient cohesion, a critical degree of polymerization must be reached, which usually lies between a molecular weight of 500 and 1000. Above this value, a sharp increase in cohesive strength and a significant improvement in the mechanical properties of the corresponding adhesives can be observed. In the case of semi-crystalline thermoplastics, the internal strength additionally depends on the degree of crystallization, which plays a major role in the application of hotmelt adhesives.

3.4.1 Summary Wetting, Adhesion, and Cohesion

The strength of an adhesive bond depends to a large extent on the adhesion generated on the substrate surface and the cohesion within the adhesive layer. In this context, good wetting of the material surfaces is a mandatory prerequisite for good adhesion. However, to achieve a high-quality adhesive bond, both adhesion and cohesion must be optimized by suitable measures and a good balance between the two must be set. An adhesive bond with excellent internal strength of the adhesive layer will fail if the necessary adhesive forces cannot develop on the substrate surface, and the same applies vice versa, if the molecular interactions within the adhesive layer are not strong enough. Optimizing adhesion and cohesion supports the overall goal of the development team to ensure that the forces encountered can be adequately transferred by the adhesive and that the bonded component will not fail in real-world applications. This can be achieved by selecting a suitable adhesive, a surface treatment appropriate for the material and suitable processing parameters.

References

1 Young, T. (1805). An essay on the cohesion of fluids. *Transactions of the Royal Society of London* 95: 65–87.

2 Krüss GmbH, Hamburg (2014). Surface Treatment for Polymers Application. *Report AR 272*. Industry Section, Hamburg, pp. 3–4.

4

Adhesive Bonding in Industry and Craft

4.1 Introduction

Adhesive bonding can be described as *the joining technology of the twenty-first century*. This way of joining materials is capable of saving time and energy in many places and bonding substrates that can almost only be connected using this method [1]. These include joint components containing only plastics and their combinations with glass or metal, for example. Even sensitive materials can be connected gently using adhesives, such as glass for facades with silicone adhesives or special double-sided adhesive tapes. In lightweight construction, such as in bicycles, cars, or aircraft, adhesives enable fiber-reinforced plastics (FRP) or light metals such as aluminum to be bonded. So, in the case of bonded car doors, suitable adhesives cushion the impact and protect the occupants from injury. The manufacture of components joined with adhesives requires suitable facilities for both the process and the facilities in which the bond is prepared. Since the adhesion and cohesion properties of the bonded joint are, in principle, determined in the production process, the corresponding parameters have a significant influence on the quality of the bond. This represents the decisive difference to classic bonding processes such as screws and rivets, where the quality of the bond is hardly influenced by the manufacturing process.

4.1.1 Role of Adhesive Bonding in Production Today

Particularly for the modern production of components in industry and trade, adhesive-bonding technology is becoming increasingly important. The strong need for adhesives capable of bonding different materials and withstanding various stresses and environmental influences has led to the development of a wide range of structural and elastic adhesives as well as pressure-sensitive adhesive (PSA) tapes [2]. These self-adhesive products are used to cover and protect component surfaces, mark parts, join webs in coating processes, and dampen vibrations and noise. In metal construction in particular, ever-increasing demands are being placed on construction and design. To meet these requirements, adhesive bonding is often the most suitable joining technique and, in many cases, has advantages over conventional bonding methods such as screwing, riveting, and welding. For

Adhesive Bonding in Five Steps: Achieving Safe and High-Quality Bonds, First Edition. Jürgen Klingen.

example, adhesives make it easy to join materials with different thermal expansion together or are used as high-strength adhesives for structural components as well as screw locks and shaft-hub connections. In building facades and roof elements for trucks, for example, elastic adhesives provide resistance to impact stress and thermal expansion as well as vibration damping. Moreover, they often not only join two substrates, but also perform additional functions such as sealing, corrosion protection, electrical and thermal insulation, as well as damping of vibrations and noise.

4.1.2 Limitations of Adhesive Bonding

Adhesive bonding is a joining process that does not permit non-destructive separation of the joint. Here, adhesives are used as an additional material whose properties must be strongly considered when planning the joint. Due to their organic nature, they have significantly different properties than the metallic components, such as screws and rivets, used in the classic mechanical joining processes. Thus, when planning an adhesive bond, one should be aware that in the bonded joint a polymer is doing its job that is endowed with low strength and viscoelastic behavior, which results in deformation properties that behave in a time-dependent and nonlinear manner. In addition, such materials can exhibit irreversible material damage starting already at a deformation of arround 1%, so far not allowing an exact mathematical description.

4.2 The Performance of Standard Adhesives Today

The development of synthetic polymer raw materials about 100 years ago was a significant step for the adhesive technology. This enabled the development of high-performance adhesives based on phenolic materials in the late 1920s and the invention of epoxy and PU adhesives in the 1940s. To this day, new developments in polymer chemistry continue to make it possible to create new high-performance adhesives that can bond to numerous materials and meet strength and conformability requirements. These include room-temperature curing (2K) epoxy adhesives that provide structural bonds with long-term environmental resistance. Today, such products are used as standard for many structural bonding applications in industry and craft, but also in the do-it-yourself (DIY) sector, where they securely hold together a wide variety of materials and transfer high loads on a daily basis.

4.2.1 Vespa is Held by a 1 cm^2 Adhesive Layer

The performance of such a *standard* 2K epoxy adhesive, which is recommended for a wide range of industrial, craft, and DIY applications, is illustrated in Figure 4.1. The adhesive with an adhesive strength (overlap shear strength) value of 17.5 MPa is theoretically capable of lifting a Vespa 300 GTS motor scooter weighting 158 kg off the

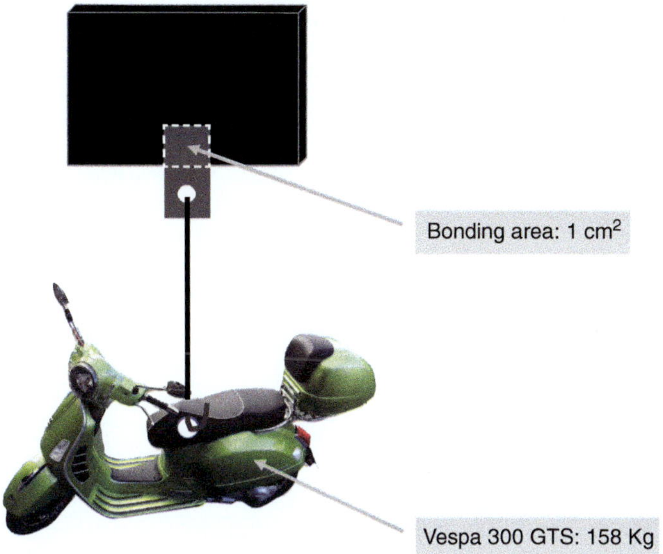

Figure 4.1 Demonstration of the performance of a *standard* 2K epoxy adhesive. Source: Dr. Jürgen Klingen, Aboso-Consulting.

ground with a bonding area of only 1 cm^2. The adhesive strength – also called overlap shear strength – is measured in the shear mode and is the stress (force divided by area), which is needed, to destroy a single overlap joint.

However, the fact that such loads can be transferred in practice over a longer period of time depends to a large extent on the design of the bonded joint, the substrates and their surface treatment, as well as the additional loads acting on the bond. These points will be discussed in detail later in the book.

4.3 Advantages and Disadvantages of Adhesive Bonding

Adhesive bonding is a so-called *substance-bound* method that works via atomic and molecular interactions, which are responsible for ensuring that there is no displacement in the bond and that forces can be transferred. This is different from classical mechanical bonding techniques, where the use of screws creates a force-fit connection (works by applying normal forces which cause friction) and the use of rivets creates a form-fit connection (works by interlocking the materials involved) (Figure 4.2). Unlike classic bonding techniques, adhesive bonding does not create material defects due to holes in the substrates, which are necessary when joining parts with screws or rivets. This method allows for uniform stress distribution and excellent fatigue strength of the joint, resulting in excellent utilization of the material properties.

However, to assess suitability for appropriate bonding applications, the advantages and disadvantages of this technique must be discussed and considered before

Figure 4.2 Form-fit (transverse to the rail) and force-fit (longitudinal to the rail) joints on a rail line Source: Dr. Jürgen Klingen, Aboso-Consulting.

planning the bonding operation [3]. When deciding which joining technique to use, the required temperature resistance of the components often plays a major role. For example, where high-temperature requirements are involved, adhesive bonding is often combined with classic bonding techniques, with the user taking advantage of the benefits of each method. One example is the combination of spot welding and adhesive bonding, which is successfully used for many bonding applications in the automotive industry. In this case, the spot welds provide the heat resistance and stability of the joint during curing at high temperatures and the adhesive, on the other hand, is responsible for good corrosion resistance, uniform stress distribution, crash resistance, as well as vibration/noise damping.

Compared to traditional bonding techniques, adhesive bonding has the following advantages:

- *No heat load on the substrates* – Since no or only a small amount of heat is required during bonding, there is no reduction in material strength. Deformation of the components – as often occurs with welding and soldering – can be easily avoided.
- *Bonding of different material combinations* – Bonding is almost independent of the nature of the substrates. Materials with different temperature resistances and surface topographies can thus be bonded.
- *Uniform stress distribution in the components* – This feature is based on the fact that there is no material weakening by holes in the substrates as created in joining processes using screws and rivets (Figure 4.3).
- *High dynamic strength and vibration damping* – Under alternating loads, adhesive layers exhibit considerable elastic deformation, allowing very high dynamic loads

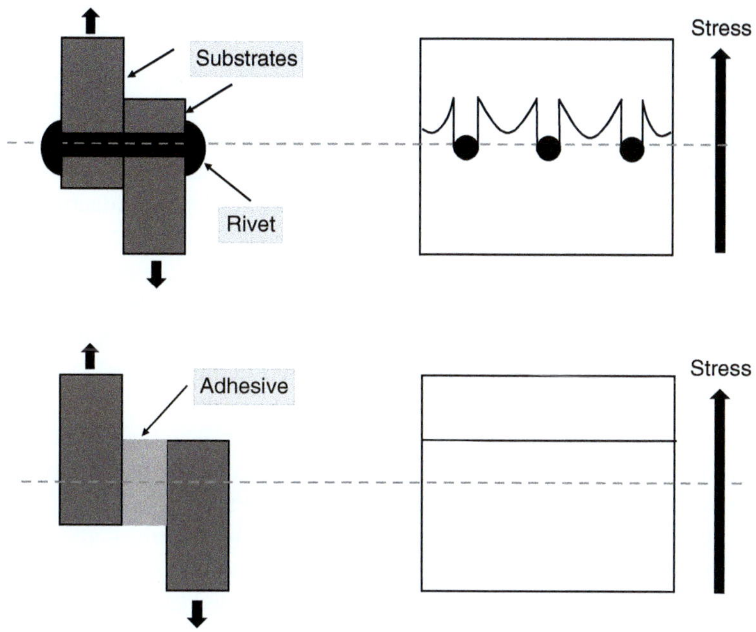

Figure 4.3 Stress distribution in riveted (non-uniform) and adhesive bonded joints (uniform).

to be applied to the components and excellent vibration damping in the bonded parts.

- *Bonding of very heat-sensitive materials* – Without the use of adhesives, bonding of very heat-sensitive plastics and components in the electronics industry would not be possible.
- *Easy bonding of thin substrates* – By using adhesives, large areas are bonded in a single operation without deformation of the substrates, enabling, for example, the application of large-area decorative film applications and the bonding of honeycomb constructions.
- *Reduction of weight* – Adhesive bonding enables constructions with high rigidity and at the same time uniform loading, resulting in a favorable ratio of substrate thickness to specific weight of the material.
- *Electrical insulation* – Because of the insulating effect of adhesive layers, there is no direct metallic contact between substrates, and thus the risk of corrosion is significantly reduced when bonding metals with adhesives.
- *Increased strength in hybrid bonding techniques* – Using adhesives in hybrid-bonding techniques – e.g. in combination with spot welding – adhesives perform the function of sealing and thus preventing corrosion, which enables components with high long-term strength, even when exposed to aggressive media.
- *Automation* – The application of adhesives to substrates can be easily handled by robotic systems. This has been demonstrated for many years in the automotive industry, e.g. in the assembly of windshields into the car body by PU adhesives and the application of door seals to car doors using PSA foam tapes.

The disadvantages of the adhesive-bonding process are:

- *Surface cleaning and treatment is required* – In cases where high durability is required for the bonded joint, appropriate surface treatment is necessary. In any case, intensive cleaning of the surfaces must be carried out before bonding.
- *100% control of the manufacturing process* – Since the user cannot detect any defects during the running process and also afterwards, the manufacturing process used must be 100% controlled.
- *Limited thermal stability of the bond* – Adhesives are present in the component as viscoelastic polymers that deform non-linearly at greatly increased temperatures. This can lead to the destruction of the bonded joint. Therefore, at high temperatures, they are inferior to the metals used in mechanical joining processes.
- *Delayed formation of the bonded joint* – Adhesive bonding takes more time than alternative joining techniques. Most systems build up their full final strength only after a certain time. Exceptions are hotmelt adhesives and PSAs.
- *Aging of the bulk adhesive and the boundary layer* – Depending on the ambient conditions, a time-dependent negative influence on the properties of the adhesive or the boundary layer may occur. This is associated with a decrease in adhesive strength.
- *Difficult recycling* – When it comes to recycling, the mechanical joining processes have an advantage, since the joined parts can be easily recycled after separation into parts. This is not possible with adhesive composites. However, thermal disposal of bonded components is possible without any problems.

4.4 Structural and Elastic Bonding

The adhesives market today offers a wide range of elastic, hard, and tough-elastic (toughened) construction adhesives in a broad strength and elongation range for the versatile bonding of materials in industry and craft (Figure 4.4). These are usually liquid one-component (1K) and two-component (2K) adhesives offered by numerous suppliers and suitable for bonding metals such as steel, aluminum, copper, or brass and many plastics such as PE, PP, polycarbonate (PC), PVC, polymethyl methacrylate (PMMA), acrylonitrile-butadiene-styrene copolymer (ABS), or glass-fiber-reinforced plastic (GFRP), but also for glass, ceramics, and wood.

Special 2K adhesives have been available for several years to bond low-energy surfaces such as PE and PP without surface treatment, which was previously not possible. They achieve structural strengths and meet requirements that for a long time were fulfilled exclusively by mechanical joining. As a result of the immense research and development (R&D) efforts of recent years, high-performance self-adhesive foam tapes are increasingly being used for the construction of components. Such so-called acrylic foam tapes have become a safe, reliable, and efficient alternative to liquid adhesives.

Figure 4.4 A limited range of adhesives for small bonding applications in industry and craft. Source: Dr. Jürgen Klingen, Aboso-Consulting.

This is due to the following characteristics of these innovative products:

- simple dosage form for immediate use,
- very easy handling,
- good immediate adhesion,
- no curing required,
- possibility of direct subsequent processing,
- tension-free bonding on a wide variety of materials, and
- relatively good temperature resistance.

An initial rough classification of adhesives can be made according to their strengths and use in the various applications:

- aerospace: 30–40 MPa,
- industry and craft, structural: 15–25 MPa, non-structural: 5–10 MPa,
- elastic sealing/bonding with PUs and silicones: <5 MPa, and
- elastic bonding with acrylic foam tapes: <1 MPa.

For applications in industry and craft, an unmanageable variety of adhesive systems is available, which often differ significantly in their strength, flexibility, resistance to environmental influences, dosage forms, and processing properties. An overview of the adhesive classes used there for structural, semi-structural, and elastic bonding applications focusing exclusively on adhesive strengths and elongations is given in Figure 4.5.

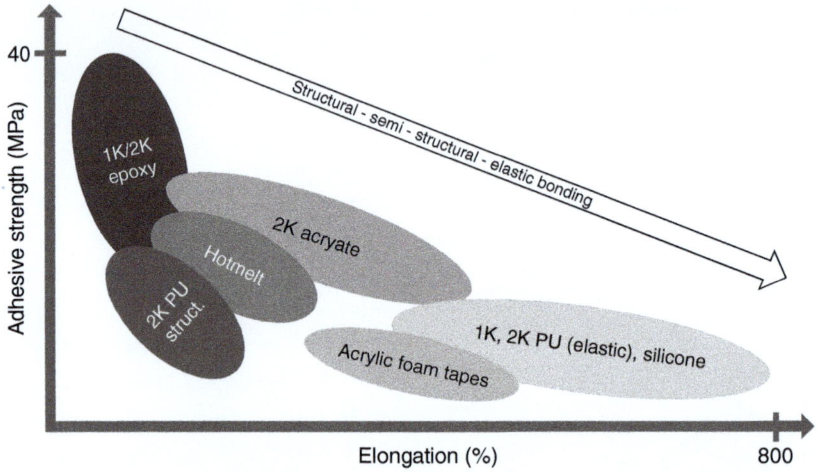

Figure 4.5 Adhesive strengths and elongations of adhesive classes used in industry and craft (rough representation).

In addition to strength and deformation properties, adhesive products are often differentiated according to characteristics that are important for the application and help the user select the right adhesive for the task at hand:

- *customer application* – such as assembly, labeling or metal adhesives, wood glue, wallpaper paste, or masking tape,
- *type of adhesive* – such as pressure sensitive, hotmelt, contact, solvent, or 2K adhesive,
- *process temperature* – such as cold glue, room temperature, or hot-curing adhesives,
- *curing mechanism* – such as anaerobic, moisture cure, UV, reaction or hotmelt adhesives,
- *form of delivery* – such as adhesive film, tape or sheet, adhesive powder or solvent adhesive, or
- *chemical base* – such as epoxy, acrylic, cyanoacrylate, silicone, or PU adhesive.

4.4.1 Structural Bonding

The main task in structural bonding is to transfer relatively high forces from one substrate to another. For this purpose, a wide range of high-performance hard or tough-elastic (toughened) *structural adhesives*, mostly based on epoxies and acrylates, is available on the market today. These adhesives usually form thin, strong layers with high adhesive strength and relatively low deformation. Adhesive strengths in the case of structural bonding are usually above 10 MPa (often above 25 MPa, sometimes reaching up to 40 MPa), and the thickness of the adhesive layer is usually between 100 and 200 μm.

4.4.1.1 Limits of Structural Bonding

In structural bonding, especially of metals, the strength of the substrates has a decisive influence on the load-bearing capacity of the bonded joint. The reason for this is that the high-strength adhesive systems commonly used cause deformation of the joined parts beyond their elastic range when subjected to very high loads. The result of these inelastic deformations is undesirable increased stresses at the overlapping ends of the bond, which lead to a reduction in adhesive strength (the force divided by area required to destroy a bonded joint, measured on single lap joints in shear mode). Stresses outside the elastic range of the materials must therefore be avoided at all costs, which is an important focus in the design of the bonded joint. An important feature of structural bonding of metals is therefore that higher loads cannot be transferred above the elastic range of the substrates to be bonded by increasing the overlap length.

4.4.2 Elastic Bonding

As already mentioned, the main task in structural bonding is to transfer high forces from one substrate to another and thus, adhesives used here are designed for very good adhesive strength of at least over 10 MPa, but often over 25 MPa, with top products reaching up to 40 MPa. In elastic bonding, the emphasis is different. The adhesives used here must also ensure force transmission to a certain degree and can therefore also be used for the construction of components. In addition, however, they must offer flexibility, for example, to overcome tolerances and compensate for different thermal expansions of the substrates. Therefore, their elongation of around 300–800% is significantly higher than that of structural adhesives of around 5–150%. However, their adhesive force at around 1–10 MPa is much lower than that of structural adhesives, and their adhesive layer thickness – in the range of around 1 and 3 mm – is usually significantly higher than that of structural adhesive layers with around 0.2 mm (Figure 4.6). Elastic adhesives are thus mainly chosen for their flexibility and there are a variety of adhesives that meet these requirements, including 1- or 2K polyurethanes (PUs), silicones, and acrylic foam tapes [4].

The main tasks in elastic bonding are:

- leveling out for tolerances in the component,
- compensating for different thermal expansions of the materials,
- avoiding stress peaks at the overlapping ends,
- enabling noise and vibration damping, and
- sealing the bonded joint to protect it from environmental influences.

4.4.2.1 Elastic Bonding for Heavier Loads

As discussed, the focus of elastic bonding is not the transfer of large loads but, for example, to compensate for tolerances and different thermal expansions of materials. The question is whether this method of bonding also allows the transfer of larger loads or whether this should be left exclusively to structural adhesives, and the answer is, it is possible because elastic bonds can be sized accordingly.

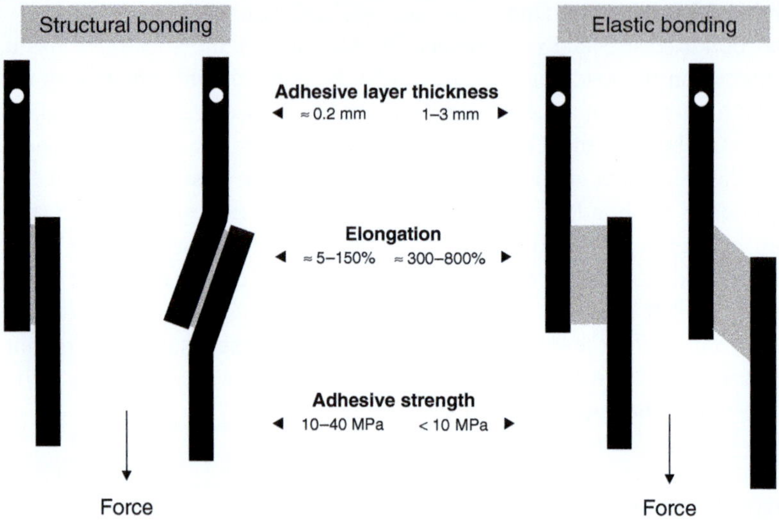

Figure 4.6 Deformation behavior of adhesive joints in structural and elastic bonding.

Unlike structural bonding, the strength of the substrates does not have a decisive influence on the load-bearing capacity of elastic bonds. The reason for this is that the strengths of the adhesives used for this purpose are generally orders of magnitude below the strength of the bonded parts and cannot stress the substrates involved beyond their elastic range. Thus, even at relatively high loads, no or only low stress concentrations occur at the overlapping ends, and the stress distribution is very uniform. This behavior causes a significant difference between structural and elastic bonding. Unlike high-strength structural adhesives, where there is a limit to the overlap length from which the bond strength is reduces (as described above), there is no value from which the bond strength decreases in elastic adhesive bonds. Here, increases in length lead to a linear steady increase in load-bearing capacity. This special feature of elastic bonding enables the user to compensate for the relatively low-bond strengths of the adhesives used by simply increasing the overlap length; doubling the bonding area allows doubling the mechanical load acting on the bond. Thus, the use of elastic adhesives also becomes possible for applications where higher loads occur. Again, in structural bonding this is only possible to a limit, namely to the point where the substrate is still just within its elastic deformation range. From this point, the adhesive strength decreases and higher loads cannot be applied.

4.5 Selected Adhesive Systems used in Industry and Craft

Depending on the focus of the application, the adhesive systems are specially selected in industry and craft. The criteria for the selection of structural and elastic bonding were described above. However, there are additional things, which are important for the use selecting the adhesive.

These include, for example:

- speed of curing,
- time to further processing,
- size of the area to be bonded,
- application range,
- necessity of the surface treatment,
- ease of use, and
- bonding–debonding options.

Below, there is a brief description of some adhesives that can meet one or even more of the above criteria and are therefore used in industry and crafts with a focus on specific applications.

4.5.1 "High-Speed" Cyanoacrylate Adhesives for Small Bonding Areas

These adhesives are very suitable for joining small and flat surfaces and are applied for bonding elastic materials. They cure quickly at room temperature and are characterized by high yield and relatively good bond strengths to a wide range of materials. These include metals and plastics such as ABS, PVC, PC, PA, elastomers - examples include ethylene-propylene-diene rubber (EPDM), styrene-butadiene rubber (SBR) and fluor rubber (FPM) – as well as glass, ceramics, porcelain, wood, leather, and cork. The cyanoacrylates form thermoplastic adhesive layers and exhibit good aging resistance as well as relative high-temperature and chemical resistance. The addition of suitable activators to these systems enables for example extremely fast curing or improved adhesion to critical material surfaces. Typically, these adhesives are dispensed dropwise, manually or with semi-automatic dispensing equipment. Light contact pressure during joining of the substrates forms a thin film, which – initiated by moisture – cures in seconds. Cyanoacrylate adhesives are offered for numerous applications, e.g. there are types with different viscosities and versions with special suitability for certain material groups such as metals, plastics, or elastomers (Figure 4.7) [5].

4.5.2 "Easy-to-Use" Hotmelt Adhesives for Rapid Further Processing

The great advantage of hotmelt adhesives is that the strength of the adhesive layers is achieved after only a few seconds, and the bonded components can then be further processed without loss of time. Therefore, these adhesives are often used for fast bonding of components, but – due to their thermoplastic behavior – for applications with low requirements on the thermal resistance of the bond. They wet surfaces in the molten state and develop good adhesive forces to many materials such as wood, cardboard, paper, cork, leather, rubber, felt, textiles, plastics, and foams, as well as ceramics, glass, and metals. Special manual application guns or automatic systems are used to extrude or spray the hotmelt adhesives onto the appropriate surfaces (Figure 4.8).

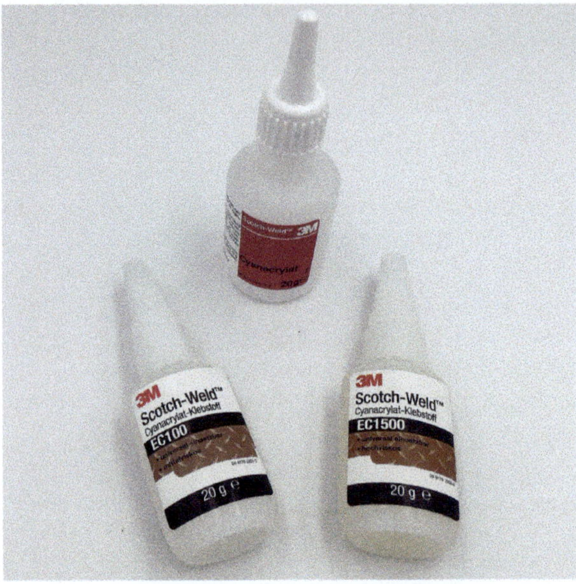

Figure 4.7 Cyanoacrylate adhesives as used in small bottles with different viscosities. Source: Dr. Jürgen Klingen, Aboso-Consulting.

Figure 4.8 Manual hotmelt adhesive gun with the adhesive in the form of dispenser sticks. Source: Dr. Jürgen Klingen, Aboso-Consulting.

Reactive, mostly PU-based hotmelt adhesives have been developed for higher-temperature resistance of adhesive composites. After joining the substrates, these adhesives are crosslinked by moisture in the air. The adhesive layers are flexible and impact resistant and have high immediate adhesion as well as high final structural strength, which results from post-crosslinking.

4.5.3 "All-Purpose" Adhesives for Numerous Applications

Solvent and dispersion adhesives have good wetting properties, are easy to apply by various processing methods, and achieve good strengths shortly after the solvents evaporate. They can be used on a wide range of materials such as metals, high-energy plastics, rubber, leather, cork, felt, wood, cardboard, textiles, glass, ceramics, and concrete. Therefore, these adhesives are often referred to as *all-purpose* adhesives. There are two methods processing these adhesives. In *wet bonding,* the liquid adhesive is applied to both substrates and, after a short time, the partially dried adhesive layers are pressed together. If at least one of the substrates to be bonded is porous, the wet adhesive can also be applied to one side, after which the other substrate is laminated directly to the first. Applying *dry bonding,* the liquid adhesives are applied to both substrates and then the material surfaces are dried. After the adhesive surface is no longer tacky, the two substrates are joined by pressing them together. This method is also called *contact adhesive bonding*. Dispersion adhesives are also processed by the wet- and dry-bonding method, have similar properties and applications compared to solvent adhesives, but the adhesive is dispersed in water rather than dissolved in organic solvents. Due to the high solids content, these systems have high yields and because they are non-flammable, they offer advantages in storage and production. The disadvantage of these water-based adhesives of having longer flash-off times can be compensated for by the use of heat drying (e.g. by infrared radiation or hot air).

4.5.4 The "All-Rounders" for Structural Applications

If relatively small-volume structural bonding applications with manageable batch sizes are to be realized in industry and craft, adhesives that do not require complex metering, application, and mixing equipment are preferred. For these applications, easy-to-handle two component (2K) epoxy and acrylic adhesives with relatively high reactivity are very popular. These are applied with very user-friendly on-demand mixing and dispensing systems using double cartridges in which the two resin components are isolated until they are needed. Mixing occurs as the materials are dispensed through the motionless mixing nozzle (static mixer) and is accomplished with hand tools or air guns. Such systems allow metering, mixing, and application of the adhesive (as dot or bead) to the material surface in only one step (Figure 4.9). After joining the adhesives are then cured at room temperature, although the curing time can be shortened by increasing the temperature.

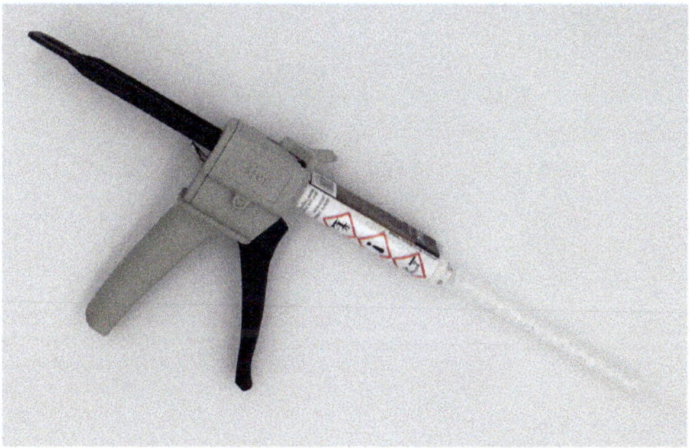

Figure 4.9 User-friendly on-demand mixing and dispensing system with double cartridges for manual operation. Source: Dr. Jürgen Klingen, Aboso-Consulting.

If structural bonds with high strength and very good aging resistance are required, 2K epoxy adhesives are the adhesive of choice. After curing, the adhesive layers are present as thermosets with relatively high cohesive strengths and relatively brittle-to-elastic properties. These adhesives exhibit very good adhesive properties on most of the commonly used bonded parts. When the focus of the user is on preparing more elastic adhesive bonds with high strengths and fast further processing options are required, 2K acrylic adhesives are used. Their very good compatibility with oily or greasy surfaces is also often highly valued, especially in the automotive industry. In many demanding bonding applications, the characteristic performance spectrum of 2K acrylic adhesives gives them a competitive advantage over high-strength but relatively brittle epoxy and very elastic but low-strength PU adhesives.

Due to the demand for low-weight components in modern manufacturing, plastics such as PE and PP are increasingly used in industrial applications. While such materials can be welded together, their low-energy nature makes them very difficult to bond with adhesives. Thus, for a long time, bonding of such materials was only possible after complex treatments, e.g. by corona or plasma. Today, after intensive research on this subject, novel tough acrylate-based 2K construction adhesives enable excellent bonding of materials such as PE or PP – but also thermoplastic elastomers (TPE) – without any treatment. Immediately after the adhesive is applied to the low-energy substrate, a chemical reaction begins at the wetted material surface, creating a bonding-friendly, high-energy surface. Thus, the new 2K acrylic adhesives enable structural strengths in the range of the inherent strength of the bonded PE and PP materials and rapid further processing of the bonded parts after 60 minutes.

4.5.5 "User-Friendly" Conformable High-Performance PSA Foam Tapes

Mechanical fastening techniques such as screwing, riveting, and welding are increasingly being replaced by modern high-performance acrylic foam tapes (Figure 4.10) [6].

Figure 4.10 A roll of acrylic foam tape as used in numerous applications. Source: Dr. Jürgen Klingen, Aboso-Consulting.

These self-adhesive products are very popular with customers because of their:

- ease of handling,
- good immediate adhesion,
- good vibration and noise dampening properties,
- tension-free bonding properties on a wide range of substrates,
- direct processing possibilities, and
- option of using application-optimized parts that are pre-applied.

Since the 1980s, they have proven themselves in countless elastic bonding applications. This is due, among other things, to their ability to absorb dynamic forces much better than mechanical bonds and avoid stress peaks at the overlapping edges of substrates. Thus, such viscoelastic acrylic foam adhesive tapes can compensate for differences in the linear expansion of joint substrates up to three times their thickness and are therefore particularly suitable for joints between materials with widely differing coefficients of thermal expansion, such as those frequently used in modern lightweight construction. Acrylic foam tapes also have excellent resistance to plasticizers and oils found in soft PVC and EPDM-based rubber-like materials. As a result, users in the automotive industry very often use these adhesive systems for all types of PVC-based exterior parts and EPDM-based door seals that are applied to car doors. Here, bonding with these special foam tapes enables sealing, noise and vibration damping, and at the same time withstand a variety of loads over many years (Figure 4.11).

Acrylic foam adhesive tapes enable highly demanding applications, especially when sealing and dampening properties as much as adaptability to different surfaces, high initial adhesion, and ease of use are required. These features can be provided, because these products consist of a closed-cell adhesive core in the center, which forms an inseparable unit with the two outer functional adhesive layers. This continuous viscoelastic adhesive structure makes it possible, in contrast to conventional foam adhesive tapes, to produce a permanent, stress-free bond with the substrates involved, with very good resistance to heat (in some cases even up to 260 °C for short periods), UV radiation, and chemical stresses.

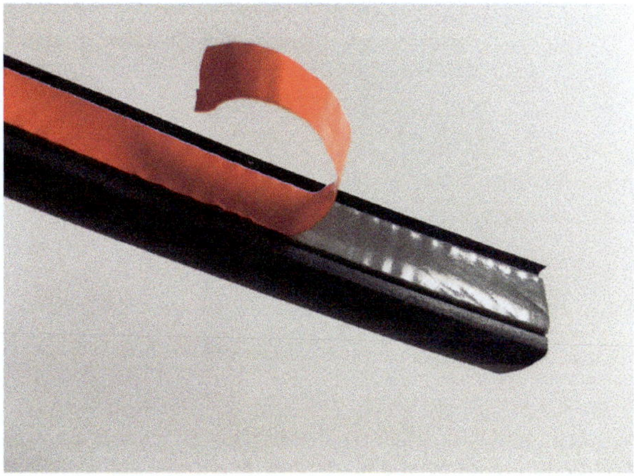

Figure 4.11 Use of acrylic foam tape for the attachment of self-adhesive EPDM seals to car doors. Source: Dr. Jürgen Klingen, Aboso-Consulting.

4.5.6 PSA Tapes for "Quick Connect and Disconnect"

These self-adhesive products are used in a wide variety of applications when, in addition to suitable and secure fastening, regular trouble-free separation and subsequent repeated joining is also required. For applications in industry and craft, there are used:

- *high-strength systems* – for carrying relative high loads requiring less frequent reconnection, and
- *low-strength systems (hook-and-loop fasteners)* – for low loads requiring very frequent reconnection.

4.5.6.1 High-Strength Fasteners

The relatively good holding power of these self-adhesive re-closable fastening systems makes it possible to replace classic mechanical fastening systems in many industrial and craft applications, e.g. for fastening wall and ceiling panels in trains and buses. This means there are no more holes in the substrate materials, corrosion is avoided, and access for maintenance work is achieved in seconds. Such systems consist of a self-adhesive polyolefin film with small mushroom heads on short stems (Figure 4.12). By joining two such structured films, which are previously applied to the substrates involved, a stable, removable closure is created. Such adhesive structures can be opened and closed hundreds of times without hesitation, with a clearly perceptible noise during the closing process indicating perfect anchorage each time.

4.5.6.2 Hook-and-Loop Fasteners

When a low-strength connection needs to be closed and re-opened continuously, self-adhesive hook-and-loop fasteners are used. These products are applied in numerous applications in everyday life, such as clothing, handbags, wallets, sports shoes and pads in motorcycle, bicycle, and work helmets. In what is probably the largest application in terms of volume, some years ago, hook-and-loop fasteners were introduced to improve the convenience of handling baby diapers and thus very successfully replaced the adhesive tapes long used for this application.

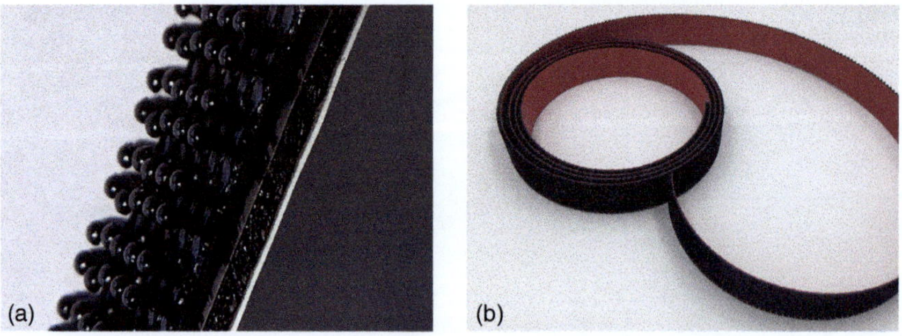

Figure 4.12 Structure of a high-strength re-closable fastening system consisting of a polyolefin film with small mushroom heads on short stems, which is laminated to a high-performance acrylic foam tape (a) and the delivery form of the associated tape (b). Source: Dr. Jürgen Klingen, Aboso-Consulting.

4.5.7 "The Strong PSA Tape" for All Cases

There are several types of PSA tapes available based on different materials for a variety of applications that go far beyond simple fastening. *Duct tape* is one of them. It is one of the most popular and versatile tapes around due to its ideal combination of different properties: it is water repellent, strong, easy to tear, flexible, and usually removable without leaving any residue. This makes it particularly suitable for quick repairs or for protecting workpieces – in fact, for all jobs where an extremely strong tape is needed. What the name was originally is not really clear and there are some conflicting stories about it. One theory is that *Duck Tape* was the original name because the tape is water repellent just like duck feathers. This may have later become *Duct Tape* as sealing heating ducts became one of many known applications.

Today's version of a Duct Tape is little different from the original version and is still just as useful. It consists of the same basic materials: a cotton fabric for the necessary tensile strength and ease of tearing, a PE coating or similar plastic coating for protection against liquids or damage, and a PSA for a strong bond to a variety of surfaces. Duct Tape has become very popular over the decades and is often used in the construction industry for example to seal joints, for minor repairs, or to protect surfaces from abrasion. In addition, there are numerous applications for this tape in many other industries and trades as well as in almost every home where they are used as an all-purpose weapon.

However, these tapes are also indispensable tools at NASA, where the fabric tape even takes part in every space mission and has already saved lives in one or two crisis situations. For example, in April 1970, when an oxygen tank in the Apollo 13 space shuttle exploded, Duct Tape acted as a lifesaver for the three astronauts. They had to retreat from the damaged command module into the lunar module. But the lunar module's air purification system was overloaded and had to be modified to use the command module's CO_2 filter. Not an easy task, because the various filters were round and square – in other words, not compatible. An adapter had to be built. But duct tape made the impossible possible and thus saved the astronauts' lives (Figure 4.13).

Figure 4.13 Duct tape saves lives by helping to adjust the CO_2 filter of the Apollo 13 space shuttle. Source: NASA.

References

1 Habenicht, G. (2005). *Kleben - Grundlagen, Technologie, Anwendungen*, Berlin: Springer-Verlag.

2 Satas, D. (1989). *Handbook of Pressure Sensitive Adhesive Technology*. New York: Van Nostrand Reinhold.

3 Brockmann, W., Geiß, P.L., Klingen, J., and Schröder, B. (2009). *Adhesive Bonding – Materials, Applications and Technology*, 1–3. Weinheim: Wiley-VCH.

4 Pröbster, M. (2013). *Elastisch Kleben – aus der Praxis für die Praxis*. Wiesbaden: Springer Vieweg.

5 Brockmann, W., Geiß, P.L., Klingen, J., and Schröder, B. (2009). *Adhesive Bonding – Materials, Applications and Technology*, 83–84. Weinheim: Wiley-VCH.

6 Brockmann, W., Geiß, P.L., Klingen, J., and Schröder, B. (2009). *Adhesive Bonding – Materials, Applications and Technology*, 298–300. Weinheim: Wiley-VCH.

5

The 5 Steps for Achieving Safe and High-Quality Bonds (Overview)

5.1 Introduction

In the previous chapters 2, 3 and 4, we learned about the history of adhesive bonding, the basics of bonding such as wetting, adhesion, and cohesion, as well as a first impression of what adhesive bonding can and cannot do when it comes to joining in industry and craft. Now is the time to start the project for the development of a safe and high-quality adhesive joint. For this, we must first discuss the requirements for the bonding process – according to ISO 9001 called the *special process* – and get to know the overall framework of the *5-step management system*. This will be accomplished in this chapter. Then, in Chapter 6, we start the actual project work with Step 1, the planning step.

5.2 Requirements for the *Special Process*

Once the decision has been made to develop a new bonding process, it is necessary to design the path that will lead to safe and high-quality bonded components, which meet the application requirements and that can be produced in the required quantities. According to the globally recognized quality standard ISO 9001, the bonding process is a *special process*, which expresses that its success cannot be verified without destroying the resulting products, i.e. the bonded components. Another feature of this process is that the adhesives used in the manufacturing process are not in their final (cured) form when they arrive at the manufacturing site. Thus, the customer using the adhesive to make the joint is almost exclusively responsible for the quality of the bonded joint. This is in contrast to the alternative classical mechanical joining techniques, where the *fasteners* such as screws or rivets are present in the bonded component in the exact form and quality as delivered by the supplier. The production of high-quality adhesive bonded components is a very complex process in which many action steps must be coordinated and responsibility must be assumed by employees from different disciplines and levels of the company. Therefore, careful and comprehensive quality management is required – from the initial inspection of raw materials to the final completion of the bonded components – which enables

Adhesive Bonding in Five Steps: Achieving Safe and High-Quality Bonds, First Edition. Jürgen Klingen.

the required orderly flow, meets quality requirements, and ensures good communication at the appropriate levels.

5.3 Quality Standards

The ISO 9001 quality standard is designed to eliminate defects and ensure defect prevention in complex processes. According to this standard, the bonding process is considered a *special process*, whereby 100% inspection of the bonded components for any defects that may be present is not possible. Also, the properties of the adhesive layers produced are not only determined by the specified raw material properties, but are generated to a very large extent during the production of the bonded components under the responsibility of the user. It is known that approximately 90% of the quality defects occurring during bonding are due to errors in the adhesive manufacturing process on the user side. The reasons for this are manifold and can be attributed, for example, to contamination of the surfaces, incorrect surface treatment, the use of unsuitable adhesives, or unsuitable curing conditions. In contrast, the adhesives used today are quality products manufactured in well-controlled processes at the supplier. Such raw materials should allow *zero-defect production* of bonded components. However, as mentioned above, this is often not the case, leading to the conclusion that to achieve the required quality, every effort must be made to control the manufacturing process to a level approaching 100%. This is where the philosophy of ISO 9001 comes into play. In essence, this standard is characterized by a prevention of errors, which is ensured by a comprehensive and lived quality management. Applying the ISO 9001 philosophy to bonding processes enables the user to develop a well-controlled process in which the execution of the sub-processes is described and the factors influencing quality are controlled and documented. However, although ISO 9001 is the basis for quality assurance in bonding, it is too general for the specific requirements of the processes used there. To serve these, the new *DIN 2304* standard *Quality requirements for adhesive-bonding processes* was published in February 2016 [1]. This new standard, introduced for the first time in Germany, supplements ISO 9001 in that it defines the specific requirements for processes to produce high-quality structural (load-bearing) adhesive bonds and describes the corresponding measures. In addition, it requires appropriate knowledge and regular training of the relevant employees, so that their qualification is now a mandatory component. Companies that manufacture in accordance with *DIN 2304* are thus signaling that they are working in accordance with the current state of the art in adhesive-bonding technology. This initiative, which is still focused on Germany, has a clear European and global dimension in perspective.

Courses to fulfill the requirement of DIN 2304 to have well-trained employees available for the relevant tasks and to provide regular further training are offered by appropriate institutes in Europe and comprise the following three levels [2]:

- European Adhesive Bonder [3],
- European Adhesive Specialist [4], and
- European Adhesive Engineer [5].

5.4 The 5-Step Project Management System

5.4.1 Introduction

As mentioned earlier, for the constant production of safe and high-quality bonded components, the bonding process must be controlled close to 100%. The development of such a bonding process should be very systematic throughout the entire process chain – from the idea to the production of the bonded component – and should be carried out by a very well-controlled cross-functional effort. For this purpose, the author has developed a special *5-step project management system* tailored to the needs of bonding. In this process, quality and bonding-related execution is ensured by the requirements specified in *DIN 2304* and tools and quality techniques of the *Six Sigma* methodology are used for optimum planning and execution (Figure 5.1).

The *5-step project management* system is designed:

- for the planning and execution of the individual development steps using proven tools and quality techniques of the *Six Sigma* methodology,
- to coordinate the tasks in the correct sequence for all functions involved, such as process development, application engineering, sales, manufacturing, and finance,
- to ensure that the necessary work in a step is completed before the next step begins,
- to ensure a high-quality and adhesive specific execution of the process steps and design of the environment according to *DIN-2304*, and
- to minimize the risk in the development of the adhesive process from the idea to the start of the production.

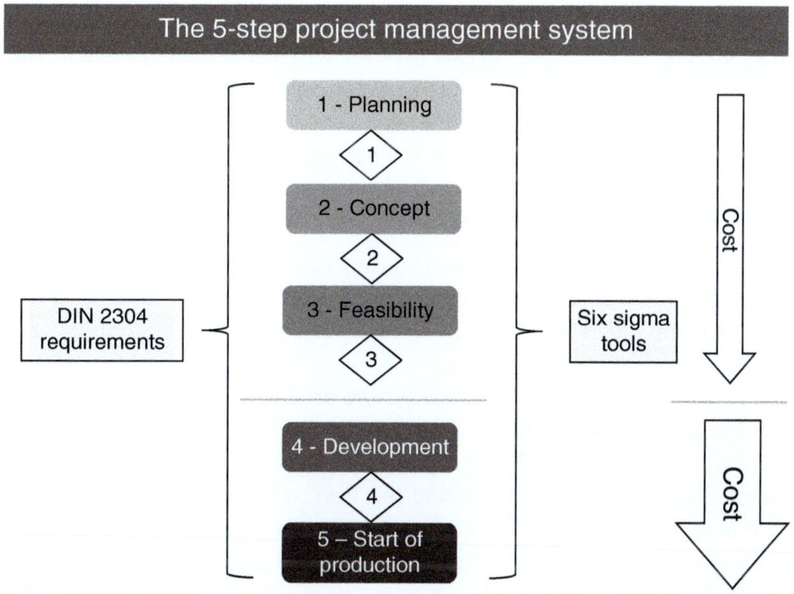

Figure 5.1 The 5-step project management system for the development of a bonding process.

As already discussed to ensure the adhesive-specific design and execution of the sub-processes, the management system is supported by *DIN 2304* and quality techniques of the *Six Sigma* methodology are applied with the following role descriptions:

- The *5-step project management system* helps to create the project plan for the development of the adhesive-bonding process, defines the individual phases from idea to start of production, the tasks to be completed, the time sequence of the project steps, the necessary approval procedures for the phase transitions and the responsibilities for the project.
- *DIN 2304*, on the other hand, ensures that the proper bonding-compatible execution of the individual steps is taken into account and provides guidance for setting up organizational structures for an appropriate manufacturing operation.
- The *Six Sigma* methodology provides the tools and quality techniques for planning and executing the corresponding steps of the 5-step system, supports the preparation of the project contract and the creation of flow charts in all phases of the project.

5.4.1.1 Process Map

Six Sigma is designed to guide through complex processes and for this purpose, it breaks down a project into sub-steps, representing the project flow, the work to be done in the corresponding stages, and the results to be achieved. When using the *5-step project management system*, a process map should be created in the early stages of the project. This provides a detailed definition of the required actions (inputs) and the corresponding results (outputs) that must be available when the actions are completed. Through the process map, the project team is supported by guidance at all stages on what needs to be done and what outcomes need to be achieved. In addition, this tool is an essential part of gate reviews, where it is used to evaluate the project and decide whether to continue or stop the project. Figures 5.2 and 5.3 show the process maps for the two parts of the management system and the deliverables that must be present at the end of each step as a prerequisite to enter the next step.

5.4.2 Planning, Concept, and Feasibility (Part 1)

The planning of the project, the definition of concepts for bond preparation, and the feasibility of the preferred solution are worked out in the first part of the *5-step system*. Here, the investments in personnel, equipment, and material are still relatively small, which changes significantly in Part 2.

At the end of Part 1, the validated concept for the production of the bonded joint is completed and ready for the development phase. Up to this point, the project has been evaluated three times in so-called gate reviews after the planning, concept, and feasibility steps. After each phase, a decision has been made whether to move the project to the next phase or cancel it. Gate 3, which concludes Step 3, is of great importance for the adhesive project and must therefore be prepared and carried out with particular care. This is because the investments and production-related activities in the development step (Step 4) cause significantly higher costs of the project compared to the activities before entering this step (Figure 5.2).

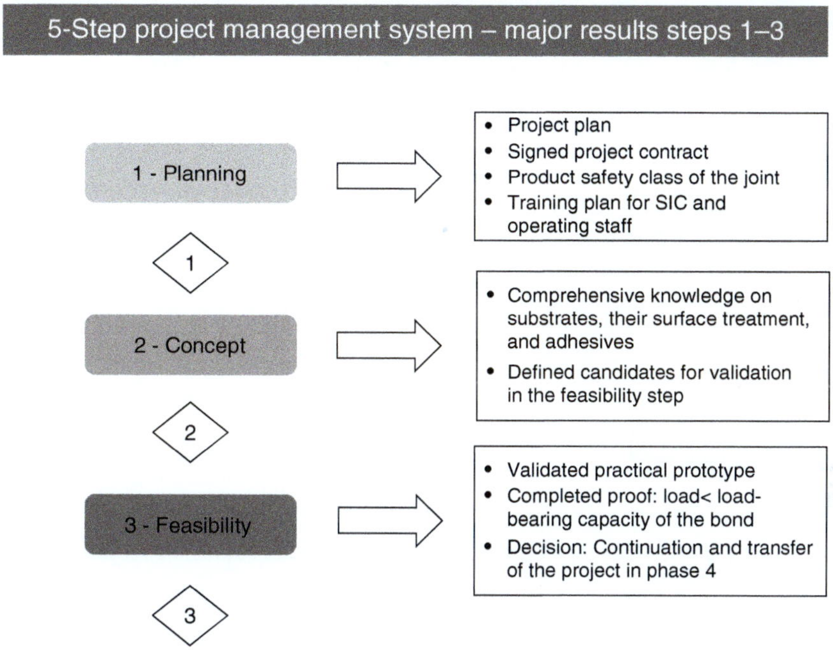

Figure 5.2 Part 1 of the management process with major results obtained after the planning, concept, and feasibility step.

5.4.3 Development and Launch (Part 2)

The second, much more elaborate part of the *5-phase system* deals with the development of the validated concept into a robust bonding process up to the launch. Before the start of production, the project has been evaluated in a gate review after the development step, and the team decides whether to introduce the developed bonding process into production (Figure 5.3).

5.5 Gate Reviews

The *5-step management system* provides for a total of four project assessments at the following gates:

- *Gate 1* – after the project is planned and the contract signed,
- *Gate 2* – after the bonding concepts have been defined,
- *Gate 3* – after validation of the preferred concept, and
- *Gate 4* – after the development of the robust production-ready process.

To move the project to the next step, it is necessary to pass through the corresponding gate. For this purpose, so-called gate reviews are performed to approve these transfers. Only after their completion, the next project step can be started, provided that all tasks have been completed. If this is not the case, the necessary work must

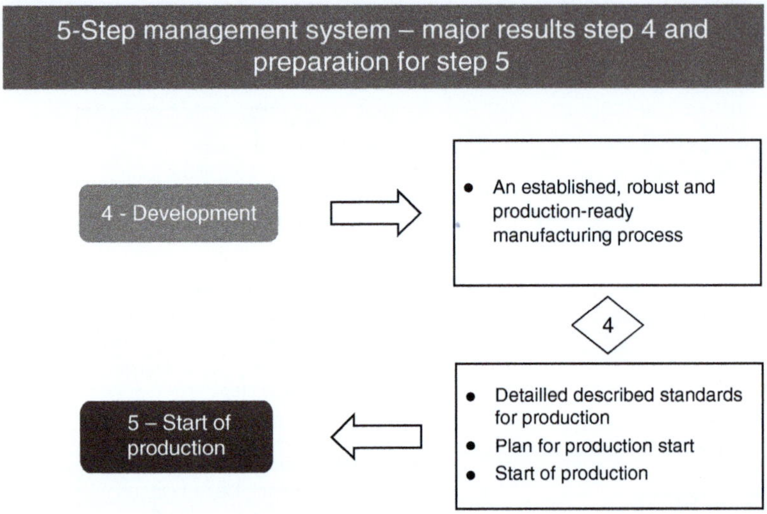

Figure 5.3 Part 2 of the management process with major results obtained after the development step and preparation for start of production.

be completed as quickly as possible, but this delays the phase transfer. In the gate reviews, projects are evaluated in detail and a decision is made whether to continue or cancel them. The gate reviews also serve to inform the decision makers about the status of the projects. They receive information about what has been achieved so far, the risks of continuing the project, critical issues, and the important open questions. In preparation for the final gate decision by the gatekeeper, the project team presents the status and assessment of project progress from their perspective. Participants in the gate reviews include the gatekeepers, champions, project managers, the for adhesive bonding responsible supervisor (supervisor in charge [SIC]), and experts to answer critical technical questions, if necessary.

5.5.1 The Gatekeeper

The gatekeeper has the task of representing the functions involved in the project, such as development, sales, production, and finance, in terms of overall decision-making in the gate reviews. The gatekeeper is usually a member of senior management, is well integrated into the project, and focuses on strategic and critical project issues. The gatekeeper should be very well informed about the project status before the gate reviews begin, be results-oriented, and have a thorough understanding of the potential of the tools used in the project. The gatekeeper is present at all gate reviews, makes the gate decision, and is responsible for demonstrating the importance and priority of the project internally and externally.

5.5.2 Color Coding

The color-coding method should be used in gate reviews and also in *high-level* status reports to senior management to demonstrate project progress or status. This method provides a clear and efficient way to show the strengths and weaknesses of the project and enables a short and precise update of the project status for both the project team and senior management. It also avoids the need to present and understand information that is not important for decision making.

The following colors are used to evaluate the project:

- *Green* – The most important questions have been answered and all tasks have been completed. The project is under control and can be moved to the next phase.
- *Red* – One or more key questions are not answered or/and not all tasks have been completed. The project cannot be moved to the next phase, the remaining tasks must be completed, and the gate review must be repeated.
- *Yellow* – One or more questions have not been answered or/and not all tasks have been completed. However, answering the questions or completing the tasks can be accomplished in a relatively short period of time and with a high probability of success. The project can be moved to the next phase and the completion of the outstanding tasks must be done in a defined time frame and documented.

5.6 Quality Requirements According to DIN 2304

DIN 2304 quality standard is designed for all structural (load-bearing) bonded joints and defines the requirements for high-quality execution of the bonding process along the process chain. Also, the standard provides guidance for setting up the organizational structure of a company involved in bonding and defines product safety classes. According to DIN 2304, a structural bonded joint is a composite material produced by the support of an adhesive that is mainly responsible for the transfer of mechanical loads. It is part of a technical design that contributes to the adequate functioning of the component over its entire life cycle and can withstand all loads that occur, regardless of whether they are static, dynamic, or combined loads with or without exposure to environmental influences. With the help of DIN 2304, the user is more involved in the planning, evaluation, or decision-making processes and is thus called upon to create a technically competent environment, to train the personnel, and to take responsibility for the design of the bonding process in the production facility.

The standard contains three core elements (Figure 5.4):

- *Part 1* – Classification of bonded components according to product safety classes.
- *Part 2* – Appointment of a qualified bonding supervisor and production personnel.
- *Part 3* – Proof that the load of the bonded joint is less than the load-bearing capacity of the adhesive layer.

5.6.1 Part 1 – Classification of Adhesive Components into Safety Classes

The selection of the requirements for the quality-compliant manufacture of structural, load-bearing bonded joints is made by classifying the planned bonded components into safety classes. For the purposes of this standard, this refers exclusively

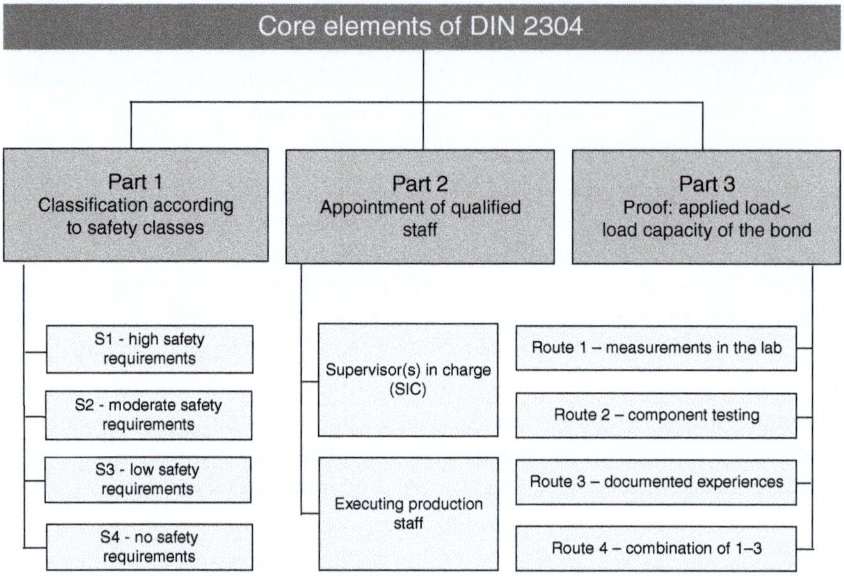

Figure 5.4 Core elements of DIN 2304.

to the expected consequences of a possible failure of the bonded joint. The task of the responsible designer or component manager is therefore to analyze what will happen if the bonded joint fails.

***S1 – High-safety requirements* – The failure of the bonded joint**

- leads directly or indirectly to an unavoidable danger to life and limb,
- leads to a loss of functionality, the effects of which are very likely to lead to an unavoidable danger to life and limb.

***S2 – Moderate-safety requirements* – The failure of the bonded joint**

- can lead to a danger to life and limb,
- leads to a failure of the functional capability, the effects of which are very likely to lead to a hazard to persons or significant damage to the environment,
- leads to a failure of the functional capability, the effects of which are very likely to lead to significant financial losses.

***S3 – Low-safety requirements* – The failure of the bonded joint**

- results in a failure of functionality, the effects of which are unlikely to result in harm to persons or significant environmental damage,
- results in a failure of functionality, the effects of which are at most associated with a loss of comfort or performance,
- results in a failure of functionality, the effects of which are not likely to result in major financial loss.

***S4 – No-safety requirements* – The failure of the bonded joint**

- results in a failure of functionality, the effects of which are not expected to result in harm to persons or the environment under foreseeable conditions,
- results in a failure of functionality, the effects of which are associated solely with a loss of comfort or performance,
- results in a failure of functionality, the effects of which are not associated with major financial losses.

5.6.2 Part 2 – Assignment of the Responsible Supervisors (SIC)

According to the standard, a responsible supervisor in charge (SIC) must be designated who is responsible for adhesive-bonding activities and related tasks in the user company. His qualifications and expertise (acquired through training, education, and/or relevant experience) must be proven and documented. The SIC is the main contact for all quality issues relating to the so-called *special process* of bonding – from planning and production to maintenance/repair. According to *DIN 2304*, a user company must provide a sufficient number of trained personnel for the relevant tasks and document their appointment. Depending on the safety class of the bonded joint (described in Part 1), the qualification required to be appointed as SIC varies and may include vocational education and on-the-job training.

5.6.3 Part 3 – Proof of Applied Load < Load-Bearing Capacity of the Adhesive Layer

An adhesive joint must be dimensioned so that the actual applied load is always less than the load-bearing capacity of the adhesive layer. In addition, *DIN 2304* requires – with the involvement of the SIC – a traceable and documented proof of this requirement: applied load < load-bearing capacity.

This can be carried out in four ways by:

- measurements,
- component testing,
- documented experience, and
- combinations of 1–3.

5.6.3.1 Measurements

The load on the bonded joint is determined from tests, calculations, standards, real data, or a combination of these. The maximum load-bearing capacity must be determined by extensive practical tests – including statistics – taking into account aging, test media, and combinations of several effects. It must be ensured that the laboratory results correspond to the results under real production conditions. The tests can be performed on test specimens or on real components, and the stresses and strains obtained are calculated for each bonded joint, followed by comparison with the allowable stresses/strains. It must be clearly documented how the data were determined.

5.6.3.2 Component Testing

For this purpose, an entire component or a part of a component may be tested under real conditions or conditions simulating reality. The test conditions must have similar effects on the bonded joint as on the entire system. When testing a subsystem, the mutual influence between the subsystem and the overall system must also be taken into account, and it must be ensured that the results are not distorted by this effect.

5.6.3.3 Documented Experience

Verification based on experience with applications that have already been carried out is also possible. For this purpose, the transferability of the bond in operation to the planned bond in terms of service life, stress conditions, and properties of the bonded joint must be demonstrated. This must be documented in detail.

5.6.3.4 Combinations of 1–3

If the verification is carried out by a combination of the methods described above, it must be ensured that the consideration is suitable to evaluate all requirements in an appropriate way and that the individual components of the verification fit to each other. The process must be documented in a comprehensible manner.

References

1 Groß, A. and Lohse, H. (2015). New DIN 2304 standard and its use in practice. *Adhesion Adhesives and Sealants* 4: 12–17.

2 Quintino, L., Fernandes, I., and Assuncao, E. (2013). European harmonized system for training and qualification of adhesive bonding personnel. *Applied Adhesion Science* 1: Article 2.

3 Fraunhofer IFAM (2021). EWF-European Adhesive Bonder (EAB). https://www.weiterbildung.ifam.fraunhofer.de (accessed 11 June 2021).

4 Fraunhofer IFAM (2021). EWF-European Adhesive Specialist (EAS). https://www.weiterbildung.ifam.fraunhofer.de (accessed 11 June 2021).

5 Fraunhofer IFAM (2021). EWF-European Adhesive Engineer (EAE). https://www.weiterbildung.ifam.fraunhofer.de (accessed 11 June 2021).

6

Planning (Step 1) – Preparing the Plan and the Project Contract

6.1 Introduction

Planning the activities to develop a safe and high-quality bonding process is the first step in the *5-step management system*, with the main deliverables being the project plan and the project contract (Figure 6.1). Prior to this, there has usually been a lot of communication and many discussions about the idea of developing a bonding process with potential internal customers and others relevant to the success of the project, and as an important outcome there is a broad consensus to start such a project.

6.2 The Project Plan and Contract

The project plan essentially contains the following elements:

- the goal of the project,
- the project description,
- the project outline,
- time frame of the overall project,
- dates for the step completions,
- dates for gate reviews,
- the safety class of the project according to *DIN 2304* [1], and
- the available budget.

The classification of the planned adhesive bond according to *DIN 2304* in safety classes 1–4 should be part of the project plan, because it will significantly influence the extent of development work to be done, the degree of training needed for the supervisors, and the budget. It is important that the persons relevant for the success of the project, in particular the sponsor, the project manager, and the project champion, are already clearly identified and assigned at this stage.

One of the main tasks of the planning phase is the preparation of the project contract, which is one of the most important tools of the project management system. The contract is created by the project manager with significant input from the project team and should be structured so that no changes are likely to be required during

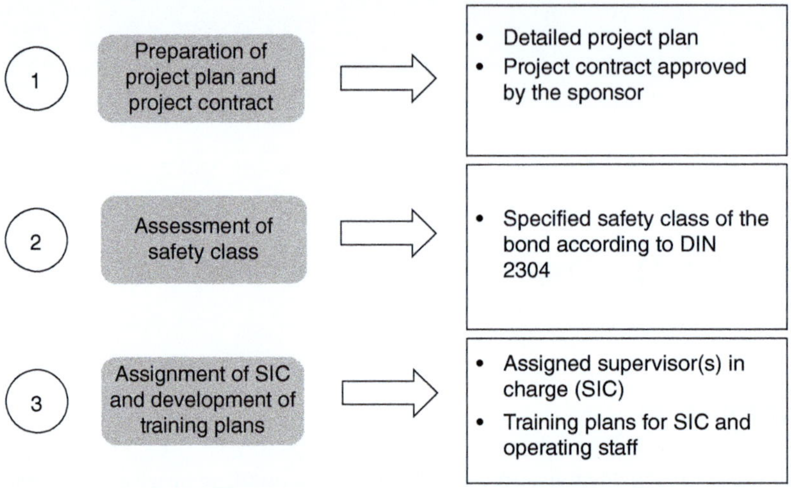

Figure 6.1 Major tasks and accomplishments of the planning phase.

the project. It is approved and the project can be started not before the contract is signed by the sponsor. Thus, the approved project contract formally gives the project manager the authority to use the planned resources and communicates to all stakeholders what can and cannot be expected of the project. At the end of the planning step, the project contract and the plan for the concept phase are discussed in a gate review and formally approved by the project team, allowing the project to move to the concept step (Step 2).

6.3 Roles of Project-Relevant People

Part of the planning phase is to identify and assign the following individuals relevant to the success of the project:

- project sponsor (client),
- project manager (planner),
- project champion (advocate),
- supervisor in charge (SIC), and
- operating staff.

6.3.1 Project Sponsor (Client)

The sponsor of a project is the main project stakeholder who signs the project contract, finances the project, and makes the final decision about launching of the project. After signing the contract, the sponsor hands over the project to the project team for implementation. Since they pay for the project, they are significantly involved in the creation of the requirements, goals and objectives, which they finally sign off on. In any case, the sponsor is a member of the project's steering committee, where they usually have the power to make the final decisions.

6.3.2 Project Manager (Planner)

The project manager is responsible for planning all tasks and work steps as well as for the overall management of the project. The manager supervises and coordinates all work processes from the planning step to the implementation.

At the beginning of the project, especially in the planning phase, the main tasks are to work out

- the precise description of the project, including the goals,
- the concrete project goals with the sponsor,
- the resource and budget planning,
- the role definition of the persons involved in the project,
- the integration of the project into the existing corporate structure,
- the structure and definition of the team,
- the definition of the communication structure, and
- the documentation of the approved project contract.

Since conflicting departmental interests and thus conflicts often arise, the project manager is largely responsible throughout the project for balancing different interests, bridging cultural differences, and forming a powerful, well-functioning project team. Another task of the project manager is project controlling, which can also be delegated to another person on the project team. This is where project performance, deadlines, and costs are monitored, as well as their adherence to the defined project goals. The project manager is also responsible for ensuring that the relevant management is regularly informed and that project results are regularly discussed.

6.3.3 Project Champion (Advocate)

The success of any project requires a champion who strongly supports the project by removing barriers. Through his power and influence, he should be able to obtain appropriate project resources such as personnel, equipment, and permits. The champion is also responsible for communicating project status with senior management. In project meetings, the champion acts as a coach, especially in discussions about criteria relevant to the success of the project. Periodically he meets with the project manager to discuss critical issues and status of the project. In gate reviews, the champion acts as an advocate for necessary decisions by the project team. It is important to the success of the project that he establishs good connections with top management to make the project visible here.

6.3.4 Supervisor In Charge (SIC)

According to the new *DIN 2304* standard, at least one person must be assigned as SIC when setting up projects with a focus on adhesive bonding. The person is responsible for the bonding-related projects and activities in the company. His qualifications and knowledge must be proven by training and/or appropriate experience. He supports the definition of quality requirements for the related technical tasks in the planning phase of the project. The SIC must hold a European Adhesive Bonder qualification or equivalent that meets the requirements of the task. For safety class S2 projects, the SIC must hold at least a European Adhesive Specialist qualification or equivalent. S1 safety class adhesives require a qualification as EWF European Adhesive Engineer or a comparable qualification. The SIC must undergo external training at least every two years.

6.3.5 Operating Staff

DIN 2304 also requires trained production personnel who have at least a European Adhesive Bonder qualification or a comparable qualification corresponding to the requirements of the task. For adhesive bonding projects of safety class S3, the SIC may, in exceptional cases, also approve employees without this qualification, who are, however, only used for specific and limited bonding projects or activities. The SIC must instruct such employees in these bonding operations and comprehensibly justify the basis on which the approval was given. The operating staff must be trained at least every two years.

Reference

1 Groß, A. and Lohse, H. (2015). New DIN 2304 standard and its use in practice. *Adhesion Adhesives and Sealants* 4: 12–17.

7

Concept (Step 2a) – Substrates, their Surfaces, and their Treatment

7.1 Introduction to Chapters 7 and 8

With the transition of the project to the concept step, an approved contract and a detailed plan for the project sequence are available and the persons responsible for the success of the project have been identified and assigned. The project team is in place, the planned composite component is classified according to *DIN 2304* safety classes, and both the SIC and the manufacturing personnel with the required qualifications have been announced. In the gate review at the end of the planning step, the handover of the project to the concept step is approved and there is a plan for the upcoming tasks in the concept step.

In Chapter 3, the basic knowledge of adhesive-bonding technology, such as wetting, adhesion, and cohesion, has already been discussed as a basic prerequisite for the production of good bonded components. The concept step now includes a comprehensive discussion of the substrates and adhesives used for bonding.

For this purpose, the concept step is divided into two chapters where to find this detailed information:

- *Chapter 7 (Step 2a)* – on the characteristics and physical properties of the substrates, their surfaces, and the surface treatment methods used prior to bonding, and in
- *Chapter 8 (Step 2b)* – on characteristics and physical properties of adhesives, their applications in industry and crafts, loads acting on bonded joints, and criteria for selecting adhesives for the appropriate application.

Thus, in chapter 7 substrates and the treatment of their surfaces are discussed, and then Chapter 8 provides the reader with detailed information on adhesives, their use and the stresses acting on the bonded joint. Based on all this knowledge gained in Chapters 7 and 8 and information about the task at hand, the user has to select 1–3 candidates at the end of Chapter 8, which will be further processed in Step 3. For this purpose, the exact specification of the adhesive(s) and their surface treatment(s) must be determined and the bond must be outlined in its structure.

As an overview, the main tasks and corresponding deliverables of Step 2a are shown in Figure 7.1.

Adhesive Bonding in Five Steps: Achieving Safe and High-Quality Bonds, First Edition. Jürgen Klingen.

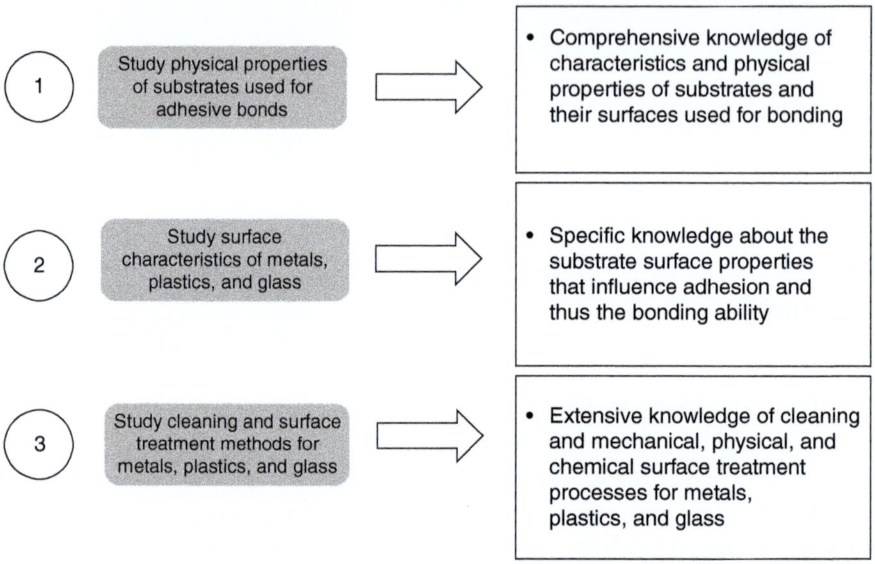

Figure 7.1 Major tasks and accomplishments of Step 2a.

7.2 Substrates and their Surfaces

7.2.1 Introduction

In addition to the adhesives used for bonding, the material properties of the substrates decisively determine the quality of a bonded component, whereby the chemical, physical, and structural properties of the material surfaces are of great importance.

The following material classes will be covered in sub-chapter 7.2:

- metals,
- plastics,
- glasses, and
- wood.

7.2.1.1 Metals

A significant proportion of all bonding applications in industry and the trades are carried out with metallic materials and it can be assumed that, apart from intermolecular interactions in the boundary layer area, no chemical reactions take place with the adhesive and the metal surfaces. Furthermore, the metal surfaces are generally not affected by the adhesive in any way. Since the dawn of civilization, metals have been used in a variety of applications. In the solid state, they are crystalline substances that are usually not used in their elemental form, but almost exclusively as alloys. The cohesion in metals is based on metallic bonds, from which the four characteristic properties can also be derived, such as

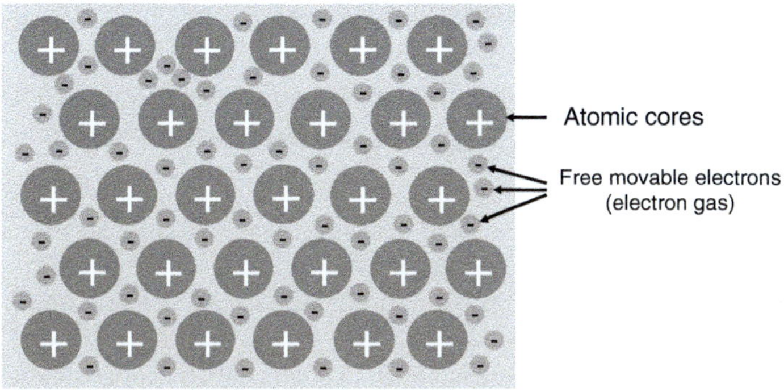

Figure 7.2 Atomic structure of metals.

- high electrical conductivity,
- high thermal conductivity,
- ductility (malleability), and
- luster.

According to their density, metals are divided into two groups:

- *light metals* with a density of < 5 g/cm^3 and
- *heavy metals* with a density > 5 g/cm^3.

Another important characteristic of metals is their high-melting point, and metals with a melting point above 1750 °C are referred to as high-melting. Most metals are base metals and occur naturally on the earth's surface in the form of compounds. In their elemental form, they are always eager to react chemically with nonmetals. Therefore, many metals very easily form compounds with aggressive liquids, gases, water, and aqueous solutions, they corrode. In metals, the atoms are in a regular ordered arrangement and the electrons move in this matrix, forming a so-called *electron gas* (Figure 7.2). Their mechanical properties depend essentially on their structure, grain size, and grain shape. Therefore, when processing them, precise knowledge of the internal structures of the metals is crucial for specifically influencing their properties. One of the most important properties of metals is their high strength combined with good formability.

Steel, the alloy of which is essentially iron and carbon, has the greatest importance in the assembly of metals in industry and craft, but the use of aluminum alloys has increased significantly in recent years due to the demand for modern lightweight structures.

7.2.1.2 Plastic Materials

In contrast to metal surfaces, which are generally not affected by the adhesive in any way, additives present in the plastic material can migrate into the adhesive layer and change the material properties. Also, organic solvents that come into contact with plastics can alter their surfaces through dissolution or swelling processes. Since welding of plastics is limited to thermoplastic materials, adhesive bonding is often the only applicable joining method in the assembly of plastics and is therefore widely used. However, due to the different structure of plastics compared to metals, other material- and process-specific factors must be taken into account when designing the components and preparing the bonding process.

7.2.1.3 Glasses

This material consists mainly of silicon dioxide and can be described as a melt solidified without significant crystal formation. Most of the glasses produced today are soda–lime glasses, where the network is formed mainly from silicon dioxide (SiO_2). Other oxides such as alumina or various alkali oxides are added to obtain alumina or alkali silicate glasses. From an adhesive-bonding perspective, glasses have some unique properties that must be considered when selecting adhesives and bonding processes. These include their brittle and elastic behavior, which does not allow plastic deformation and bending at the overlapping ends of the bonded joints. A specific feature of glass is that stress peaks caused by damage to the glass surface can adversely affect the strength of the bonded component.

7.2.1.4 Wood

The dry mass of wood consists of cellulose, hemicellulose, lignin, and small amounts of other constituents such as resins, fats, peptides, tannins, and dyes. Depending on species, location, and felling time, freshly harvested wood contains 50–150% water. A distinction is made between *free* water, which is stored in cell cavities, and *bound water*, which is directly associated with the cellulose fibers. For adhesive bonding, wood is a grateful substrate because the inherent pore structure, which results from the evaporation of the capillary water in the drying process, provides the possibility of a mechanical anchorage of the adhesive film in the structure. Rich opportunities for the formation of hydrogen bonds and even covalent chemical bonds in the adhesion layer result from the chemical structures of the main components of the wood matrix (cellulose and lignin), which contain hydroxyl and other polar functional groups to a large extent. If the basic requirements for bonding are well met for this material, however, some special features must be taken into account. For example, the behavior of the adhesive at the boundary layer is greatly influenced by the wood density, which depends on the wood species, but within the wood species also on whether it is early or late wood. Furthermore, the adhesive used must have a certain compatibility with water and water vapor, both during application and in continuous use, since wood always has a residual moisture content of 6–15% in practical use. Wood is subject to moisture-related and thus climate-related dimensional fluctuations, resulting in high shear stresses at the joint interfaces, which must either be absorbed by the elasticity of the adhesive or held by sufficiently high adhesion and cohesion forces. Since wood surfaces are never absolutely flat, even after sanding, wood adhesives must be able to fill these unevennesses and gaps very well.

Figure 7.3 Heavily corroded iron parts from the First World War (1914–1918), found in the Dolomite Alps. Source: Dr. Jürgen Klingen, Aboso-Consulting.

7.2.2 Ferrous Materials

7.2.2.1 Introduction

Pure iron is a silvery white metal and is resistant in dry environments. In moist air and in water containing oxygen or carbon dioxide, it readily oxidizes to form iron oxide hydrate and thus rust forms on the surface (Figure 7.3). When iron is heated in dry air, a thin, strongly colored, jet-black layer of iron oxide Fe_3O_4 forms. The melting point of iron is 1537 °C and the boiling point is 2750 °C. The density of ferrous materials is about 7.9 g/cm^3. Originally – since about 1500 before Christ – the first iron was produced in furnaces by the reduction of iron ore with charcoal at about 1300 °C. Much later, it was discovered that much better properties, in particular lower brittleness, were achieved by contacting the pig iron with air and thus reducing the carbon content.

When the mass content of iron is not exceeded by any other element in the alloy, these materials are called ferrous materials. The most important classification is the distinction between cast iron and steel (Figure 7.4). Here, an iron–carbon alloy is called steel if the carbon content is less than 2%. Steel can be forged well, but is comparatively difficult to cast. If the carbon content is above 2%, the material is called cast iron. Here, the reverse is true: casting works well, forging becomes difficult. Cast steel is produced by casting an iron–carbon alloy with less than 2% carbon into a mold without subsequent forming processes such as forging or rolling.

Although the increase of carbon content in ferrous materials contributes to a significant increase in strength, it also causes a decrease in ductility and toughness, the material becomes brittle. Thus, the increase of carbon content in the ferrous alloy leads to a negative impact on the forging and welding properties, elongation at fracture, and impact energy. Ferrous materials are the most widely used and versatile metals because their physical properties can be widely varied.

Additional alloying elements can be added for improvement:

- *chromium* – for hardness and corrosion resistance,
- *cobalt and manganese* – for wear resistance,
- *molybdenum* – for tensile strength,
- *niobium and vanadium* – for elasticity, and
- *tungsten* – for heat resistance.

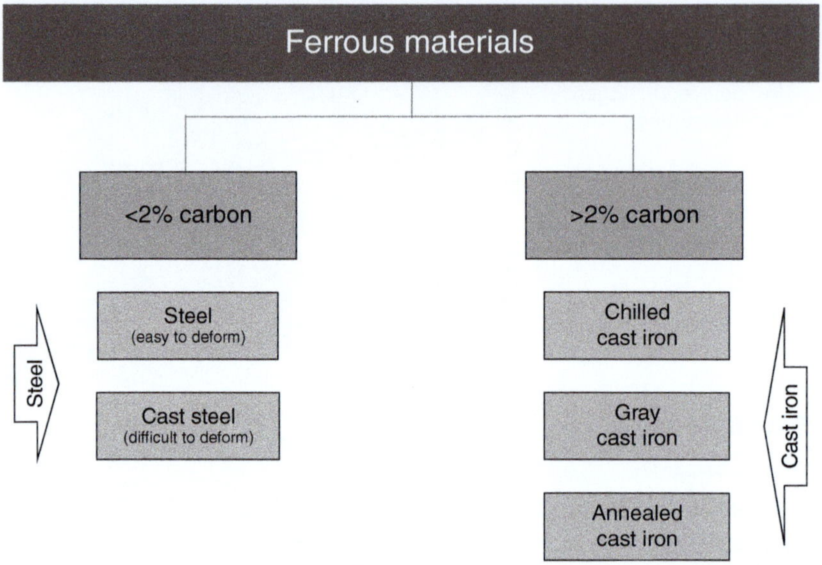

Figure 7.4 Classification of ferrous materials.

For example, steels can be made very ductile and formable as tinplate for cans, but also with high strength and relatively brittle behavior for use in building construction. Current developments are focusing on steels with very high strength and at the same time good formability. Such properties are particularly needed in modern lightweight construction, e.g. automotive engineering.

7.2.2.2 Mechanical Properties of Ferrous Materials

Steel is a very high-strength but also relatively soft material, while cast iron is hard and brittle. Therefore, components made of steel can transmit very high forces, and before the material breaks, there is usually significant deformation. In contrast, cast iron breaks without prior deformation. Thus, steel has a kind of safety margin, which allows the material to be used in the manufacture of safety-relevant vehicle parts such as steering, chassis, and drivetrain.

Common unalloyed structural steels used, for example, in the manufacture of vehicles, machinery, ships, buildings, and furniture, have a tensile strength R_m- the maximum stress a material can withstand in strain or tension before breaking - of about 350–500 N/mm^2, which can be improved with increasing carbon content and the use of other alloying elements. Young's modulus of these materials is about 210 GPa and the elongation at break is about 20–25%. These materials have a pronounced yield strength $R_{p0.2}$ of about 180–350 N/mm^2. $R_{p0.2}$ is the maximum stress at which the material begins to deform plastically and for practical reasons measured at 0.2% plastic deformation. In application, the structural steels can be loaded up to this value, but for processing by rolling and forging, the yield strength must be exceeded. High-strength tool steel can achieve tensile strengths

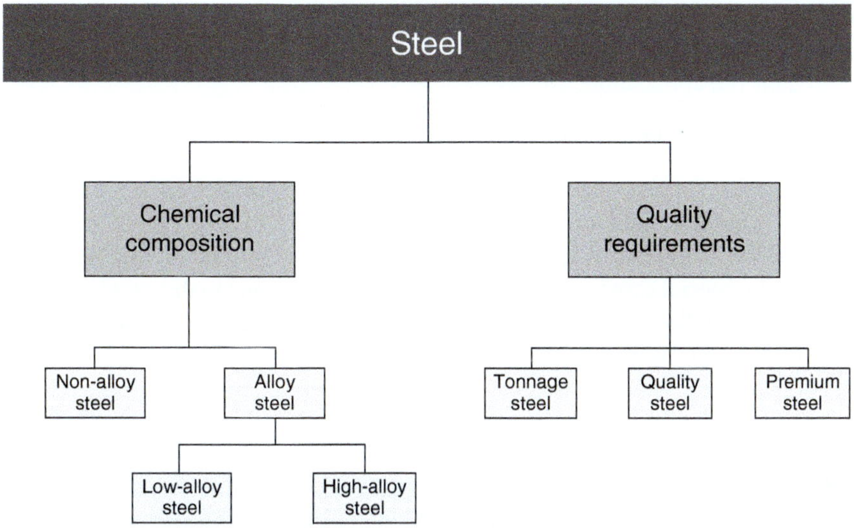

Figure 7.5 Classification of steel according to chemical composition and quality requirements for its use.

of 800–1500 N/mm^2 and have a modulus of elasticity of 210–250 GPa. Compared to materials such as wood, concrete, and stone, steel has very high strength and excellent density-related specific strength. For example, steel has a tensile strength about 10 times higher than wood. Thus, for a given load-bearing capacity, steel structures are much lighter than those made of wood, concrete, or stone.

7.2.2.3 Classification of Steel

Steel can be classified according to its chemical composition and the quality requirements for its use (Figure 7.5).

Classification according to their chemical composition:

- *Non-alloy steel* – Steel is called *non-alloyed* if the additional constituents do not exceed a certain level. Thus, non-alloyed steel is not free of other components, but contains foreign components only in very small quantities. The properties of this easy-to-process material can be adjusted in many ways, especially by varying the heat treatment and carbon content. This relatively inexpensive metal, which accounts for about 70% of world steel production, has proven itself over a long period of time in industry and craft, where it is used for numerous applications with relatively moderate performance requirements.
- *Low-alloy steel* – Here, an alloying element does not exceed a 5% content. These materials have similar properties to unalloyed steels and are used when their special advantages such as better high-temperature strength or better cold toughness are needed.
- *High-alloy steel* – In this steel, at least one alloying element exceeds 5%, which results in a completely different structure and properties of the material. Due to

the high price, they are taken only when, for example, special corrosion resistance, high temperature resistance, good cold strength, or magnetic properties are required for the application.

Classification according to the quality requirements:

- *Tonnage steel* – Tonnage steel is unalloyed and has no special service properties. This steel is not intended for heat treatment.
- *Quality steel* – As the name implies, higher demands are placed on quality steel, which is why special care is also required in its manufacture. Quality steel belongs to either the unalloyed or alloyed steel group, has a fine-grained structure and a purity of no more than 0.045% content of phosphorus and sulfur (so-called iron companions). The steel is subject to precisely defined requirements, such as toughness, weldability, and formability and used for the production of sheets, tubes, and profiles.
- *Premium steel* – Premium steel can be alloyed or unalloyed and is demonstrating very high quality. It has a particularly high degree of purity and their chemical composition is very well controlled. Thus, its phosphorus and sulfur content must not exceed 0.025%. Contrary to what is often assumed, a premium steel is not necessarily a stainless steel, just as a stainless steel is not necessarily a premium steel. Premium steels are used as heat-resistant steels for the manufacture of tools, machines, and container structures and in applications, where chemical resistance of components is required.

In industry and trade, steels are also very often classified according to their special field of application:

- *Structural steel* as standard steel for industrial applications where, for example, vehicles, machines, ships, buildings, and furniture with moderate requirements must be built,
- *tool steel* as a high-grade steel where particularly good material properties, especially in terms of hardness and toughness, are required for the manufacture of cutting and forming tools as well as drills and molds, and
- *stainless steel* as a very resistant steel, where components must be particularly resistant to aggressive media, e.g. containing chlorine, and salt water over a long period.

Structural steel – As the name suggests, this standard material in steel and mechanical engineering is used to build things, namely vehicles, machines, ships, buildings, and furniture. These low-cost steels with a carbon content between 0.05% and 0.60% are relatively low in carbon and can be cold-formed very well at room temperature. The mechanical properties – the tensile strength of structural steels is usually below 500 N/mm^2 – are moderate, but sufficient for many applications.

Tool steel – Here, too, the designation refers to the area of application addressed. High-quality tool steel, which must have a carbon content between 0.60% and 2.10%, belongs to the premium steels and shows high hardness, strength, toughness, and wear resistance. Thus, it is used for the production of, for example, cutting and forming tools as well as for drills and casting molds. Depending on the application, tool steels are offered as unalloyed and alloyed steels.

Stainless steel – Invented in 1912 by Friedrich Krupp AG in Essen, this material contains at least 10.5% chromium, which forms a protective and dense passive layer of chromium oxide on the surface of the material. The addition of other alloying elements such as nickel, molybdenum, manganese, and niobium provides even better corrosion resistance and/or mechanical properties. Due to the good formability of the corresponding sheets and especially due to the long-term resistance of the manufactured articles to environmental influences, stainless steel is very popular in industry and trade, despite the lower tensile strength of the material. For 100 years, the designation V2A (abbreviation of experimental melt 2 austenite) has been in use for stainless steel. This steel grade contained 18% chromium and 8% nickel, but is no longer produced in this composition. The designation V2A is still frequently used in industry today, as is the designation V4A for a chromium-nickel steel with a molybdenum addition of 2%, which is also no longer available today. In the case of V4A, the addition of molybdenum has made the material more resistant to corrosion in the presence of chloride-containing media. Today, alloys similar to V2A and V4A are used very successfully in applications with aggressive media, such as in swimming pools, in the chemical and construction industries, in the offshore sector, on tankers, and in the food and beverage industry.

7.2.2.4 Surface Improvement of Steel

To improve the properties of the steel surface, modification is often carried out by depositing an additional metal, either by electrochemical deposition or by physical processes. Electrochemical deposition can be done either by an external circuit (galvanic metal deposition) or electroless processes (chemical metal deposition). Physical metal deposition is carried out either by immersion in molten metal (as in hot-dip galvanizing, for example) or by various mechanical processes such as hammering, rubbing, rolling or also by applying the metal in a vacuum. Very often, the aim of such treatments is to improve corrosion resistance; however, when depositing chromium on the steel surface, the aim is to increase wear resistance.

The following surface modification processes are used to improve steel surfaces:

- zinc deposition,
- chromium deposition, and
- phosphate deposition.

7.2.2.4.1 Zinc Deposition

This surface treatment ensures corrosion protection of steel in numerous applications. In this process, the zinc virtually sacrifices itself to protect the iron from attack by aggressive media. If it comes into contact with such corrosion-promoting media, the zinc is preferentially dissolved and thus prevents the destruction (dissolution) of the steel. This effect is due to the lower position of zinc (−0.76 V) in the electrochemical voltage series compared to iron (−0.04 V). The deposition of zinc on steel

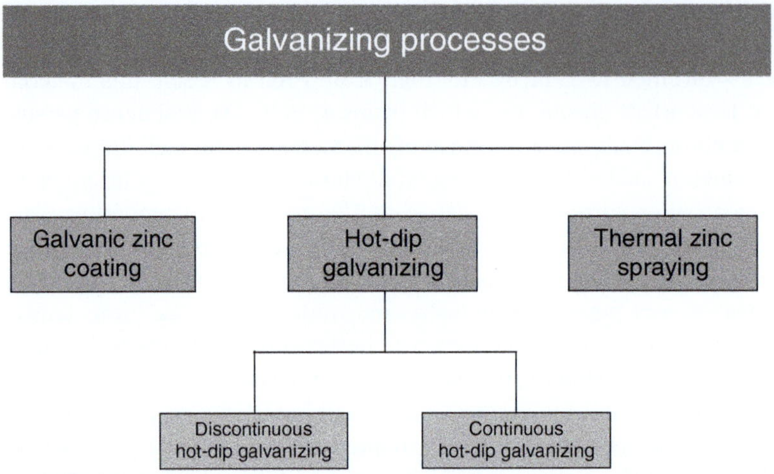

Figure 7.6 Galvanizing processes for the treatment of steel surfaces.

Table 7.1 Galvanizing processes and applications for galvanized steels.

	Thickness (μm)	Alloy with steel	Typical applications
Hot-dip galvanizing (discontinuous)	50–150	Yes	Outdoor metal construction (steel beams, balcony constructions)
Hot-dip galvanizing (continuous)	7–25	Yes	Steel sheets for indoor (cable ducts) and car bodies in automotive
Thermal zinc spraying	80–150	No	Steel construction, parts with complex structures
Galvanic zinc coating	3–25	No	Small articles for indoor use

is usually referred to as *galvanizing* and is carried out by various processes, which result in different properties of the zinc coating. The most important processes are continuous and discontinuous hot-dip galvanizing, electrogalvanizing, and thermal spraying of zinc (Figure 7.6, Table 7.1).

Galvanic Zinc Coating Here, the metal is deposited onto the steel surface with the aid of an electric current. The resulting zinc layers are significantly thinner than in hot-dip galvanizing, at approximately 3–25 μm, and depend on the required corrosion resistance. Unlike hot-dip galvanizing, electrogalvanizing does not allow alloy formation between the zinc coating and the steel. Therefore, this method is mainly used for temporary corrosion protection tasks in weakly corrosive environments and suitable for both large and small parts, such as screws or rivets, which are processed in baskets or drums.

Hot-dip Galvanizing This process is carried out by immersing the cleaned and hydrochloric acid–etched parts in a liquid zinc alloy bath at a temperature between 430 and 640 °C. The melting temperature can be reduced by small additions of foreign metals. This process allows the deposition of 50–150 μm thick zinc layers and thus complete corrosion protection of steel components such as staircase structures, which are first assembled and then galvanized. The hot-dip galvanizing process can also be carried out continuously, in which case the steel strip is immersed in the molten zinc, resulting in thinner layers of approximately 7–25 μm. In this case, the galvanized steel is a preliminary product, which subsequently takes the final shape by forming, punching, and cutting. A disadvantage of this method is that these process steps destroy the protective zinc layer at the edges. Because of their difference in corrosion protection, discontinuous and continuous (strip) hot-dip galvanized steel materials have different applications. Strip-galvanized steels are mostly used in slightly corrosive mostly interior areas, of which cable ducts or air-conditioning elements are typical examples. Discontinuously galvanized steel is used in outdoor areas, where protection times of several decades must be achieved. Here, this process has proven to be an extremely durable, robust, and maintenance-free corrosion protection.

Spray Galvanizing This variant of galvanizing is often used for steel parts with complex structures that cannot be galvanized using the other processes. Here, zinc is melted in a spray gun and sprayed onto the surface of the steel part. Unlike hot-dip galvanizing, spray galvanizing also does not allow any alloy formation between the zinc layer and the steel.

7.2.2.4.2 Chromium Deposition

This method is often used to improve the wear and corrosion resistance of heavily loaded steel materials and allows the electrochemical deposition (in aqueous phase) of about 0.2–1.0 μm thick chromium layers on the steel surface (cathode). The result is very hard steel surfaces, as required for rollers in the paper industry or metal processing or for tools in the textile industry. The layer thicknesses depend on the requirements of the parts in terms of wear. Due to the formation of a stable and inert chromium oxide layer, which is formed in air and passivates the surface, these materials have very good corrosion resistance. However, the presence of this newly formed protective layer is rather disadvantageous for the bonding process, since the non-reactive character of these surfaces allows only partial wetting and thus hinders the molecular interactions necessary for good adhesion formation.

7.2.2.4.3 Phosphate Deposition

In this process, phosphates are deposited from an aqueous phase on the steel surface by means of the electrochemical process of anodization. The resulting microscopic crystals increase the surface area of the metal and thus improve the adhesion of polymer materials – and thus also adhesives – to these surfaces. A major application of this surface coating is found in the automotive industry. Here, for the purpose of optimal temporary corrosion protection, the oil absorption capacity of steel sheets used for add-on parts such as doors and trunk lids is greatly improved by the unique topography of the phosphate-treated surfaces. At the same time, it is possible to produce very good adhesive bonds with these oiled sheets without removing the oil.

7.2.3 Aluminum and their Alloys

7.2.3.1 Introduction

Aluminum is a silvery-white light metal and the third most abundant element in the earth's crust after oxygen and silicon, accounting for about 8% by weight. However, because of its base nature, aluminum occurs almost exclusively in bound form, with the largest proportion found in aluminosilicates, which are constituents of clay and granite, for example.

7.2.3.2 Properties of Aluminum

In industrial applications, pure aluminum with at least 99.0% aluminum, ultra-pure aluminum with at least 99.7% aluminum, and especially aluminum alloys are used in numerous applications. Pure aluminum conducts electricity well, with electrical conductivity reaching about two-thirds that of copper, which allows this material to be used in high-voltage power lines. The coefficient of linear thermal expansion of aluminum is relatively high, reaching about twice the value of unalloyed steel. Thus, an aluminum structure expands twice as much as a steel structure when the temperature increases. The high thermal conductivity of aluminum allows the use of this metal in the manufacture of heat exchangers.

7.2.3.2.1 Strength of Aluminum

Pure, unalloyed aluminum has a density of about 2.7 g/cm^3, making it a typical *light metal* with a strength far below that of steel. Aluminum alloys, however, are capable of achieving mechanical strengths comparable to those of steel, but at only onethird of its density. Because of this favorable density-to-strength ratio, aluminum is an excellent material for today's requirements in modern lightweight construction, which is currently experiencing a major boom, especially in the automotive, aerospace, and marine industries. Commercially available unalloyed aluminum has a tensile strength of about 90 MPa, a yield strength of about 35 MPa, and an elongation at break of about 35%. However, by adding small amounts of alloying elements such as silicon, iron, copper, manganese, magnesium, zinc, titanium, and chromium, it is possible to significantly improve the mechanical and chemical properties of pure aluminum without losing its good conformability properties. For example, an increase in the yield strength ($R_{p0.2}$) of pure aluminum Al99.5 from 40 MPa to about 120 MPa for $AlMg_3$ and to about 300 MPa for $AlCuMg_2$ can be achieved (Table 7.2).

7.2.3.2.2 Resistance of Aluminum

Although aluminum is a base metal with a standard potential of −1.67 V, it is very resistant to many corrosive media. The reason for this is a very thin oxide layer, up to about 0.05 μm thick, on the surface of the aluminum, which forms spontaneously

Table 7.2 Physical and mechanical properties of aluminum.

Property	Unit	Al 99.5	$AlMg_3$	$AlCuMg_2$
Solidification range	°C	650–660	610–640	640
Tensile strength	MPa	65–145	190–350	400–450
Yield strength $R_{p0.2}$	MPa	20–85	80–130	270–310
Elongation at break	%	3–35	10–20	10
E-Modulus	GPa	70	70	73
Heat conductivity	W/mK	210–220	140–160	120
Coefficient of linear thermal expansion	$10^{-6} \cdot 1/K$	23.5	23.9	23.1

in the presence of oxygen. This passivation layer ensures that aluminum is highly resistant to corrosion under normal environmental conditions, the degree depending on the thickness of the oxide layer. Because of their excellent resistance to corrosive media, aluminum and its alloys are widely used in the construction, chemical, automotive, and shipbuilding industries.

7.2.3.3 Applications of Aluminum

According to the method of processing, aluminum alloys are divided into casting and forging alloys. Because of their low viscosity in the melt and their low shrinkage, casting alloys are commonly used in the automotive industry for the production of, for example, engine blocks, pistons, or wheel rims. In contrast, forged aluminum alloys produced by forming processes are often found as strips, sheets, tubes, rods, and profiles. Pure aluminum, with a tensile strength of about 65–145 MPa, is weldable and highly corrosion resistant. The material is used as aluminum foil, in the chemical industry for tanks and pipes, and in applications where high electrical conductivity is important. Aluminum–magnesium alloys such as $AlMg_3$ with a tensile strength of 190–350 MPa are medium- to high-strength weldable materials used in shipbuilding, transportation, pressure vessels, bridges, and building construction. However, alloys containing more than 3% magnesium are not suitable for applications at temperatures above 65 °C, as they are prone to stress corrosion cracking. Aluminum–copper alloys such as $AlCuMg_2$ have high tensile strengths in the range of 400–450 MPa and are therefore used in aircraft construction and aerospace engineering. However, these materials are susceptible to corrosion and are generally difficult to weld.

7.2.3.4 Surface Improvement of Aluminum

The up to about 0.05 μm thick oxide layer on the aluminum surface already provides good corrosion resistance. However, if higher corrosion resistance is required, this can be achieved by an anodic treatment process. Here – in contrast to the galvanic coating processes – the protective layer is not built up on the surface, but is created by converting elemental aluminum in the uppermost metal layer into aluminum oxides and aluminum hydroxides. The resulting 5–25 μm thick layer is much harder than the underlying untreated aluminum and thus protects the material against mechanical or weathering influences, but only until no cracks appear in this layer. The adhesion of this layer to the underlying elemental aluminum is exceptionally good, since the newly formed protective layer is the result of transformation of the basic aluminum material. However, not all aluminum alloys are suitable for treatment with this process. On pure aluminum or aluminum alloys containing magnesium or zinc, stable and uniform oxide layers are formed, while the presence of the alloying elements copper and iron negatively affects the quality of the coating. Relatively thick (20–25 μm) protective coatings are produced on the base aluminum material for applications in the construction industry and for items used in automotive, aerospace, and household. Colored and colorless thinner coatings from 8 to 20 μm serve both decorative purposes (e.g. in door hardware assembly) and for better heat dissipation in heat sinks.

7.2.4 Plastic Materials

7.2.4.1 Introduction

Today, it is impossible to imagine our daily lives without plastics, which are used in countless applications. Many products have proven their worth over decades and are used in a variety of ways, for example as extremely inexpensive *plastic articles* for everyday use or as *high-performance plastics* (possibly fiber-reinforced) for *high-tech* applications with very good temperature resistance and high strength. Plastic materials have a very different structure from metals, glasses, and ceramics, resulting in different properties and applications. The discussions of the characteristics and properties of plastic materials in this subsection refer to both plastic substrates and the adhesive used in bonding technology.

7.2.4.2 Properties

Due to their relative loose structure, plastics have a very low density of 0.9–1.4 g/cm^3 and compared to metals much lower strengths moduli of elasticity (Table 7.3). However, by stretching the material or reinforcing it with fibers, the mechanical properties can be significantly improved, resulting in lightweight materials with very good strength-to-density ratios. Compared with other materials, plastics have excellent damping properties in the event of mechanical vibrations. In general, the temperature resistance of plastics is rather low, some fail already at 80 °C or become soft; others, however, are long-term resistant at 200 °C. For all plastics, there is a decomposition temperature at which the long polymer chains break and the mechanical properties are greatly reduced. At this point, the material no longer meets the requirements of the application.

If we consider the influence of deformation on the strength of plastic materials, they can be divided into brittle, tough-hard, and soft-elastic (Figure 7.7). They are electrical insulators, but can be made electrically conductive by incorporating metal powders or carbon black into the polymer matrix. However, the thermal conductivity of such modified plastics is significantly lower than the conductivity of

Table 7.3 Mechanical properties of plastic material compared to metals.

Material	Density (g/cm^3)	*E*-Modulus (GPa)	Tensile strength (MPa)	Elongation at break (%)
High-density PE (HDPE)	0.95	0.7–1.4	18–35	100–1000
Low-density PE (LDPE)	0.92	0.2–0.5	8–23	300–1000
Polystyrene (PS)	1.02	3.2–3.3	45–65	3–4
PVC (non-plasticized)	1.38–1.55	1–3.5	50–75	10–50
PA 6	1.13	1.4	70–85	200–300
Construction steel	7.80–7.85	210–215	240–830	18–26
Aluminum	2.65–2.80	60–78	45–710	2–60

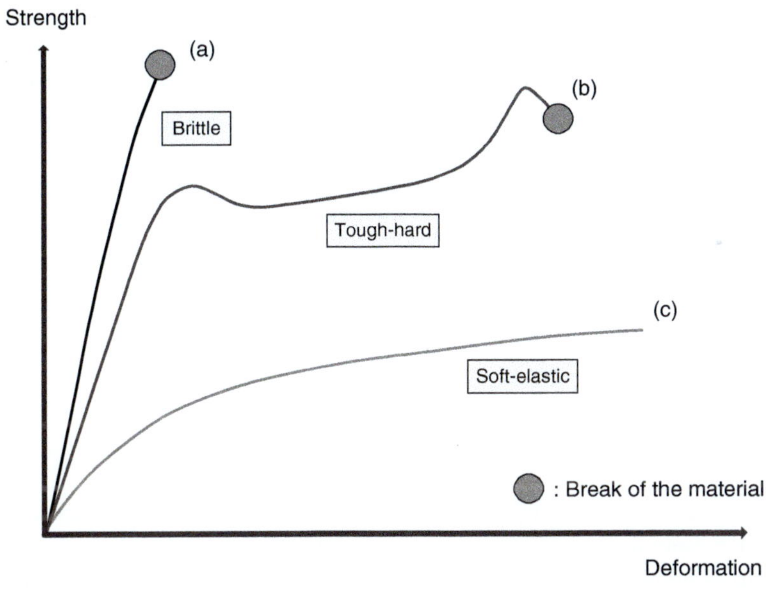

Figure 7.7 Influence of deformation on strength in the case of brittle (a), tough-hard (b), and soft-elastic (c) plastic materials.

metals. A characteristic feature of many plastics is their swelling capacity after coming into contact with organic solvents or water. This reduces the bonding forces between the molecular chains, which has a negative effect on the mechanical properties. However, this effect is used positively in the plasticization of relatively hard plastics, such as PVC, in which special low-molecular organic substances – so-called plasticizers – are added to the polymer. The swelling behavior of plastics is very often an advantage when bonding them with adhesives and, by exploiting this special property, often makes it possible to bond these materials in the first place.

7.2.4.3 Chemistry of Plastic Materials

Plastic materials are synthetic products formed by a chemical reaction called polymerization. The resulting products – the polymers – are the basis for plastic materials. The word polymer is composed of the ancient Greek words poly (= many) and meros (= part). A polymer is therefore something that consists of many parts and can be described as a chain of individual molecules called monomers. Such a chain can consist of several thousand molecules.

If such *macromolecules* consist of only *a few* monomers, they are called oligomers, which are described in more detail as:

- *dimers* – consist of two,
- *trimers* – consist of three, and
- *tetramers* – consist of four units.

The basic requirement for successful polymerization is that the molecules (monomers) can be bonded together. This can only be fulfilled with three elements: carbon (C), silicon (Si), and phosphorus (P); in practice, however, only carbon and silicon play a role. In addition to the *chain-forming* elements carbon and silicon, a polymer designer usually only needs hydrogen (H), oxygen (O), nitrogen (N), sulfur (S), chlorine (Cl), and fluorine (F) in the laboratory to produce a great variety of polymers. Characteristic of polymers, which form the basis for plastics, is their high-molecular weight. They can be linear, branched, or cross-linked and can be disordered (amorphous) or ordered (crystalline). Chemically, there are many possibilities of variation.

7.2.4.3.1 Synthesis of Polymers

Three methods have been developed for the synthesis of polymers:

- polymerization,
- polyaddition, and
- polycondensation.

Polymerization One of the most important types of reaction through which plastics can be formed is polymerization. This is a sequence of reactions – known as a growth chain reaction – in which simple starting materials (monomers) are combined with one another via reactive double bonds. In the strongly exothermic reaction, no by-products are split off. In this method, which is correctly called *chain polymerization*, the starting material is a monomer whose carbon atoms are linked by a double bond, such as in the case of the monomer ethylene. This double bond is broken with the aid of an initiator, often a thermally unstable peroxide, which breaks the double bond of the ethylene molecule, for example, and its single free electron forms a stable electron pair with one electron of the double bond. The other electron of the double bond looks for the next monomer and breaks the double bond there, and so on. Thus, theoretically, the reaction is not complete until there are no more monomers with double bonds. The polymerization method is called a chain growth reaction because the free, unreacted monomers are added to an already-existing chain. The disadvantage of this method is that long macromolecules hinder each other's growth and shield the monomers that are still free. Consequently, the reaction is stopped even though significant amounts of free monomers are still present. Important plastics produced by polymerization are PE, PP, PVC, PS, PTFE, PMMA, and polyacrylonitrile (PAN).

Polyaddition In polyaddition, monomers containing two or more functional groups react with each other to form macromolecules, with a hydrogen atom being rearranged during bond formation. It can also happen that two already-formed longer polymer chains combine to form an even longer one. Thus, in the polyaddition method, molecules come together – they add up – which is in contrast to polymerization, in which more and more monomers are added to an existing chain. In contrast to polycondensation, no reaction by-products are formed in this process. Important polymers produced by polyaddition are epoxy resins and polyurethanes.

Polycondensation Just like polyaddition and polymerization, polycondensation is an important type of reaction used to produce plastics. Besides, this reaction has great importance in nature. In the polycondensation reaction, macromolecules are formed from many monomers by splitting off smaller molecules, usually water. Because we are dealing with a so-called equilibrium reaction and the synthesis of the polymer should be as effective as possible meaning that as much plastic as possible should be produced with the monomers used, the released water is continuously removed during the synthesis. This constantly shifts the equilibrium reaction to the right side of the equation and thus maximizes the amount of material produced.

7.2.4.3.2 Structure of Polymers

Polymer chains can be built up from only one type of monomer, such a substance is then called a homopolymer. However, they can also be built up from two (or even more) different monomers, which are then called copolymers or heteropolymers. For example, the plastics PP, PP, and PVC are homopolymers and ABS or styrene-acrylonitrile (SAN) are copolymers.

The monomers can be arranged in the copolymers as follows:

- *statistical copolymer* – the two monomer species are irregularly distributed,
- *gradient copolymer* – along the chain, the number of monomers of one species increases and the other decreases,
- *block copolymer* – each monomer species remains among itself; one section in the polymer chain consists only of one species, the other section only of the other one,
- *alternating copolymer* – the two types of monomers are evenly alternated, and
- *graft copolymer* – one type of monomer is grafted and attached to the chain consisting of monomers of the other type (in itself forms a homopolymer).

The polymers without side chains – such as PE – are called linear polymers and the polymers with side chains – such as PP – are called branched polymers. The ethene monomer is symmetrical and thus the corresponding PE chain is also symmetrical. However, the propene monomer contains a CH_3 group at one position instead of a simple hydrogen atom and after polymerization, this group is positioned off the PP chain, resulting in a non-symmetrical polymer. This CH_3 group can now be positioned in three different ways in the polymer chain. The generic term for this variety is tacticity and in some plastics, only one of the three forms is present.

However, there are other plastics, such as PP and PS, which have all of the following three forms:

- *atactic* – the side groups are randomly distributed,
- *isotactic* – the side groups are all on one side of the main chain, and
- *syndiotactic* – the side groups are arranged alternately on both sides of the main chain.

Tacticity has a significant effect on the properties of the polymer. The more evenly the side groups are distributed, the more likely the polymer will be semi-crystalline. Otherwise, it becomes amorphous. In the crystalline regions, the molecular interactions between the polymer chains act more strongly than in the amorphous regions. Polymers with atactic side groups are practically always amorphous and generally transparent. They are softer and more elastic and begin to flow at lower temperatures than semi-crystalline polymers. Syndiotactic and isotactic polymers are harder and stiffer, but also more brittle and become liquid at higher temperatures. Therefore, they can withstand a higher temperature of use. However, there are also some disadvantages. For example, pure isotactic PS takes too long to crystallize and would therefore not be useful for industrial use. PP, on the other hand, can only be used technically in isotactic form, because only in this form it represents a plastic material with suitable strength.

7.2.4.4 Thermomechanical Properties of Plastic Materials

The most interesting features of a material, from which a component is to be made, are certainly the thermomechanical properties. Constructing such a component it is essential to know how much pull or pressure does the material withstand and how does the component deform when an external force under temperature load is applied. Depending on the structure and arrangement of the polymer chains and the degree of crosslinking in the material, different chemical, physical, and mechanical properties arise. The essential criterion here is the mechanical behavior under temperature loads.

According to their thermomechanical properties, plastic materials can be classified into three groups (Figure 7.8):

- *thermoplasts,*
- *thermosets*, and
- *elastomers*

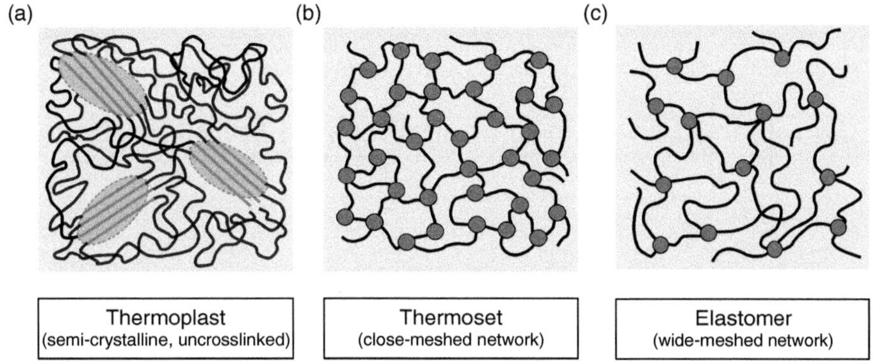

Figure 7.8 Structure of thermoplasts (in this case semi-crystalline) (a), thermosets (b), and elastomers (c).

7.2.4.4.1 Thermoplasts

These polymers are not crosslinked and the chains are held together by weak intermolecular forces. These are usually based on van der Waals forces, which are composed of dipole–dipole, induced dipole, and dispersion force components. Such linear or branched polymers, which can appear in their amorphous and semicrystalline form are liquefied relatively easily by heating and after subsequent cooling, the van der Waals forces are re-activated, and the material solidifies. Thus, these polymers are able to go through reversible state changes. Because this type of polymer becomes plastically deformable when heat energy is added, it is called a thermoplastic. Depending on the arrangement of the polymer chains, thermoplastic polymers are amorphous, as in PS, PMMA, and PC, or semi-crystalline, as in PE, iso- and syndiotactic PP, PA, and PET.

7.2.4.4.2 Thermosets

If the polymer chains are chemically bonded together via many small molecular bridges, a strong, hard network is formed. Such a strongly cross-linked polymer is called a thermoset, indicating that this material cannot be melted once these bridges are formed in the polymer. Because of the highly limited mobility in the polymer structure, they are present in a rigid, sometimes brittle, amorphous state. Unlike thermoplastics, the thermosets are insoluble in virtually all organic solvents.

7.2.4.4.3 Elastomers

If the polymer chains are only slightly cross-linked (wide-meshed), they can still move among themselves very well. There are only a few networking points, which are relatively far apart. The same applies here as with thermosets: once the network has formed, it can no longer be dissolved. Such materials are called elastomers, which can be deformed reversibly regardless of temperature. However, they are not able to flow below the decomposition temperature.

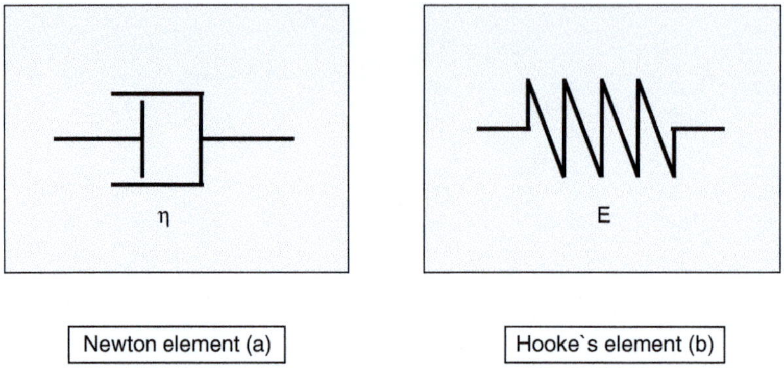

Figure 7.9 Viscous behavior represented by the Newton element (a) and linear-elastic behavior represented by the Hook element (b).

7.2.4.5 Viscoelastic Properties of Plastic Materials

In general, plastic materials creep under load. For example, a plastic rod becomes longer and longer even though the force pulling on it remains constant. However, the loaded plastic part often returns completely to its original shape after the load is removed. Thus, the material behaves a bit viscous and a bit elastic.

7.2.4.5.1 Viscous Behavior

This feature of a material is caused by fluid friction and here the relationship between deformation and force is velocity dependent. The faster the deformation, the greater the resistance. The energy expended is completely *dissipated*. Since energy cannot be destroyed, of course, it is rather dissipated and converted into heat. For the user, it is lost, and after relief, the plastic part remains as it was deformed. Fluid friction is symbolized by a shock absorber, as used in motor vehicles. When an external load is applied, the piston is pushed into the cylinder and the oil in the cylinder space is forced through small holes in the piston into the space behind it. In honor of Isaac Newton, who was the first to describe the mathematical relationships of fluid friction, this type of friction and its symbol (Newton element) were named after him (Figure 7.9a). Characteristic of fluid friction is the dynamic viscosity η, where the rate of deformation depends on the external stress applied to the part. This motion lasts as long as the force acts and the body continues to deform. If the force no longer is applied, the part retains the deformation and there is no return to the original shape.

7.2.4.5.2 Linear-Elastic Behavior

This behavior of materials is also called Hooke's behavior. In this case, the deformation is proportional to the applied force. The higher the force, the more the

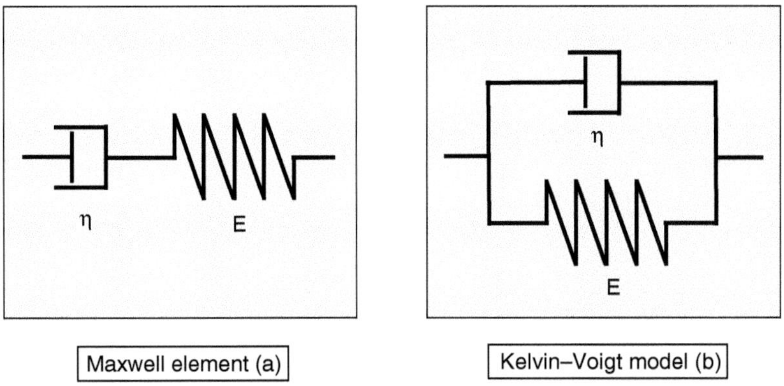

Figure 7.10 Viscoelastic behavior represented by the Maxwell model (a) and the Kelvin–Voigt model (b).

body deforms. If the force is completely removed, the deformation also disappears completely. This behavior is symbolically represented by a highly simplified helical spring called the Hook element (Figure 7.9b). The measure of Hooke's behavior is the modulus of elasticity *E*. In such systems, there is a linear relationship between deformation and applied force. The energy expended for deformation is fully released during recovery, and the loaded part returns completely to its original shape. This is the usual behavior of metals as long as one remains in the linear-elastic range of deformation.

However, plastics do not behave linearly elastic like metals, but mostly viscoelastic. This behavior is usually described by the models according to:

- *Maxwell*,
- *Kelvin–Voigt*, and
- *Burger*.

Maxwell Model This is the simplest combination conceivable for illustrating viscoelastic behavior consisting of a series connection of a Newton (damper) and a Hook element (spring) (Figure 7.10a). The damper is not able to follow a short-time, fast-load application and behaves like a quasi-rigid body in this situation. The deformation work is completely absorbed by the spring and can be completely released again. However, if the force is applied for a longer period of time, the shock absorber begins to dip and the body continues to deform, even though the load is not increased. If the load is then removed, the body cannot fully recover its original shape. In summary, when the force is applied for a longer period of time, the Maxwell element deforms and does not return to its original position when the load is removed. The energy used for the viscous flow (η) is completely dissipated.

Kelvin–Voigt Model Unlike the materials represented by the Maxwell model, there are also materials that may very well return completely to their original shape after the load is released. For this reason, Kelvin and Voigt developed a different model,

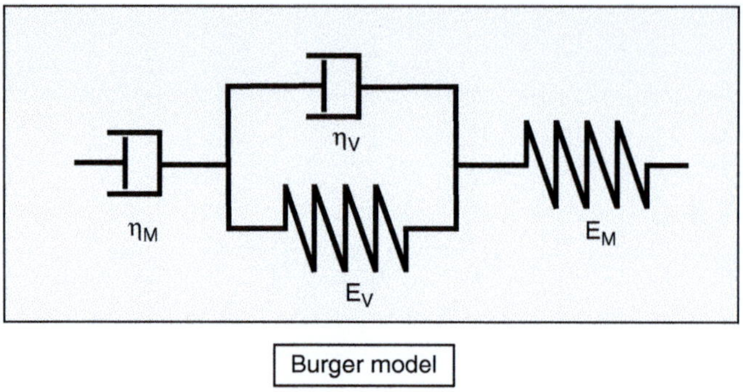

Figure 7.11 Viscoelastic behavior represented by the Burger model.

called the Kelvin–Voigt model or just the Voigt model (Figure 7.10b). Here, the damper and spring are connected in parallel, and the force stored in the spring can push the damper element back to its original position after the load is released. This model also shows how the suspension of cars works. The spring provides the elastic deformation and the damper absorbs the vibrations. Since the damper is connected in parallel with the spring, the faster an external load is applied, the stiffer the Kelvin–Voigt model behaves. If the speed of the force application is very high, this model behaves quasi-rigidly and does not deform at all. Such materials also exist, but most of them definitely deform no matter how fast the force is applied. In summary, when the force is applied for a longer period of time, the Kelvin–Voigt model deforms and returns to its original position when the force is removed. The energy used for the viscous flow (η) is completely regained.

Burger Model There are also materials that deform no matter how fast the load is applied, but do not completely return to their original position after the load is removed. To understand and calculate this behavior, Burger connected the Maxwell model in series with the Kelvin–Voigt model, which is basically a Kelvin–Voigt model connected in series with a spring and a damper (Figure 7.11). When a load is applied quickly, the Maxwell spring (E_M) deforms and the rest behaves in a quasi-rigid manner. When the model is unloaded, it returns completely to its initial position. If the load lasts longer, the two dampers start to move and the material begins to creep. If the model is subsequently unloaded again, the damper in the center, the Voigt damper (η_V), is pulled back to its initial position by the parallel spring E_V. The Maxwell damper (η_M), on the other hand, remains in its deformed position even after unloading. In summary, when the force is applied for a longer period of time, the Burger model deforms and does not return to its original position when the force is removed. The energy used for the viscous flow is partly regained (η_V) and partly dissipated (η_M). This behavior most closely matches the real-world behavior of polymeric materials and must be strongly considered in the selection of polymeric substrates and adhesives used in bonded components.

7.2.4.6 Plastic Materials Used in Industry and Craft

7.2.4.6.1 PE

The polymer is semi-crystalline, non-polar and by suitable choice of the synthesis process, catalysts, etc. it is possible to produce not only plastics, but also waxes, fats, and even oils.

According to density, PE is divided into

- *low-density (LDPE)* with density from 0.91 to 0.93 g/cm^3, 45% crystallinity, and melting point of 110 °C,
- *linear-low density (LLDPE)* with a density of 0.93–0.94 g/cm^3 and low crystallinity, and
- *high-density (HDPE)* with a density of 0.94–0.96 g/cm^3, 70% crystallinity, and a melting point of 135 °C.

PE in the original form is not UV resistant and can be made resistant by adding carbon black. Strength, hardness, and stiffness are relatively low, but ductility and impact strength are very high. The material exhibits good frictional behavior and is therefore relatively resistant to abrasion. The continuous service temperature is 60–85 °C, but strong creep deformation already starts at room temperature under continuous load. However, creep can be reduced by fillers such as short fibers. PE expands strongly at elevated temperatures, and above 290 °C the material begins to decompose. The polymer is flammable and continues to burn after ignition; the flame does not extinguish. PE is characterized by low water absorption and low water vapor permeability, but high permeability to oxygen and carbon dioxide. It is resistant to mineral lubricants, aliphatic hydrocarbons, gasoline, weak and strong acids and alkalis, as well as perchloroethylene, acetone, alcohols, and hot water. PE is moderately resistant to aromatic hydrocarbons, oxidizing acids, and trichloroethylene.

7.2.4.6.2 PP

This material is so successful because it is relatively inexpensive and also very easy to process. It is the polymer with the lowest density (0.90–0.92 g/cm^3), which allows more parts to be made from the same mass of material. PP can be blended with mineral fillers such as talc, chalk, or glass fibers, which greatly expands the range of its mechanical properties. As a result, PP is increasingly displacing more expensive plastics such as PAN, ABS or PA, making it the plastic whose market share is growing fastest. Chemically, PP is very similar to PE. It is also non-polar and is mainly used in the semi-crystalline form. The polymer can be divided into *isotactic, syndiotactic,* and *atactic.* In *isotactic PP*, the side groups are all on one side of the main chain, so the polymer chains are closely linked. As a result, *isotactic PP* has a high degree of crystallinity (60–70%) and its melting point is between 160 and 165 °C. Due to its better properties, *isotactic PP* accounts for the largest share of the total production volume of PP and only this type can be processed into fibers. This material is slightly harder, stiffer, and more heat stable than PE. The operating temperature is up to 100 °C and the material has the highest hardness of all polyolefins. Like PE, however, it offers no UV resistance and is not weather-resistant, which is a limitation for outdoor use. An improvement is possible here by adding 2–2.5% graphite to the base polymer. The thermal expansion of this *isotactic PP* is very high, but lower than that of PE.

Additional features of *isotactic PP* are:

- high ductility,
- good abrasion resistance,
- flammability (continues to burn after ignition, the flame does not go out),
- excellent dielectric properties,
- low water absorption,
- good resistance to mineral lubricants, gasoline, aliphatic hydrocarbons, weak and strong acids and alkalis, acetone, alcohols, and hot water, as well as
- conditional resistance to aromatic hydrocarbons, oxidizing acids, trichloroethylene, and perchloroethylene (PP softens on contact with these substances).

Syndiotactic PP is slightly less crystalline and therefore less rigid than *isotactic PP*, but is no longer available on the market. *Atactic PP* is amorphous and has a lower density. It is rubbery, tacky even at room temperature, and becomes liquid at 127 °C. This polymer is used in niche applications such as potting compounds for electrical components, sealants or insulating materials, and as additives for bituminous compounds.

7.2.4.6.3 PVC

This material is amorphous and – without the addition of plasticizers – hard and brittle. After polymerization, it is present as a porous, white powder. Its processing temperature, e.g. during injection molding, is higher than the decomposition temperature of the material, so that in high temperature processing of PVC suitable thermal stabilizers must be added. For softening, the polymers are compounded with plasticizers, resulting in two types of PVC: *unplasticized polyvinyl chloride* (PVC-U), with a plasticizer content of 0–12% and *plasticized polyvinyl chloride (PVC-P)*, with a plasticizer content of <60%. PVC can be translucent to opaque or crystal clear. The material is resistant to light and UV radiation, flame retardant, and self-extinguishing when no longer flamed. It is resistant to salt solutions, seawater, acids, alkalis, ethanol, greases, oils, and gasoline. PVC is not resistant to acetone, ether, benzene, chloroform, concentrated hydrochloric acid, and aromatic hydrocarbons. It is also not resistant to hydrolysis. *PVC-U* is processed, for example, into window profiles, roller shutters, and pipes, and also credit cards are made from this material. In foamed form, it is used as display panels for advertising purposes. *PVC-P* is used to make roofing films, cable sheathing, and flooring. Other applications include cover films for agriculture, swimming pool linings, or transparent swing doors in supermarkets. As thin films, PVC is used for tablecloths, shower curtains, and self-adhesive films.

7.2.4.6.4 PS

An important property of PS is that it has very low thermal conductivity. However, the material does not like heat and accelerated aging begins already at 55 °C. PS is resistant to alkalis and mineral acids, but is not resistant to non-polar solvents such as gasoline. It does not show good UV resistance, and embrittlement also occurs under the influence of visible light. The polymer is physiologically safe and is the only plastic material approved for the storage of raw meat or fish.

There are other features of PS that strongly recommend the material for food packaging:

- allows stability of odor and taste,
- shows resistance to fats,
- demonstrates very low water absorption, and
- provides dimensional stability of the parts up to 80 °C.

This thermal stability of the material makes it suitable for hot-filling processes in food packaging, as well as for transporting hot coffee filled in cups.

Basically, there are four different types of PS on the market:

- *general-purpose polystyrene (GPPS),*
- *foamed polystyrene (PS-E),*
- *impact-resistant polystyrene (PS-I),* and
- *cross-linked polystyrene (PS-X)*

General-purpose PS (GPPS) GPPS is not foamed, amorphous, and therefore transparent. It is characterized by good electrical insulation and low water absorption as well as high stiffness and low impact strength. The material is mainly used for short-life products such as packaging materials or disposable tableware. However, the production of these mass-produced items benefits from the fact that the GPPS shows very good melt characteristics and can therefore be injection molded very well. Thus, its efficiency in processability is thus unmatched by any other plastic. In addition, the high stiffness enables low wall thicknesses and thus low material consumption as well as low weight of the packaged parts. Since GPPS is highly resistant to gamma radiation, medical devices, e.g. syringes that need to be sterilized, are often packaged with this material.

Foamed PS (PS-E) The best-known application of PS is probably in foamed form, with the largest amount used in construction for good thermal and sound insulation. However, PS-E is also used as a load-distributing sub-base for roads on non-load-bearing soils. As thermal insulation, PS competes with mineral wool and PU foams, with PS insulating about as well as mineral wool but worse than PU foam. This is because the air in the pores of PS foam and mineral wool conducts heat better than the cyclopentane gas present in PU foam. Other applications of PS-E can be found in the construction industry, where the material is used for insulating plasters, lightweight mortars, and concrete. Here, the PS-E is mixed into the base compound in the form of small particles. Widely known and used are applications in packaging, not only in the form of chips for filling cardboard boxes, but also tailor-made for packaging of televisions and computers, for example. The reason why PS-E is used as a packaging material is that it can absorb a great deal of impact energy. This property can even save lives, for example, the padding layer in bicycle and motorcycle helmets is made of PS-E, as are the seat shells for children in cars. Because foamed PS is very lightweight, there are other life-saving applications as a filler for life jackets and life rings, as well as buoys, surfboards, and buoyancy aids.

Impact-resistant PS (PS-I) If somewhat higher-quality components are to be made of PS, PS-I is used, with about 40% of these applications being in the packaging sector. However, the interior walls of refrigerators are also made from PS-I, as are injection-molded articles for electronics and office supplies.

Cross-linked PS (PS-X) Cross-linking PS produces a thermoset that is insoluble in almost all solvents due to the dense network. This makes PS-X the ideal material for ion exchangers, e.g. for water treatment, as it does not dissolve in the electrolyte.

7.2.4.6.5 PET

This polymer is polar and therefore there are high bonding forces between the polymer chains. These chains are linear in structure and therefore partial crystallinity is possible.

There are two types of PET available in the market:

- *APET*, which is amorphous and transparent, and
- *CPET*, which is semi-crystalline (up to 70%) and white.

CPET has high strength and dimensional stability up to 80 °C, but low-impact strength. Both grades are characterized by very low friction and very good wear resistance.

PET provides the following properties:

- low shrinkage and thus high dimensional stability,
- high flammability,
- good resistance to dilute acids, oils, greases, and alcohol,
- no resistance to strong inorganic acids, e.g. sulfuric, nitric, or hydrochloric acid, ketones, hot water, and alkalis, and
- low permeability to oxygen and carbon dioxide.

The latter characteristic property of PET is the reason why bottles for carbonated beverages are made of PET. Originally, PET was only processed into fibers and the textiles made from it are crease-resistant, tear-resistant, and weatherproof. Since they absorb very little water, PET textiles are very popular for sportswear and, because of their weather resistance, are very suitable for awnings, tarpaulins, tents, or flags. Technically, they are used for conveyor belts, drive belts, tire cord, fire hoses, ropes, and safety belts.

PET films are mainly used for food packaging, but also as a carrier for various films. More technically demanding applications include films for radiography and membranes as well as metallized films used as rescue blankets. These keep the casualty warm, but are much lighter and take up much less space than a wool blanket.

Perhaps the best-known use of PET is as a bottle material, where the polymer competes with PVC and PC. Compared to PVC, PET has two outstanding advantages: like PVC it does not release substances that are harmful to health and it is easy to recycle. Compared to PC, it is more resistant to breakage. To ensure that PET not only allows little gas to pass through, but also keeps beverages fresh, an additional barrier layer, usually made of silicon dioxide, is applied. PET is also used to manufacture parts by injection molding such as bearings, rollers, gears, cam discs, plugs, bolts, and nuts. The polymer is selected for this purpose because of its low friction, wear resistance, and high dimensional stability.

7.2.4.6.6 PC

Due to its relatively high price, PC is little used. The transparent and colorless polymer is classified as amorphous because the degree of crystallinity is less than 5%.

PC demonstrates the following properties:

- resistant to UV radiation and weathering,
- impact resistant and tough down to −40 °C,
- can be used up to 120 °C,
- has a tendency to stress cracking,
- does not continue to burn after ignition,
- has good resistance to aliphatic hydrocarbons, weak mineral acids, and weak organic acids, and
- is not resistant to aromatic hydrocarbons, gasoline, oxidizing acids, weak and strong alkalis, trichloroethylene, perchloroethylene, acetone, and hot water.

Immediately after polymerization, PC is extruded from the solution into fibers or films; for injection molding, it must first be processed into pellets. The material can be welded, printed, bonded, and also coated with metal and offers an even more scratch-resistant surface than the other polyesters. It is also crystal clear, which is why it is used for CDs, DVDs, and glasses such as lenses for binoculars and microscopes. When used for safety glasses, the material offers the advantage of high-impact resistance and thus break resistance. Also, the diffuser of car headlights and the colored windows in traffic lights are made of PC and for aircraft windows, underwater housings for cameras, flash units for photography and projectors the material allows good light transmission. Because there are no excess monomers left in the polymer after polymerization (unlike PVC), PC is food-safe and accordingly, microwave dishes, baby milk bottles (also because of breakage), or coffee filters are made from it. In mechanical engineering, PC is used for sight glasses, protective hoods, and also cam discs.

7.2.4.6.7 PVAC

The material is amorphous and therefore transparent. It offers high light and weather resistance, is combustible but flame retardant. Because its resistance to chemicals is low, PVAC is often copolymerized with other monomers for improvement. Over time, the polymer converts to polyvinyl alcohol, making it more hydrophilic and sensitive to water. PVAC is insoluble in water, butanol, aliphatic hydrocarbons, and soluble in lower alcohols, ketones, esters, ethers, and aromatic and chlorinated hydrocarbons.

In the form of a solution in organic solvents or as a dispersion in water, PVAC is used as a binder for the following applications:

- glues for wood,
- general-purpose adhesives (as a 40% solution of PVA in acetone/methyl acetate),
- paper coatings,
- textile impregnation, and
- carpet backing coatings.

PVAC is also used to modify concrete and as a component of chewing gum.

7.2.4.6.8 ABS

This amorphous thermoplastic terpolymer is colorless to gray and consists of a SAN matrix into which butadiene rubber particles are incorporated. Thus, the SAN provides stiffness and hardness, while the butadiene provides impact strength.

The proportions in ABS vary as follows:

- *acrylonitrile:* 15–35%,
- *butadiene:* 5–30%, and
- *styrene:* 40–60%.

ABS is resistant to mineral lubricants, gasoline, weak acids, weak and strong alkalis, and hot water, and is not resistant to strong mineral acids, oxidizing acids, and trichloroethylene. The plastic is suitable for processing by injection molding, extrusion, and thermoforming. It can be easily painted and printed, welded, bonded with MEK, and coated with metals as well as other plastics. ABS is used in the automotive and electrical industries for many applications and also for household products, toys (Lego bricks), motorcycle helmets, and as a material for 3-D printing.

7.2.4.6.9 PMMA

This plastic material is used to a relatively large extent for LED flat-panel displays. It is amorphous and therefore transparent (clear), has high scratch resistance, and can be used as a glass substitute. It even transmits light better than clear glass, but absorbs UV light, an important feature of PMMA sunglasses. The continuous-use temperature is 0–80 °C and the glass transition temperature is about 105 °C. This means that – at room temperature – PMMA *works* below its glass transition temperature and thus in a *frozen* state. PMMA becomes soft and plastically deformable above 100 °C, a feature allowing the production of curved surfaces for hoods, for example, without the need for an expensive mold. The polymer is resistant to moderately concentrated acids and alkalis as well as gasoline and oils, but not to polar solvents such as acetone or alcohols. After its market launch in 1933, this material found its first large-scale application for windows installed in cockpits of military aircraft.

Today, PMMA has found a much wider use, for example:

- for household items, e.g. bowls, salad servers, salt and pepper mills,
- in the automotive industry for turn signal and taillight lenses, but also for A-, B-, and C-pillar cladding,
- in the building industry for glazing,
- in aviation for windows, but also for headlight covers,
- in mechanical engineering for protective hoods, and
- in medicine for ear molds, bite splints, and eyeglasses.

Other applications include displays for illuminated advertising, photo coatings for the manufacture of printed circuit boards and watch glasses.

7.2.4.6.10 PA

This engineering polymer occurs in both the semi-crystalline and amorphous variants and exhibits significantly better mechanical and thermal properties than previously discussed standard thermoplastics. The two types PA 6 and PA 6.6 are frequently used in industrial applications, which differ chemically. PA 6 is derived from ε-aminocaproic acid and PA 6.6 from adipic acid and hexamethylene diamine. Thus, only one type of monomer is used in the polymerization to PA 6, while PA 6.6 uses two types of monomers: hexamethylenediamine and adipic acid. The number 6 indicates the number of carbon atoms in the monomers involved. Physically, there are no differences between the two versions. Both exhibit excellent strength and toughness, but absorb a great deal of water (up to 3.7%). Other polymers remain well below 1%.

Polyamides are polar and generally show good resistance to weathering and media such as dilute alkalis, aliphatic and aromatic hydrocarbons, fuels as well as alcohols, esters, ketones, fats, and oils. They are not resistant to strong acids and oxidizing chemicals such as hydrogen peroxide. They are processed by injection molding and extrusion and can then be welded, bonded, and painted. Since 3–4% humidity is required for optimum impact strength, PA parts that are very dry after thermal processing must be stored in warm water or a humid atmosphere, which can take several weeks to months. High-strength fabrics made of PA fibers are used for motorcycle protective clothing, workwear, backpacks, parachutes, and sails. Other products often based on PA fibers include ropes, fishing lines, and guitar strings. Because of its excellent strength and toughness, PA is often used as a construction material. Products such as dowels, bearing shells, gears, and pulleys benefit from the excellent properties of this plastic material. Applications also exist in electrical engineering for power tool housings, as well as insulating parts and cable ties. PA is also resistant to lubricants and fuels and is therefore used in automotive engineering for oil pans, fuel lines, and engine covers. Other automotive applications include accelerator and clutch pedals, lamp housings, and bumpers.

7.2.4.6.11 Polyurethane (PU)

PUs are extremely versatile plastics and are therefore used in many different areas. The material's success is due to the wide range of possibilities for varying its properties. PUs come in their linear and cross-linked forms, allowing the production of all four types of polymers:

- *thermoplastics,*
- *thermosets,*
- *elastomers,* and also
- *thermoplastic elastomers.*

The degree of crosslinking determines which polymer is present. In thermoplastic PU, the polymer chains crosslink via hydrogen bonds. This relatively weak chemical bond can be reversibly dissolved by heat and reconnected by cooling. The material then returns to its elastic behavior. Due to the variety of PUs on the market, the properties can only be given on an application-specific basis. The density of the solid PU is 1.0–1.25 g/cm^3. The material is not resistant to strong acids and alkalis, aromatics, alcohols, and hot water.

PU Foams The main application form of PU is as foams with densities of 0.005–0.04 g/cm^3 for flexible and 0.03–0.09 g/cm^3 for rigid foams. For their production, it is necessary that gases are generated during polymerization, which can be carried out either chemically or physically. These gases form small bubbles in the polymer network, creating the foam. In chemical foaming, water is added during polymerization and reacts with the monomers to produce carbon dioxide. Physical foaming is usually done by adding cyclopentane, which evaporates due to the heat released during exothermic polymerization. This process is mainly used to produce rigid, closed-cell foams that are used for heat or cold insulation, such as for refrigerators or hot-water tanks. In the construction industry, rigid foam applied from cartridges is used for the assembly and insulation of facades.

The open-cell and thus air-permeable flexible foam is used in a variety of applications, e.g.:

- mattresses,
- upholstered furniture,
- household sponges,
- winter clothing,
- coating of carpets,
- dust filters,
- insulation, and
- packaging.

PUs are much more than just foam and there are other interesting applications, including:

- coatings that adhere well to many surfaces but are not resistant to strong acids and alkalis, aromatics, alcohols, and hot water,
- adhesives that are very suitable for large area and flexible bonding, as required for bonding wood (laminated wood panels) and also shoes,
- casting resins, which are popular in prototype construction because they can simulate the mechanical, thermal, and optical properties of the production part very well,

- fibers, which are used in extruded form as the starting material for elastic textiles, and
- elastomers, which are thermoplastic and used for a variety of applications such as lightweight water-repellent footballs, outer skins of bowling balls, high-quality rubber boots, treads of skateboard wheels, or shoe soles.

7.2.5 Glass

7.2.5.1 Introduction

Glass is a collective term for a group of inorganic amorphous solids consisting mainly of silicon dioxide (Figure 7.12). The material can be described as a melt solidified without significant crystal formation and, because it is composed of various chemical compounds, cannot be described by a single chemical formula. Most of the glasses produced today are soda–lime glasses, which belong to the group of quartz glasses. All glasses in this group have in common that their network is formed mainly from silicon dioxide (SiO_2). By adding other oxides such as alumina or various alkali oxides, the alumina or alkali silicate glasses are formed. The decisive factor for classification is which oxide is contained second most frequently in the silicate base glass in terms of quantity. A silicate glass without other components – pure SiO_2 – is called quartz glass, which also has an amorphous structure. Quartz also occurs as a single crystal and then naturally shows a crystalline structure (Figure 7.12). There are thus a very large number of glasses of different compositions, which are of economic or scientific interest depending on their properties.

The tensile strength of glass is in the range of 30–80 N/mm^2 and significantly below the theoretical value. This is due to microscopic cracks and notches that can be found in any glass material. These *weak spots* in the material may have occurred both in the manufacturing process and as a result of mechanical or corrosive stress during use. Glass fibers, on the other hand, have fewer cracks and notches and thus achieve much higher tensile strengths of about 1000–5000 N/mm^2.

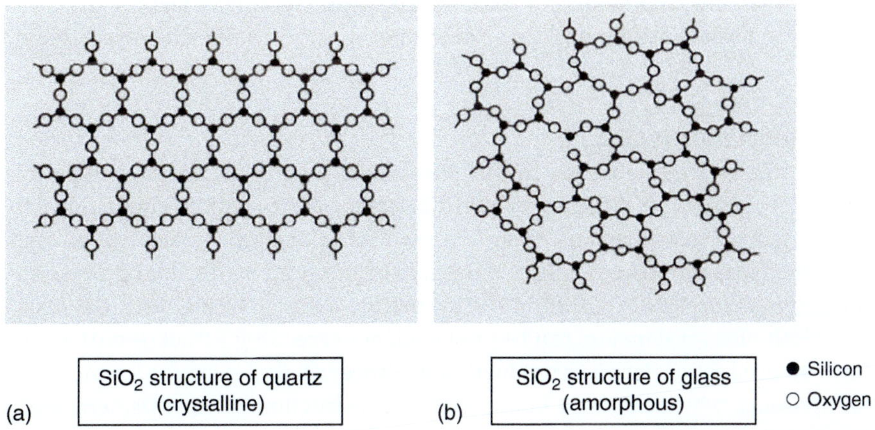

Figure 7.12 Structure of quartz crystal (crystalline) (a) and glass (amorphous) (b).

7.2.5.2 Classification of Glass

According to chemical composition, glass can be divided into three main groups, which account for about 95% of glass production:

- *soda–lime glass*,
- *borosilicate glass*, and
- *lead glass*.

7.2.5.2.1 Soda–Lime Glass (Standard Glass)

Soda–lime glasses represent the strongest group within glasses and consist of about 70–75% sand (silicon dioxide, SiO_2), about 10–15% lime (calcium oxide, CaO), about 15% sodium oxide (Na_2O), and a few percentages of other aggregates. This glass is also known as soft glass and is characterized by good light transmission and smooth, non-porous surfaces. Soda–lime glass is used for the production of bottles, drinking glasses, and also flat glasses. It is suitable for products that must withstand short-term chemical exposure, but are not subject to high thermal stress.

Borosilicate Glass Compared to soda–lime glass, borosilicate glass has a higher proportion of silica and owes its name to the boric oxide that is part of the formulation. Due to the low coefficient of thermal expansion of approximately $3.3 \cdot 10^{-6}\ K^{-1}$ and the associated low sensitivity to temperature changes as well as the good resistance to chemical influences, the borosilicate glass is preferred for glassware in the laboratory and in chemical processing. For corresponding applications in the household, the borosilicate glass is used as a so-called *cold refractory glass*. Due to its high resistance to abrupt temperature changes, this material is also used as fire-proof glass.

Lead Glass A high percentage of lead oxide is used instead of calcium oxide to produce lead glass. These glasses melt at relatively low temperatures, have a higher density than the other glasses, and because they achieve particularly high light refraction, they are well suited for glass cutting. Therefore, lead glasses are used mainly for polished drinking glasses, decanters, vases, ornamental objects, and lenses.

7.2.5.3 Manufacture of Glass

Several processes are established in the glass industry to produce various types of glass. Hollow glasses are made by pressing, blowing, suction, and combinations of these processes. Glass fibers are spun and used for optical fibers, glass wool, and also for fiberglass-reinforced plastics. Flat glass is produced by the float process, in which the molten glass is passed into a tin bath to solidify. In this way, flat glass with a width of several meters can be produced, not exceeding a thickness of about 2 mm. The product of flat glass production is a sheet of glass used for a wide variety of products, which are used in automotive construction, as mirrors, tempered or laminated glass. Toughened safety glass is a thermally or chemically tempered glass in which the glass is heated to about 600 °C and then rapidly cooled by cold air.

As the core cools more slowly than the outer areas, stresses are created in the glass material. The result is a glass material that is particularly sensitive to mechanical stresses and can therefore no longer be processed. Laminated safety glass consists of alternating layers of glass and a plastic film made of polyvinyl butyral (PVB). This material prevents glass fragments from flying around and causing accidents in the event of breakage. Safety glass with a thickness of about 2.5 mm is called bulletproof glass and is used, for example, for shop windows, display cases, and car windows.

7.2.5.3.1 Surface Finishing of Glass

Often the surface of glass is finished by chemical and physical vapor deposition. Thin metal coatings are applied that reflect infrared light (heat radiation), for example, in the case of window and automotive glass. This leads to improved comfort, as rooms in houses and car interiors heat up less from sunlight in summer. To improve light scattering, the surface of the glass can be roughened by sandblasting or by treatment with hydrofluoric acid, after which the glass appears milky and is no longer transparent. This surface modification of glass is used, for example, for lampshades or for opaque windows to provide privacy.

7.2.6 Wood

7.2.6.1 Introduction

Wood is one of the oldest and most popular materials of mankind and, with the exception of desert regions, is available everywhere and in large quantities. Compared to metal or stone, wood is relatively easy to extract and process, and since it is a renewable resource, the eco-balance of wood and wood-based materials can be described as excellent. Because the material demonstrates very good insulating properties to make it feel warm, most people feel comfortable when their living and working environment is made entirely or partly of wood. Because wood is able to absorb and release moisture, it regulates the indoor climate and is therefore particularly valuable from a building biology point of view. Another special feature is the acoustic properties of wood. Both the propagation and the insulation of sound waves behave characteristically differently from other materials, a property that is not only used in musical instruments, but is also often used in the furnishing of concert halls.

7.2.6.2 Properties of Wood

From a technological point of view, wood is a complex material. Its distinctive structure means that most physical properties, especially mechanical properties, are strongly anisotropic. For example, depending on the wood, the tensile strength in the direction orthogonal to the grain (transverse tensile strength) is only 10–15% of the tensile strength in the direction of the grain, which in the case of beech wood with 12% moisture content and a density of 0.7 g/cm^3 gives 120 N/mm^2. If this is compared with the corresponding values for structural steel with a density of 7.8 g/cm^3, which are 310–630 N/mm^2, this means that wood allows lighter structures if the component is designed so that the main forces that occur act in the direction of tension or bending. When considering this, however, it should be noted that not all wood is the same, but reflects the natural diversity of the tree species from which it is obtained. Thus, depending on the tree species, but also on the place of growth and its climatic history, a great variety of wood densities, strengths, colors, and grains results. Because of these possibilities, wood is very attractive for the builder and the designer.

7.2.6.3 Wood Surfaces

The adhesion build-up on the porous wood surfaces is due to both the physicochemical interactions and the mechanical anchoring of the adhesive layer in the material, which is ensured by the relatively large effective surface area per joint cross-section, and also voids in the material. Wood is processed by sawing, planing, filing, sanding, brushing, or cutting. After these processing steps, the so-called face surface is usually more large-pored, which is due to the fact that the cell cavities, like the intercellular spaces, run in the direction of the fibers. Thereby, surfaces after sawing are more uneven than planed or sanded surfaces. While planing or slicing cleanly splits the cell walls, sanding creates *debris layers* of cellular material that negatively affect the quality of the bond. Also, the wood density, which depends on the species, but within the species also on whether it is early or late wood, plays a significant role for the behavior of the adhesive at the boundary layer.

7.3 Surface Cleaning and Treatment

7.3.1 Introduction

To ensure high-quality bonded joints, the surface layer of the substrates to be joined should be clean and free of grease, wettable, firmly bonded to the base material and should not change its properties after bonding. To meet these requirements, surface preparation and often surface treatment of the substrates is necessary prior to the bonding process to create an adhesive-compatible surface that is free of dust and grease. The goal is to improve its chemical and/or physical properties by mechanical, chemical, or physical methods to enable high-quality adhesive bonding (Figure 7.13). To avoid contamination of the surface after treatment, the bonding process should be carried out immediately after completion of the treatment.

In particular, the treatment of the material surfaces will improve:

- wetting of the surface and resulting adhesion build-up,
- long-term durability by removing all adhesion-reducing layers, and
- reproducibility of the strength properties of the bond.

The necessary process steps for surface preparation and treatment are therefore extremely important for the creation of a successful bonded component and must therefore be given special attention during the planning and execution of the bonding process.

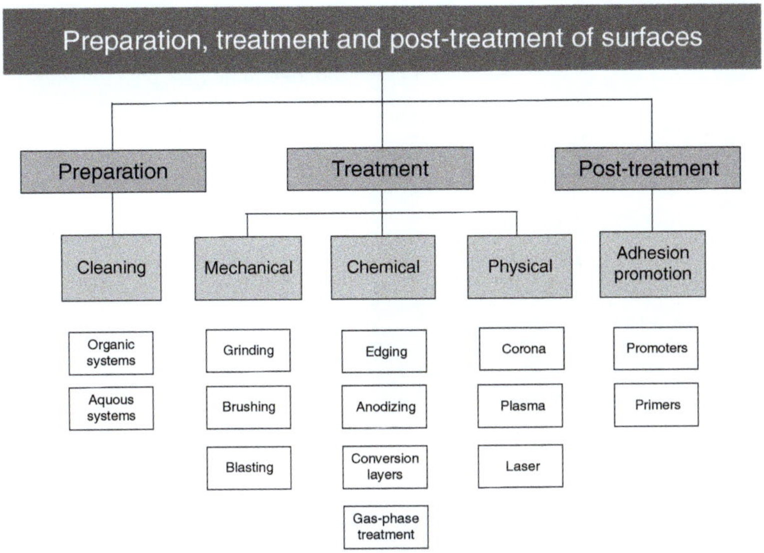

Figure 7.13 Preparation, treatment, and post-treatment of surfaces.

7.3.1.1 Importance of Surface Treatment

Why is surface treatment so important for the successful production of a bonded component? The quality of a bond ultimately depends on the adhesion between the adhesive layer and the material surfaces as well as the cohesion, which is the internal strength of the adhesive. Cohesion is determined almost exclusively by the type of adhesive, which can be easily defined at the user's site after knowing the requirements, studying supplier specifications and performing laboratory evaluations. It can also be assumed that the modern adhesive systems offered today are of the quality that ensures the cohesion required for the application. It therefore remains to establish good adhesion between the adhesive and the material surfaces, which is highly dependent on the nature of the substrate surface. Unfortunately, however, the surface condition is very often not precisely known and, in most cases, cannot be accurately analyzed. The task of surface treatment is now to bring the material surface into a defined state with the reactivity necessary for good adhesion. In addition to improving adhesion, however, the treatment should also ensure adequate aging and corrosion resistance of the joint. In any case, proper cleaning and degreasing must be carried out before the material surfaces are treated and, of course, afterwards as well. A cleaned and degreased surface is the basic requirement for the production of a successful bonded component.

The type and extent of surface treatment of the substrates depend on:

- adhesive strength requirement of the bond,
- requirement for long-term stability,
- stresses acting on the bonded joint,
- properties of the substrates,
- properties and condition of the surface, and
- properties of the adhesive used.

Figure 7.14 Adhesive (a), mixed (b), and cohesive failure (c) of a bonded joint after peel loading. Source: Dr. Jürgen Klingen, Aboso-Consulting.

7.3.1.1.1 Goal of Surface Treatment

The goal of surface treatment is to optimize the adhesion forces between the material surfaces and the adhesive layer to ensure good bonding. To achieve this, active centers must be created to enable the necessary intermolecular forces in the interfacial layer. The goal of these efforts is that if the bond is destroyed, cohesive failure will occur in any case, which is characterized by the fact that the breakage of the adhesive bond occurs within the adhesive layer and not in the interface between the adhesive and the surface. If this is the case even after the bond has been subjected to appropriate stresses, such as high temperature, humidity, and UV radiation, the objective of the surface treatment has been achieved and the substrates are well prepared. The occurrence of such a fracture in the adhesive layer thus represents a basic condition that should always be achieved during the development, optimization, and validation of a bonded component. Exceptions are applications with most PSA tapes. Specimen (c) in Figure 7.14 represents a clean cohesive fracture in a bonded joint of two steel strips joined with a structural adhesive that was peeled off after complete curing. The fracture clearly occurs in the adhesive layer, and there is no evidence of failure in the interface, which is an indicator that the surface treatment was properly performed. In this sample, the treated surface has not allowed the forces applied during the peel test to destroy the adhesion on the surfaces involved. Thus, the strength of the bonded component is determined almost exclusively by the cohesion of the adhesive. Bonded components exhibiting this (preferred) failure mode allow very good control of the bond and show excellent reproducibility of strength data. Specimens (a) and (b) in Figure 7.14 demonstrate what happens when surface treatment has not been adequately performed. Unlike specimen (c), neither specimen fails cohesively, but mixed failure (b) and complete bond failure (a) are observed. There is a good probability that the corresponding joined components would fail in real applications when the corresponding loads are applied.

Although surface treatment of the materials to be bonded is a basic prerequisite for good bonding, there are also very few areas of application in which the substrates do not need to be prepared. This is the case in automotive production, where oiled steel and aluminum sheets are bonded directly without surface treatment. However, this is only possible because specially formulated adhesives have been developed that are capable of achieving high strength and long-term resistance on these oiled sheets. In this case, application development was a collaborative effort between adhesive manufacturers, users, and oil producers.

At this point, the author would like to point out two permanently underestimated causes of adhesive failure:

- surface contaminations from silicone products, and
- unwanted depositions of fingerprints on the substrate surfaces.

Both are very often caused by the personnel performing the bonding process during the manufacture of bonded components. These contaminations are very difficult to detect visually and usually have a very dramatic effect on the quality of the bonded components. In particular, silicones on the surface of materials used in industry and trade repeatedly cause considerable problems. The reasons for this are their very low surface energy of 24 mJ/m^2 and the extremely high presence of these materials in many areas of industrial production, but also in a wide variety of things in everyday life.

For example, silicones are used as:

- lubricants for machines and bearings,
- mold release agents for injection molding,
- coatings for release papers used for pressure sensitive adhesives,
- lubricating creams for latex gloves,
- additives for plastic furniture, paints, and plastic care products such as cockpit sprays,
- additives in cosmetics such as hand creams, face creams, lipsticks, and hair sprays, and
- waterproofing agents for shoes, upholstered furniture, and carpets.

Fingerprints pose a very serious threat to the quality of the bonded joint and are very often the cause of failure of a bonded component in industry and crafts. They are generated by the transfer of sweat adhering to a finger onto a surface, which often happens when materials are handled improperly during transport, alignment, or positioning before bonding. It becomes even more dramatic when the fingerprint sweat also contains foreign substances, such as silicones from hand creams, face creams, lipsticks, and hair sprays. The fingerprint consists of sweat and the composition can vary depending on individual factors such as the ambient temperature, the physical condition, or the dietary habits of the person concerned.

Hand sweat is in the acidic range, with a pH of around 4.5. 99% of sweat is water, with the remaining percentage composed of:

- sodium, potassium, and other salts,
- urea and uric acid,
- fatty acids,
- amino acids,

- ammonia,
- sugars,
- lactic acid,
- ascorbic acid (vitamin C), and
- cholesterol.

Under normal circumstances, a person produces more than a liter of sweat per day, and especially during physical exertion, excessive sweating often occurs on the palms. However, the problem of surface contamination by sweat can be avoided very easily. To this end, it is essential that gloves be worn at all times during the transportation of components, the surface treatment of materials, and the joining process.

7.3.2 Surface Characteristics

7.3.2.1 Metals

These *adhesive-bonding-friendly* materials can be bonded with almost all adhesives on the market. Although the wettability of bare metal is usually not critical, it is considerably affected by the substances usually present on the surfaces, such as metal oxides, corrosion inhibitors, greases, or oils. If one takes a close look at metal surfaces, one finds that the surface to be bonded very rarely has the composition and/or structure of the pure material. This has already been briefly discussed in Chapter 1.

Compared to plastics and glasses, the surface structure of metals is relatively complex and consists of (Figure 7.15):

- *base metal material*,
- *interface layer (2–5 μm)* – with modified physical and/or mechanical properties, caused by the forming process during the production of the material,
- *reaction layer (1–10 nm)* – formed by chemical changes at the interface and representing the actual adhesion layer,

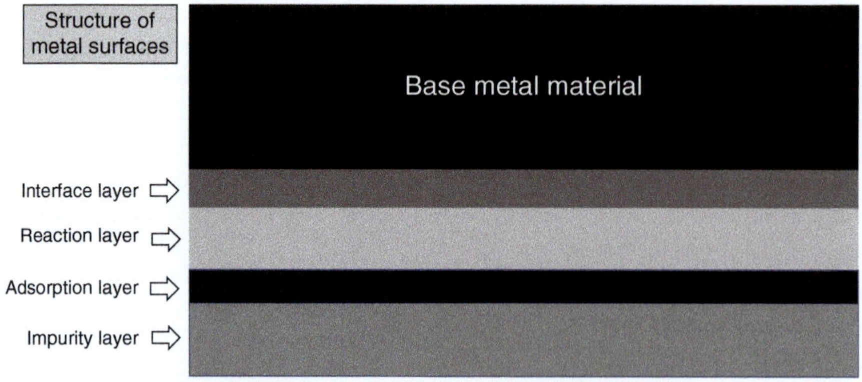

Figure 7.15 Structure of metal surfaces (schematic and not to scale).

- *adsorption layer (0.1–1 nm)* – with adhering molecules such as water and gases, and
- *impurity layer* – with solid materials such as dust or liquid contaminants such as oils, greases, or moisture.

The objective of surface treatment is to remove impurities and adsorbed substances mostly up to the reaction layer. When bonding metals, this reaction layer is usually well wetted by the adhesive and thus enables good bonding. However, in the case of inert, porous, unstable, or low internal strength reaction layers, this is difficult and then removal of this layer is unavoidable. This is necessary, for example, with low-alloy or unalloyed steels, since their reaction layers (rust) do not adhere sufficiently to the base materials. It is also often performed with aluminum substrates to remove their porous, relatively unstable oxide layers, especially if very good aging resistance of the bonded components – as required, for example, in aircraft construction – is to be ensured. Because of the extraordinary importance of the reaction layers for successful bonding, it is extremely important to deal with their formation, structure, and properties already early during the development of the bonding process.

7.3.2.1.1 Low and Unalloyed Steel

Reaction of atmospheric oxygen and water with the iron causes a porous oxide layer (rust) to form on low-alloy or unalloyed steels, which has low adhesion to the base material. This is caused by the lattice structure deviating from that of the base material. The main constituents of the reaction layer of these steel grades, which are commonly used in industry and crafts, are iron (II) oxide, iron (III) oxide, iron (III) hydroxide, and water of crystallization.

7.3.2.1.2 Stainless Steel

Stainless steel contains alloying additions such as chromium and nickel. Due to the reaction of chromium with atmospheric oxygen, a layer of chromium (III) oxide a few nm thick forms on the surface and is firmly bonded to the base material. This layer is responsible for corrosion protection and has self-healing properties. In case of a punctual destruction of the surface, chromium migrates to the surface and after reaction with oxygen in the atmosphere, new chromium (III) oxide is formed. This process is repeated with each damage and thus contributes to excellent long-term corrosion protection of the material. This feature of stainless steel is good for corrosion protection, but is critical for bonding processes with stainless steel. The reason for this is the inert character of the chromium (III) oxide at the surface, which does not allow the degree of intermolecular interactions at the surface that is required for good wetting. This often results in inadequate adhesion properties of this material.

7.3.2.1.3 Galvanized Steels

The reaction layer of elemental zinc or galvanized surfaces is formed, among other things, by reaction with CO_2 in the air and consists of a mixture of zinc hydroxide and zinc carbonate. The layer is firmly bonded to the base material, but is sensitive to acidic substances. Since fresh zinc surfaces are still chemically unstable, they often pose a challenge in terms of bonding. The crystal structure of the zinc metal is also an important factor in the quality of adhesion to zinc surfaces. Large crystallites, such as those present in hot-dip galvanized surfaces, can lead to localized stress peaks in the adhesive coatings due to their preferred directions, resulting in reduced adhesion. Electrogalvanized surfaces with smaller crystallites, however, do not show this effect and therefore exhibit good adhesion properties.

7.3.2.1.4 Aluminum

The porous oxide layer of aluminum, which is firmly bonded to the base material, is formed spontaneously by the reaction of the metal with atmospheric oxygen and consists of aluminum oxide, hydroxide, and carbonate. This naturally grown layer largely protects the underlying metal, but when it comes into contact with salt solutions, there is a risk that this reaction layer will detach from the base material. In addition, these layers are morphologically complex and do not always have a positive effect on the quality of the bond due to their brittleness and porosity. This is particularly critical for bonded components with very high requirements for long-term aging resistance, such as those required for bonding applications in the aerospace industry.

7.3.2.1.5 Titanium

The approximately 20 nm thick reaction layer of titanium exhibits good adhesion to the base material, but exhibits passivating properties. Depending on the ambient conditions, oxides with different lattice structures (rutile, anatase, or brookite) are formed. When bonding these surfaces, it should be noted that rutile can be converted to anatase at a temperature of about 80 °C, which often leads to a loss of adhesion.

7.3.2.1.6 Copper

The structure of the reaction layer of copper depends on the duration of exposure to the environment. After a few hours in air, a solid and resistant layer of brown copper oxide is formed, which greatly delays the progression of the corrosive attack. After months to years, reaction with CO_2 in the air transforms the oxide into green copper (II) carbonate – the patina – which often contains other anions such as sulfate and/or chloride. Good-quality adhesive bonds can be formed with both surfaces, the brown copper oxide and on the green patina layer.

7.3.2.2 Plastics

Due to their chemical structure, plastics have relatively low-polarity surfaces and, unlike metals, are difficult to bond with adhesives. Thus, they allow only low wetting of their surfaces by adhesives, which largely limits the formation of the necessary molecular interactions. For this reason, plastics are also referred to as *low-energy surface* materials. In general, the degree of wetting on these substrates varies greatly depending on the chemical structure of the plastic. For example, polymers such as PTFE with a surface energy of 18 mJ/m^2, polydimethylsiloxane (PDMS) with 22 mJ/m^2, PP with 32 mJ/m^2 and PE (low density) with 31 mJ/m^2 are only marginally or not wettable at all. However, PVC with 45 mJ/m^2 and PA 6

Table 7.4 Surface energies of selected low- and high-energy materials used for adhesive bonding.

Low-energy materials (<38 mJ/m^2)	Surface energy (mJ/m^2)	High-energy materials (>38 mJ/m^2)	Surface energy (mJ/m^2)
PTFE	18	PMMA	40
PDMS	22	PET	43
LDPE	31	PA 6	43
PP	32	PVC	45
		Glass	300
		Ceramics	500–1500
		Zinc	1020
		Aluminum	1200
		Copper	1850
		Titanium	2050
		Chrome	2400
		Iron	2550

with 43 mJ/m^2 exhibit relatively good wetting behavior. In contrast, the metals with surface energies of approximately 1000–5000 mJ/m^2 are in a different order of magnitude and can be wetted very well.

7.3.2.2.1 Low- and High-Energy Plastics

The surface energy can be used to assess the wetting properties and thus the suitability of the plastic material for bonding. The guide value for the surface energy required to enable a good adhesive bond is often given as 38 mN/m^2. Below this value, plastics are referred to as low-energy, above this value as high-energy. For example, PTFE, PP, and PE with a surface energy far below 38 mN/m^2 are low-energy plastics that are very difficult to bond and require surface treatment to enable a functional adhesive bond. Because PMMA, PET, and PC are all above 38 mN/m^2, they can be bonded well without surface treatment. The surface energies of metals with very high-energy surfaces of about 1020 mN/m^2 for lead and 2550 mN/m^2 for iron, allow very good wetting of course without any surface treatment (Table 7.4). When bonding plastics, not only the surface energy but also the solubility of these materials by adhesives or adhesive components should be used to evaluate the bonding behavior. For example, polystyrene with a surface energy of 35 mN/m^2 can also be bonded without surface treatment; the reason for this is its good solubility.

In addition to the low energy characteristic of plastic materials, release agents on the surface, additives in the matrix and stress cracks very often have an even more negative effect on the quality of the surface. Also, the same plastics by name often contain different types and contents of fillers, plasticizers, color pigments, and flame retardants, which makes it very difficult to clearly assess the suitability of the material for bonding.

7.3.2.2.2 Stress Cracks

Stress cracks in plastics are caused by tensile and residual stresses as well as by contact with stress-promoting substances. These are for example low-molecular alcohols and surfactants in aqueous cleaners, which dissolve low-molecular components out of the plastic. Also, adhesives, which contain stress-promoting solvents or surfactants can be the cause of stress cracks. The cracks occur spontaneously and initially show up as hairline cracks on the surface. If the plastic material is subjected to continuous static or dynamic mechanical stress, larger cracks form and eventually the material breaks.

The following stresses on the material promote the formation of stress cracks:

- high tensile stresses,
- residual stress in the material,
- rising and falling temperatures,
- prolonged contact of crack-inducing media,
- mechanical surface treatments such as grinding, blasting, or brushing, and
- chemical/physical surface treatments.

In addition, stress cracking is favored by a change in the polymer structure, which includes

- increase in the partial crystallinity of the polymer,
- increase in the orientation of the polymer chains in the tensile stress direction,
- broadening of the molecular-weight distribution,
- increase in the low-molecular-weight fractions in the polymer, and
- decrease in network density.

7.3.2.2.3 Additives

As discussed, additives very often have a negative effect on the plastic surface and thus, negatively affects the bonding performance.

In this context, the following additives, which are common in plastic materials, deserve special attention:

- *silicone additives* – used for improving surfaces,
- *plasticizers* – used to improve flexibility of plastics,
- *washing additives and lubricants* – used for the improvement of processing,
- *mold release agents* – used in injection molding processes,
- *color pigments* – used for proper design,
- *flame retardants* – used for fire protection, and
- *fillers* – used for cost reduction.

Thus, for adhesive bonding the following articles are critical:

- *injection-molded plastic parts* – due to the release agents used in the process,
- *PVC-P* – due to the possible migration of plasticizers to the plastic surface,
- *rubber compounds* – due to talc, oils, and/or plasticizers used here for a variety of reasons, and
- *EPDM profiles* – that generally contain a variety of additives, including oils.

7.3.2.3 Glasses

With a surface energy of around 300 mN/m^2, glass is a high-energy material and it should therefore be easy to make a good adhesive bond with them. However, under normal ambient conditions, thin, invisible layers of water molecules condense on the surface, hindering the creation of a good bond. Another challenge for bonding glass is glass corrosion, which is caused by the dissolution of oxides in the first layers of the glass material. This also negatively affects the quality of the bond and, in particular, the long-term stability of the bonded components.

7.3.3 Surface Cleaning

Cleaning or removing foreign matter such as dirt, grease, oil, and porous substances from the surface before joining the substrates is an absolute requirement for a well-functioning adhesive bond. Therefore, a cleaning and degreasing step must always be performed before and also after surface treatment.

Typical types of foreign substances are:

- *on metals* – oils, greases, rust, and moisture,
- *on plastics* – release agents, lubricants, plasticizers, and antistatic agents,
- *on glass* – fingerprints and moisture,
- *on tiles* – detergents, moisture, and coatings, and
- *on rubber* – talc, grease, and plasticizers.

For the cleaning process, very different cleaners are used, such as organic solvents, acid or alkaline pickling, washing-active substances, and special cleaners. In the case of light contamination, very good results are achieved with both polar and non-polar hydrocarbons. The following are used for this purpose:

- *alcohols* – such as isopropanol,
- *ketones* – such as acetone and methyl ethyl ketone,
- *esters* – such as ethyl acetate, and
- *non-polar hydrocarbons* – such as petroleum ether and heptane.

The following cleaning and degreasing processes are used in industrial and craft production of bonded components:

- wiping with solvent-soaked rags or cloths,
- degreasing with solvents in ultrasonic baths,
- steam degreasing with solvents,
- washing with aqueous alkaline solutions,
- cleaning with the aid of a high-pressure water jet, and
- mechanical processes such as brushing, grinding, or blasting.

7.3.3.1 Cleaning with Organic Solvents

Solvents and solvent mixtures used for the preparation of material surfaces must meet special requirements with respect to dissolving power, evaporation time, water solubility, flammability, and boiling limit. An important criterion is the evaporation ratio of a solvent, which is the ratio of the time in which a given substance evaporates

Table 7.5 Evaporation ratios, boiling points, and flash points of some solvents used for cleaning in bonding processes (ordered by their evaporation ratio).

Solvent	Evaporation ratio	Boiling point (°C)	Flash point (°C)
Diethyl ether	1	35	−40
Acetone	2	56	−19
Methyl acetate	2	57	−10
Tetrahydrofuran	2	66	−17
Ethyl acetate	3	77	−4
Methyl ethyl ketone	6	80	−1
Methanol	7	64	11
Ethanol	8	78	12
Isopropanol	11	82	12
Butyl acetate	12	127	25
Cyclohexanone	40	156	43
Water	80	100	—

completely to the time required for diethyl ether to evaporate. A high evaporation ratio means relatively slow evaporation, while a small number means the opposite. For example, diethyl ether with an evaporation ratio of 1 evaporates 11 times faster than isopropanol (Table 7.5).

The following classification applies to the evaporation ratio of solvents:

- *highly volatile solvents* – with an evaporation ratio of up to 5,
- *medium-volatile solvents* – with an evaporation ratio of up to 35, and
- *low-volatile solvents* – with an evaporation ratio of more than 35.

Mixtures of solvent vapors with air are often explosive. The indicator of this behavior is the flash point, which is the lowest temperature at which a liquid releases flammable gas or vapor in such quantity that a flame is immediately produced when the vapor phase comes into contact with an effective ignition source. The appropriate test is conducted at normal pressure and under prescribed test conditions.

Solvents are classified into three hazard groups according to their flash point:

- *extremely flammable* – with a flash point below 0 °C (e.g. ethyl ether and acetone),
- *highly flammable* – with a flash point below 21 °C (e.g. isopropanol and ethanol), and
- *flammable* – with a flash point between 21 and 55 °C (e.g. white spirit and turpentine).

In addition, the harmfulness of the solvents when inhaling the vapors or in case of direct skin contact must be taken into account.

7.3.3.1.1 Alcohols

Alcohols or their mixtures with water – often based on isopropanol – can be used universally for surface preparation. Their ability to dissolve grease from surfaces is limited, but they are able to dissolve polar and also ionic contaminants very well. Therefore, they are very commonly used to remove fingerprints, moisture, salts, and detergents. The advantage of alcohols is their relatively high flash point of, e.g. 12 °C for isopropanol compared to −19 °C for acetone and their fast-drying characteristic after cleaning or degreasing the surface.

7.3.3.1.2 Ketones

Ketones, such as acetone and methyl ethyl ketone, have better oil- and grease-dissolving properties than alcohols and are therefore used to remove grease and waxes from surfaces. However, they are also capable of removing polar contaminants, such as fingerprints. Compared to non-polar pure hydrocarbons, ketones dry faster but have a relatively low flash point. Because of its good cleaning action and environmental friendliness, acetone is often used as a grease solvent. However, due to its high flammability with a flash point of −19 °C, appropriate fire protection measures must be taken when preparing surfaces with this solvent.

7.3.3.1.3 Esters

Esters, such as ethyl acetate or butyl acetate, show better oil- and grease-dissolving properties than alcohols and ketones and dry faster than the pure hydrocarbons. In general, they have lower toxicity than alcohols and ketones, but are sensitive to acidic or basic impurities (hydrolysis).

7.3.3.1.4 Alkoxy Propanols

Alkoxy propanols, such as 1-methoxy-2-propanol, show better oil and grease solubility than alcohols, ketones, and esters and demonstrate very good dirt pick-up capacity with polar contaminants. This is particularly important when cleaning processes take place in solvent baths. Their lower toxicity compared to alcohols and ketones and their higher flash point (32 °C for 1-methoxy-2-propanol compared to 12 °C for isopropanol) are also advantages.

7.3.3.1.5 Aliphatic Hydrocarbons

Aliphatic hydrocarbons, such as petroleum ether, white spirit, or heptane, also have very good grease-dissolving properties, but are subject to strict fire and explosion protection regulations due to their low flash points. High-boiling aliphatic hydrocarbons are often used when oils or greases are to be removed. However, they dry slowly and are therefore not suitable as the sole cleaning agent for adhesive preparation processes.

7.3.3.1.6 Caution: *Halogenated Hydrocarbons*

When cleaning and degreasing material surfaces prior to adhesive bond preparation, the use of halogenated hydrocarbons, such as tetrachloromethane (CCl_4) and trichloroethane ($CHCl_3CH_3$), which were often applied in the past, should be avoided at all costs. These solvents have excellent oil- and grease-dissolving properties, dry quickly without leaving residues, and are non-flammable. However, they are of concern from a health and environmental perspective and may no longer be used for industrial applications. Fluorocarbons are generally banned for industrial applications in view of the environmental impact of ozone depletion.

7.3.3.2 Cleaning with Aqueous Systems

Various aqueous systems are offered for cleaning or degreasing surfaces before making an adhesive bond:

- *neutral cleaners* – with a pH of 5–9, containing alkanolamines such as ethanolamine or alkoxy alcohols such as alkoxy propanol,
- *alkaline cleaners* – with a pH > 9 containing sodium hydroxide, silicates, and phosphates, and
- *acidic cleaners* – with a pH < 5 containing salts of inorganic or organic acids.

The action of aqueous cleaners is based on various mechanisms, such as physical effects like emulsification and dispersion or chemical conversions. In the case of emulsifying and dispersing, fats and oils are finely dispersed in the aqueous phase by the action of appropriate emulsifiers or dispersants. In the second case, alkaline media enable the chemical transformation of water-insoluble fats into water-soluble soaps. To increase effectiveness, the cleaners often contain oil- and grease-binding additives such as surfactants or complexing agents. Depending on the degree of soiling, cleaning often takes place over a period of several hours. When using aqueous degreasing agents in the bath, care must be taken to ensure that the proportion of released fats or oils is not too high, as this can lead to re-greasing during the cleaning process. In any case, after completion of the degreasing process, sufficient rinsing with deionized water must be carried out to completely remove the reaction products. When substrates are degreased with caustic solutions, the substrates are often immersed in a dilute acid (usually sulfuric acid) to achieve complete removal of the cleaning agents.

7.3.3.3 Cleaning of Metal Surfaces

During the mechanical processing of metals, processing aids are often used, which, for example, support forming processes or/and prevent corrosion. For this purpose, water-soluble cooling emulsions or water-insoluble greases are used. These substances are present as a thin film on the metal surface after processing and have a negative effect on the wettability of the metal surface due to their low surface energy. In addition, porous and poorly adhering oxide layers – such as rust in the case of iron – may be present on the substrate surface, which would significantly reduce the strength of the bonded joint. For this reason, metallic components almost always have to be cleaned before bonding, where the type of cleaning medium is determined by the substance to be removed.

Generally, water-soluble substances can be removed with low-concentration aqueous cleaners, while organic solvents may often be required for water-insoluble greases. Often the cleaning effect is improved by the application of heat or

Table 7.6 Cleaning ability of organic solvents for plastic surfaces.

Plastic material	Temperature limit (°C)	Alcohol	Ester	Ketone	Aliphate
PS	50	+	–	–	+
PVC	60	+	+	o	+
PE	70	+	+	+	+
ABS	80	+	o	–	+
PMMA	80	–	o	o	+
PP	100	+	+	+	+
PA	110	+	+	+	+

+, good; o, moderate; –, bad.

ultrasound. In the case of aqueous degreasers, the corrosive effect on metal coatings can be reduced or eliminated by the addition of inhibitors. Oxide layers on the surface of metals are removed by acids or alkalis. This process – called pickling –enables the creation of a bright metal surface that is easy to join. As in automotive engineering, for example, oils and greases are often used to protect metal surfaces from oxidation, but also to assist in the mechanical forming of components. This is common practice in the processing of steel sheets, for example. If these materials are bonded after the forming process, the oils and greases used act as release agents between the base material and the adhesive. For this reason, the metal substrates must be degreased before bonding. However, there is an exception in the automotive manufacturing sector, where oily steel and aluminum sheets are bonded directly without a cleaning step before bonding. In this application, special oil-absorbing adhesives are used, which enable the production of bonded components with excellent long-term aging resistance.

7.3.3.4 Cleaning of Plastic Surfaces

Cleaning plastic surfaces with organic solvents can lead to the formation of stress cracks, which can have a negative effect on the quality of the bonded component. Dissolving or swelling processes taking place on the surface are responsible for this effect. Although there is always a risk of cracking with plastics, solvent cleaning is completely harmless if the right solvent is selected (Table 7.6). Applying aqueous cleaners, sufficient wetting of the plastic surfaces, which is necessary for effective cleaning, can be achieved by using surfactants, but stress cracking can also occur here. In general, when using aqueous systems for cleaning plastics, the moisture absorption behavior of the materials must be checked, and often considerable time must be allowed for substrates to dry back.

7.3.3.5 Cleaning and Degreasing Processes

In industrial and craft production, the very simple method of wiping the material surfaces with solvent-soaked rags or cloths is very often used. However, when higher-quality requirements must be met or production necessities must be considered, more complex cleaning or degreasing methods may be applied [1].

7.3.3.5.1 Immersion Degreasing

This cleaning method allows cleaning and degreasing by immersing the substrates in the degreasing agent. In this process, the immersion time depends on the degree of contamination. The effectiveness of cleaning can be improved by increasing the bath temperature, but air contamination must be considered when using organic solvents.

7.3.3.5.2 Vapor Degreasing

In vapor degreasing, the parts to be cleaned are placed in a closed system located above a degreasing solution that is heated to just below boiling point. The increase in temperature causes the solvent to condense on the surface of the substrate, effectively degreasing the surface. The degreasing solvent, which has just absorbed the grease, flows back into the liquid phase under the component and there – over time – an increase in concentration of the extracted fats and oils takes place. The advantage of this method is that re-greasing of the surface is excluded.

7.3.3.5.3 Ultrasonic Cleaning

To support the effect of cleaning agents, additional ultrasonic treatment has become established in a large number of bonding applications. The process enables efficient, material-friendly, and rapid removal of a wide variety of contaminants. Ultrasound-assisted cleaning and degreasing is also successful for parts with complex geometries and in hard-to-reach places such as holes, cracks, and undercuts. The process works by converting electromagnetic waves into mechanical vibrations of 20–40 kHz, which cause hydraulic shocks with significant energy densities and trigger microflows in the solvent. These microflows blow off the impurities present on the substrate surface. By supporting ultrasound in the cleaning and degreasing of material surfaces, a reduction in processing time of up to 90% can be realized.

7.3.4 Surface Treatment

After cleaning the substrate, a surface free of dirt, grease, and oil is present. However, to improve its adhesive properties, the surface must often be brought into a condition that ensures good wetting of the adhesive. This can be achieved either by modifying the existing surface or by creating a new surface that is firmly bonded to the base material. Mechanical, physical, chemical, or electrochemical processes are used for this purpose. However, for most bonding tasks in industry and trade, good bonded components with adequate strengths can be produced by cleaning and degreasing the surface only. If necessary, for higher strength and better resistance to environmental influences, additional mechanical treatment by grinding or blasting might be applied. However, if extremely high aging resistance is required – as in aircraft construction, for example – considerable effort must be expended using chemical or electrochemical processes. Such an effort can be useful in the manufacture of aircraft parts, but is not necessary for by far the most applications in industry and trade. In general, in the development of any bonded component, it must be kept in mind that the effort required for adhesive surface treatment must be in close coordination with the requirements for bond strength and aging resistance of the bonded component.

7.3.4.1 Mechanical Surface Treatment

Mechanical surface treatment both cleans the substrate by removing material and increases the effective area available for bonding. Thus, grinding, brushing, or blasting produces a change in surface topography in the form of greater roughness.

7.3.4.1.1 Grinding and Brushing

This by far the most commonly used surface treatment process in adhesive-bonding technology and is very often sufficient to produce high-quality adhesive bonds. Grinding causes material removal from the surface, which is achieved by the action of hard particles that are either firmly bonded to fibres, as in the case of abrasive fleeces, or are coated on a carrier material, as in the case of abrasive papers. The surface effect depends on numerous factors, the most significant being the type of abrasive and the grit size; the finer the grit size, the smoother the surface. Abrasives are classified according to their grit size as coarse, medium, fine, and very fine. Here, the grit size indicates how many meshes per inch were used in screening the particles for the purpose of classification. Thus, with a high mesh count, only the small particles fall through the sieve, while the large ones remain in the sieve. Therefore, the higher the mesh count, the finer the grit of the abrasive.

The following classification applies to abrasives:

- *12–80* – coarse grit with a size of about 1800 to 200 microns,
- *100–280* – medium grit with a size of about 160 to 50 microns,
- *320–600* – fine grain with a size of about 45 to 25 microns, and
- *800–2500* – very fine grit with a size of about 20 to 10 microns.

The abrasive grit is made of various materials that differ in hardness:

- *Aluminum oxide*, also called *corundum* is very often used. The material is hard, resistant, and an ideal material for abrasives.
- *Silicon carbide* is slightly harder than aluminum oxide, has very good frictional resistance and a high melting point. Therefore, it is often used in machine grinding, where high speeds and thus high temperatures occur in the grinding process.
- *Boron nitride* almost reaches the hardness of diamond, is considered the second hardest known substance, and is even superior to diamond at high temperatures.
- *Diamond* has a special crystal structure that results in extremely high strength and hardness. It is the hardest material known to us, but due to its lower heat resistance, it cannot be used where very high temperatures are generated during the grinding process.

Three-dimensional abrasive fleeces, available as hand pads, belts, or discs, are often used for cleaning and/or processing surfaces. Such products consist, for

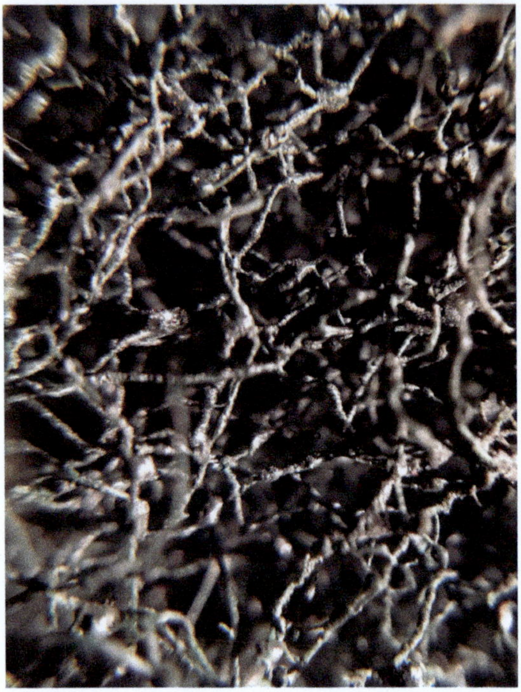

Figure 7.16 Microstructure of a three-dimensional grinding fleece with aluminum oxide particles on nylon fibers. Source: Dr. Jürgen Klingen, Aboso-Consulting.

example, of nylon fibers with aluminum oxide or silicon carbide particles on the surface embedded in resin (Figure 7.16). These products are very handy and adapt very well to the surface, reaching even hard-to-reach places. They also have a long service life and are therefore very suitable for adhesive machining of all types of metals. These abrasives are not classified by grit size, but in coarse, medium, fine, very fine, and ultra-fine.

To prevent contaminants from spreading over the entire surface or being firmly incorporated into the substrate material, the surfaces must be carefully cleaned and degreased before grinding. The reason is that later removal is difficult or impossible. When grinding, the parts to be joined should preferably be machined transversely to the direction of the acting mechanical load or crosswise. The grinding process for substrates used for bonding is carried out manually or mechanically using abrasives with an average grit size of 120–180, which best meets the requirements for bonding. For performing the grinding process by machine, belt, angle or random orbital sanders are used and subsequently, a further cleaning step must be carried out using, for example, solvent-soaked cloths to ensure that the dust on the surface is completely removed. During grinding, constituents of the grinding binder can be transferred to the surface of the joined part, but experience shows that such substances usually adhere very well to the substrates and are therefore generally not critical for the quality of the joint.

Grinding of surfaces is easy to perform and very often results in bonded components with adequate strength. However, sandblasted surfaces allow even higher performance, especially when the bonded component is exposed to critical environmental conditions over a long period of time. As with grinding, brushing of the substrate also allows material to be removed from the surface, but the elasticity of the brushes used means that roughness depths are lower. With regard to cleaning and degreasing requirements, as well as the suggested processing direction, the same recommendations apply as for grinding. The advantage of this process is that even thin substrate sheets can be machined without the risk of deflection. This is caused by lower compressive stresses on the surface compared to grinding or blasting.

7.3.4.1.2 Blasting

Blasting of material surfaces with approximately 0.2–2.0 mm metallic, mineral, or organic particles is another mechanical surface treatment process used before bonding [2]. Compared to grinding and brushing, this process enables even higher-quality surfaces, which are needed when long-term resistance of the bonded joint to critical environmental conditions is required. Depending on the process conditions, such as the duration of the impact, grain size, and hardness of the material to be processed, a highly broken surface with undercuts is produced. As a result, the adhesion buildup, which is usually mainly due to intermolecular forces, is also supported by mechanical anchoring of the adhesive in these structures. During the blasting process, the particles impinge on the material at a velocity of about 40 m/s, resulting in very high local energy concentrations that cause temperatures in excess of 700 °C. It is assumed that a short-lived high-energy plasma is subsequently formed from the particle and substrate material, which condenses on the substrate and determines the chemistry and dimension of the resulting layer. Thus, the material properties of the abrading particles also have a significant influence on the quality of the resulting surface. Macroscopically, no temperature increases of the substrates can be observed, since the plasma and the associated temperatures only occur in microareas of the surface.

Blasting Equipment Both air-driven equipment and high-speed paddle wheels are used to blast material surfaces. In air-operated blast machines, the abrasive is applied under pressure through a nozzle. This is done according to the injector principle or by a pressurized system. When using the injector principle, the compressed air or water is accelerated just before exiting a nozzle, creating a vacuum in a second feed and allowing the particles to be transferred to the surface. In the pressurized system method, the particles are stored in a chamber, usually pressurized with air. From there, they are transferred to the nozzle head and then accelerated toward the surface. Because of the large amount of dust generated, blasting with high-speed paddle wheels is usually performed in enclosed booths. For reuse of the blasting media, such plants are equipped with collection, cleaning, and recycling facilities. To avoid surface contamination, it is necessary to set up the corresponding blasting process area in a separate room from the bonding work area. The blasting processes can be performed either dry with the undesirable corresponding dust generation or wet, in which case the dust is bound but corrosion can occur on the treated substrate surface.

The materials used in blasting are between 0.2 and 2.0 mm and consist of:

- *metal particles* – of hard cast iron, steel, aluminum, or zinc,
- *mineral particles* – of corundum, glass, ceramics, or slags,
- *organic particles* – of plastic beads or particles based on nutshells or apricot kernels, or
- *frozen particles* – of liquids or gases such as carbon dioxide.

To remove all residues from the material surface, it is necessary to clean the substrates before and after blasting. This usually involves particles and carrier materials, but there may also be small amounts of oil on the surface from the compressors that pressurize the air for the blasting process.

Cryogenic Blasting This method involves blasting with dry ice (solid CO_2) in the form of granules (pellets) at a temperature of approximately −78 °C. Due to the low hardness of the pellets, even sensitive materials such as rubber, plastics, wood, glass, or polished metals can be cleaned without damage, and since no blasting media residues are produced, the method is considered particularly environment friendly. This method uses both thermal and mechanical effects to remove dirt from the surface. Thus, the different thermal expansion coefficients create stresses between the contamination and the substrate material, which loosen the layer to be removed. Complete removal of the contamination is then subsequently achieved by CO_2 gas production at the surface, resulting in an increase in CO_2 volume of approximately 700 times. CO_2 snow blasting is a simple new variant of cryogenic blasting that converts liquid CO_2 into compressed, solid CO_2 snow particles 1–100 μm thick. The particles are accelerated toward the surface by compressed air through a nozzle, creating a homogeneous jet of CO_2 snow. Depending on the nozzle geometry, a round-shaped jet with high spot cleaning power or a flat jet allowing uniform cleaning up to a width of 100 mm is produced.

SACO Process The so-called SACO (sandblasting and coating) process performs both mechanical and chemical treatments of surfaces. The process is particularly suitable for the production of bonded metal parts with excellent long-term stability, but also allows to improve the adhesive-bonding properties of plastics such as PE and PP. The process was developed in the dental industry and was originally used to improve the adhesion of dental substances to synthetic polymer materials, ceramics, and metals. Now the SACO process makes a significant contribution to improving bonded components made of mild steel, stainless steel, and aluminum. This unique blasting process uses special silane-coated corundum-based blasting particles, which allow both the removal of impurities adhering to the substrate surface and the creation of a silane coating with excellent adhesive properties. After the particles hit the substrate, impurities are detached and subsequently – after the particles have been rebound – a few μm thick layer with excellent adhesive-wetting capabilities is left on the surface. By adjusting the process parameters appropriately, a good balance between ablation and coating efficiency is achieved, ensuring the generation of a high-purity and very active surface. Prior to the application of the SACO process for metal substrates, the surfaces involved are usually blasted with corundum. For plastics, simple degreasing is sufficient as a preparatory step.

Table 7.7 Overview of mechanical surface treatment methods and their suitability for the treatment of metals and plastics.

	Grinding	Blasting	Cryo-blasting	SACO
Material	Metals, plastics	Metals, plastics	Metals, plastics, fiber-reinforced plastics	Metals, plastics
Bond strength	+	+	+	++
Aging resistance	–	o	–	++
Cleaning	+	++	++	++/–
Activation	–	+	–	++
Equipment cost	Very low	Low	Low	Low
Safety effort	Very low	Low	Very low	Low
Energy cost	Low	High	Very low	High
Environmental impact	Good	Good	Very good	Good
Special suitability for adhesive bonding	Good increase of initial strength of metal bonds, low increase of aging resistance of bonds with stainless and high-alloyed steels	Good increase of initial strength of metal bonds, moderate increase of aging resistance of bonds with stainless and high-alloyed steels	Environment-friendly cleaning of nearly all materials applied in industry and craft	Significant increase of ageing resistance of bonds with steel (including stainless-steel), aluminum, and well suitable for PE and PP

+, good effect; ++, very good effect; o, moderate effect; –, no effect.

7.3.4.1.3 Summary: Mechanical Surface Treatment Methods

- Mechanical surface treatment methods (Table 7.7) are suitable for improving the strength and to a degree aging resistance of bonded components made of metal, glass, and plastics.
- By simply grinding the appropriate surfaces, bonded components with good strength properties can be produced that meet the requirements of many industrial and craft applications.
- If higher demands are placed on the strength of the bonded joint or/and higher aging resistance is required, more complex blasting processes must be used.

7.3.4.2 Chemical Surface Treatment

For many bonding applications in industry and trade, mechanical treatment is sufficient to produce a high-quality bond. Very elaborate and time-consuming chemical and electrochemical surface treatment processes are only used when excellent long-term aging resistance of the components under the influence of environmental stresses is required in practical applications – as is the case in aircraft construction. However, due to the toxicity of the substances used, these processes must be carried out with extreme care, and environmental protection and occupational safety measures must be applied. Chemical surface treatment is carried out both by pickling the substrate surface and by creating a conversion layer, with the corresponding layers being produced by dipping, anodizing, or gas treatment processes.

7.3.4.2.1 Pickling of Substrates

This is one of the basic treatment steps for high-quality metals, especially ferrous materials. The aim of pickling is to remove oxide layers that are usually present on the surface of metallic materials, thus producing clean surfaces for bonding, coating, or making joints by welding. For steels, hydrochloric and sulfuric acids are preferably used for pickling, the effect being optimized by varying acid concentrations or working temperatures. When non-oxidizing acids – such as hydrochloric acid or dilute sulfuric acid – are used, a pure metal or metal oxide-acid reaction takes place, which causes the oxide layer to be removed from the surface, resulting in a bright, clean surface. The pickling process provides chemical cleaning of the surface and additionally roughens it submicroscopically. At the same time, the high-energy areas necessary for the formation of adhesion forces are created or exposed.

7.3.4.2.2 Generation of a Conversion Layer

In surface technology, a conversion coating is a very thin layer produced on the surface of metals by chemical or electrochemical reaction. During conversion, either metal ions from the surface react with the treatment solution or inorganic layers are deposited on the metal surface, which are firmly bonded to the underlying metal. If oxidizing acids such as nitric, concentrated sulfuric, phosphoric, or chromic acid are used – possibly with the addition of oxidizing salts such as sodium or potassium dichromate – solid oxide, phosphate, or chromate layers are produced. Thus, in this process, the surface is not only cleaned, but also oxidized. When compounds containing phosphorus or chromium are used, this process is also referred to as phosphating or chromating. The objective of creating a conversion layer on substrate surfaces is to incorporate oxygen atoms into the surface layer, which increase its polarity and thus enable the formation of intermolecular interactions with the adhesive molecules, which are necessary for good adhesion and, in particular, good long-term resistance of the adhesive bond.

Phosphating This process is widely used in surface engineering, where a conversion layer of firmly adhering metal phosphates is formed by chemical reactions of metallic surfaces with aqueous phosphate solutions containing phosphoric acid, metal salts, and additives. Phosphating is most commonly applied to non-treated steel, but can also be used on galvanized steels and aluminum. For aluminum surfaces, a variant of phosphating is proposed that results in virtually insoluble metal-phosphate layers a few micrometers thick. In this process, the materials are placed for one minute in an 80 °C aqueous 6% solution of phosphoric acid salts to form phosphate-containing aluminum hydroxide layers. The main applications for phosphating are adhesion promotion, corrosion protection, friction and wear reduction, and electrical insulation. Phosphating is also known under various brand names, including *bonderizing*.

Chromating In applying this process, complex chromic acid salts (chromates) are formed on metallic surfaces by the action of chromic acid and chromium ions, which have a strong oxidizing effect on metal surfaces. The process is used primarily on aluminum and zinc to produce relatively corrosion-resistant surface layers with a thickness of approximately 0.1–0.5 μm. In *yellow chromating* Cr (VI) compounds and in *green chromating* Cr (III) compounds are incorporated into the substrate surface. The resulting inorganic chromate layers serve to passivate or protect the corresponding surfaces. A variant of chromating is the so-called *pickling* process, which is used primarily for the production of extremely long-term resistant aluminum bonded joints in the aircraft industry. The process enables the production of highly corrosion-resistant coatings with needle- to honeycomb-shaped oxide structures about 4 μm thick. Such coatings are formed after the aluminum substrates are treated for 30 minutes with chromium (VI) ions, which are dissolved in a 60–65 °C hot, acidic, aqueous sulfuric acid solution.

Anodizing This is an electrochemical surface treatment process that is often used on substrates that are subjected to particularly high stresses. These processes accelerate the chemical reactions required to form conversion layers on substrate surfaces and provide highly reproducible results. To perform anodizing, the metallic substrate in the anodizing bath is switched as the anode. To complete the process setup, an inert cathode is also introduced into the system to enable current flow and thus the desired electrochemical reactions at the substrate surface. Through the specific adjustment of the various process parameters such as chemicals used, their concentrations, current densities, temperatures, and times, this process enables the preparation of surfaces that are optimally designed for demanding bonding applications. Relatively thick nanostructured oxides with particularly good adhesive properties are obtained, especially on aluminum alloys, but also on zinc and stainless-steel surfaces.

Gas-phase Treatment This process is used for special applications, for example, for the surface activation of plastic surfaces by oxidation with elemental fluorine gas. In this process, the surfaces are treated at room temperature for a short time with a nitrogen–fluorine gas mixture containing about 0.1–5% fluorine by volume. The resulting fluorinated layers are in the range of 10–30 nm and are very stable. Even after several weeks, these layers are able to form a very good bond with the corresponding adhesives, which allows high flexibility in production. The composites do not have to be joined directly, but can be processed even after weeks. This method is particularly suitable for PE, PP, and PET surfaces and enables the successful treatment of complex geometries as well as films, fibers, and foams. For example, gas-phase treatment with elemental fluorine gas is used for the production of self-adhesive PE foam tapes.

7.3.4.2.3 Summary: Chemical Surface Treatment Methods

- Chemical surface treatment methods enable excellent long-term aging-resistant bonded metal components under the influence of environmental stresses as required in aircraft construction.
- Due to the toxicity of the substances used, this type of treatment must be carried out with extreme care and it is imperative to provide appropriate environmental protection and occupational safety measures.
- The treatment process is carried out both by pickling the substrate surface, which basically means cleaning the surface from metal oxides, leaving bare metal, and by creating a new layer (conversion layer), which is much more suitable for bonding.

7.3.4.3 Physical Surface Treatment

Like the chemical processes, physical surface treatment also leads to a chemical change in the surface. These modern processes with very good environmental compatibility are used in particular for the treatment of metal and plastic surfaces. Depending on the choice of process, they are capable of significantly improving the initial adhesion and aging resistance of bonded components.

7.3.4.3.1 Flame Treatment

Short-term flame treatment of surfaces is used especially for thermoplastic polymers such as PE and PP, but also for other plastics. The aim of this method is to increase the surface tension of these substrates by partial oxidation of the surfaces. In this process, the flame generates a rough surface with polar carbonyl or/and carboxyl groups, which lead to a significant improvement of the adhesion properties, especially with non-polar plastics. However, due to the low stability of the modified layers, the bonding process must be carried out immediately after flame treatment. Flame treatment is carried out at temperatures of up to 1700 °C by a free-burning propane or acetylene flame with an excess of oxygen. The flame is moved rapidly over the substrate surface at a distance of 5–150 mm, where it causes a brief increase in temperature of around 200–400 °C. Due to the oxidative effect of the flame and the resulting formation of polar oxygen-containing groups, a chemical and also physical modification of the surface occurs, which enables the intermolecular interactions required for good adhesion of such low-energy surfaces.

The properties of the treated surface are determined by:

- torch spacing,
- duration of the treatment,
- flame temperature, and
- mixing ratio of the gas in the flame.

Flame treatment is not a purely thermal process, but a targeted use of the oxidation effect of a flame. The process is considered environment friendly and is used primarily for thicker-walled components and simple geometries. Since there is always a risk of damaging heat-sensitive polymer films, cooling rollers may have to be used with this process. Systems with web widths of up to 10 m and web speeds of up to 700 m/min are available for flame treatment of plastic surfaces in production.

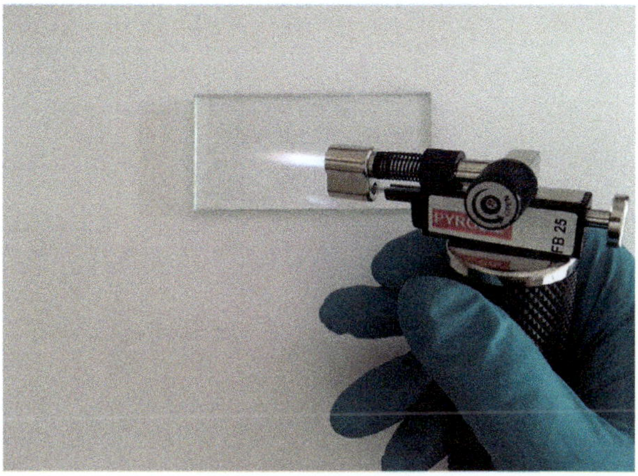

Figure 7.17 Manual Pyrosil® flame treatment on a glass surface. Source: Dr. Jürgen Klingen, Aboso-Consulting.

7.3.4.3.2 *Pyrosil® Flame Coating*

In this process, the surface is treated with a gas flame modified with silicon-containing substances such as tetra-ethoxy-silane (TEOS) or hexa-methyl-di-siloxane (HMDSO) (Figure 7.17). After treatment, these substances leave very thin, 5–100 nm thick amorphous silicon dioxide layers on the substrate surface, which are very dense and adhere very well to the substrate surfaces. The result of such treatments is significantly improved strength and aging resistance of the corresponding adhesive bonds.

The Pyrosil® process is an effective, inexpensive, and easily scalable process that can be used for a wide variety of materials. Since the short exposure time of the flame only moderately increases the surface temperature of the material, the process is suitable not only for glass, ceramics, and metals, but also for some temperature-resistant plastics. Corresponding devices are offered for both manual and automatic operations.

7.3.4.3.3 *Corona Treatment*

Corona treatment of a surface is carried out by the support of a characteristic blue glowing high-voltage discharge, which causes ionization of air molecules located between the two electrodes of the corona system and breaking of organic bonds on the material surface. Initially, the emitted electrons experience a strong acceleration, ionizing the air molecules oxygen and nitrogen, which subsequently leads to the formation of ions, radicals, ozone, and nitrogen oxides. The electrons impinging on the polymer surface split carbon–carbon bonds and carbon–hydrogen bonds, forming reactive fragments on the surface that can react with the gaseous reaction products formed from the air molecules. This creates polar structures on the plastic surface that enable the intermolecular interactions with the adhesive that are necessary for good bonding. In addition to this chemical change in the surface, the impact of the accelerated chemical species also causes a change in the surface structure, which also has a positive effect on the adhesion buildup capability of the substrate due to the resulting increase in surface area.

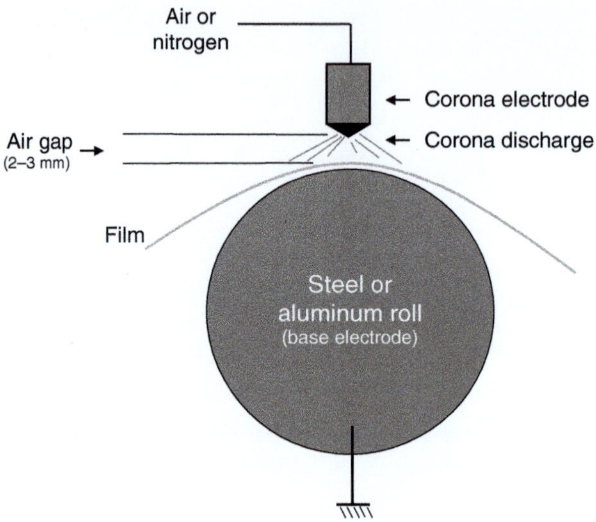

Figure 7.18 Illustration of a corona system used for film treatment (schematic and not in scale).

Despite its widespread use, the exact mechanism of corona treatment is still not fully understood. However, the main task of the treatment is to carry out oxidation processes on the surfaces, which generate polar functional groups, such as hydroxyl, carbonyl, carboxyl, or peroxide groups, depending on the plastic. Very often, the process is used in polymer film production, usually *in-line* at the end of the manufacturing process. The corona treatment is accomplished by high-voltage electrical discharge between a grounded, polished roll of steel or aluminum – the so-called base electrode – and an insulated electrode close to it – the so-called corona electrode (Figure 7.18). Since the foil is closely connected to the roller, only one side of the plastic foil facing the electrode is usually treated. However, treatment is also possible on both sides if an air gap is provided between the roll and the web. Corona treatment can increase the surface tension of a PE film, which was originally 30–32 mN/m, to 38–44 mN/m. However, the influence of the treatment decreases with time, which limits the durability of the treated materials. For example, the surface tension values after four weeks are about 10% lower than immediately after treatment.

7.3.4.3.4 Characteristics of the Plasma Process

Plasma was discovered by Irving Langmuir in 1928 and occurs in its natural form, for example, in lightning or as aurora borealis in the Arctic and Antarctic. In adhesive-bonding technology, low-pressure and atmospheric-pressure plasmas are used to clean, activate, and functionalize metal and plastic surfaces. Due to the very low use of chemicals, these processes are considered to be environmentally-friendly and, because of the low-temperature load on the surface, particularly gentle on the material. Plasma stands for a mixture of chemical substances in a gaseous state of aggregation that consists partly or completely of charged components, such as ions and electrons. Since the plasma state can be generated from the gaseous aggregate

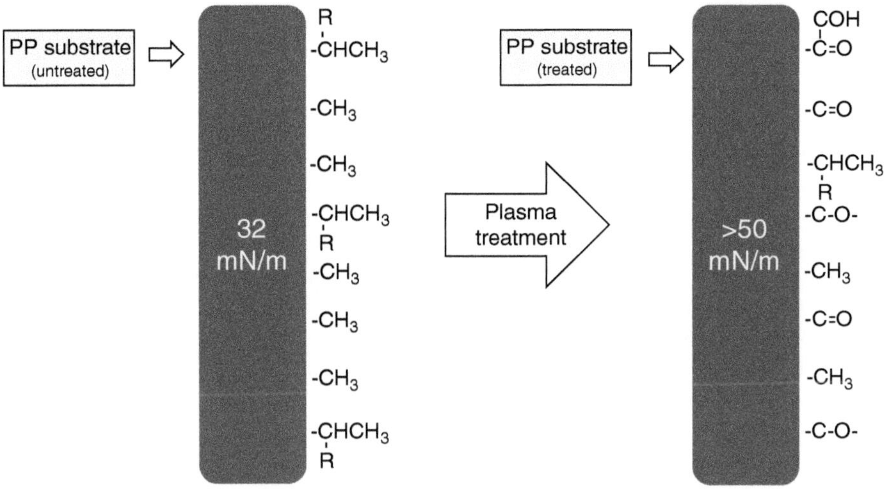

Figure 7.19 Influence of plasma treatment on the chemistry and surface energy of a PP substrate.

state by adding additional energy, it is often also referred to as the fourth aggregate state. During the generation of plasma, ions and electrons are produced, but also neutral or electrically charged, highly excited molecular fragments. The plasma thus contains a large proportion of free charge carriers, and the degree of ionization can vary from <1% to 100% (complete ionization).

7.3.4.3.5 Plasma Activation of Plastic Surfaces

If these charged plasma particles encounter, for example, plastic substrate surfaces that are in an oxygen atmosphere, polar chemical components such as carbonyl, carboxyl, or hydroxyl groups form on the surface, which significantly increase the polarity of the plastic surface (Figure 7.19). The process thus enables the generation of the intermolecular interactions required for the production of high-quality adhesive-bonded components. By varying the reactive gases and thus the resulting ions, the plasma process allows the chemistry on the surface to be tailored to match the adhesive of choice. As an environment-friendly and easy-to-use surface treatment method, the plasma treatment process is applied in numerous bonding applications involving low-energy plastics. In the case of PP, for example, it is possible to increase the surface energy from 32 (untreated) to 52 mN/m (five hours after treatment).

Low-Pressure Plasma Treatment This process variant is often used for surface treatment of parts with complex geometry and shows the advantages of:

- very effective treatment,
- very good long-term effect,
- uniform surface effect regardless of geometry,
- excellent gap accessibility (even undercuts are well reached), and
- low heat load on the material.

The process works with a chamber that is evacuated after insertion of the parts to be joined and then brought to the necessary working pressure of 0.1–5 mbar. Then a high-frequency voltage is used to start plasma generation. Since the efficiency of a gas discharge increases significantly with increasing frequency, plasma excitation by means of microwaves has proven to be particularly effective. Due to the low gas pressure applied in this process variant, a low concentration of gas molecules prevails in the chamber, causing a *cold* but still very reactive plasma. Thus, this process can also be used with thermally sensitive materials.

The low-pressure plasma process is characterized by:

- closed vacuum reactors,
- low technical vacuum requirement,
- very low treatment depth of about 1 μm,
- environmental friendliness,
- high operational and process reliability,
- possibility of a high degree of automation,
- high investment costs, and
- very low operating costs.

Atmospheric Plasma Treatment This process variant is a very efficient one for cleaning, activating, and coating plastics, metals, and glasses. Unlike the low-pressure plasma process, no separate chamber system is required. Unlike the corona surface treatment process, the beam generated in the plasma source has no electrical potential, so the substrate to be treated is not under electrical voltage.

In atmospheric plasma treatment, a pulsed arc is generated in a nozzle by a high-voltage discharge. The process gas flowing past there is excited by the discharge and thus converted into the plasma state. Subsequently, it passes through a nozzle head onto the surface of the material to be treated (Figure 7.20). Usually, the gas consists of air or oxygen, but to achieve special effects, additional reactive gases might be added shortly before reaching the substrate surface. The geometry of the emerging jet is determined by the type of nozzle head and has a length of about 50 mm and a width of about 15 mm.

During the treatment, the plasma source moves over the material surface at a distance of approximately 10–40 mm and a speed of approximately 6–600 m/min, depending on the required quality of the treatment. By using rotating plasma nozzles, it is possible to cover a width of up to 130 mm per nozzle, with treatment speeds of up to 40 m/min. For the treatment of larger areas, entire rotating systems are available that contain several plasma sources rotating at very high speeds. Depending on the diameter and arrangement of the plasma nozzles, such a system can treat areas up to 2000 mm wide. Atmospheric plasma systems can be very easily integrated into automatic processes for surface treatment, for example with robots, which is illustrated for the case of treatment of a metal component in Figure 7.21.

The atmospheric plasma process is characterized by:

- gentle treatment of the material surfaces,
- very low heat load on the materials,

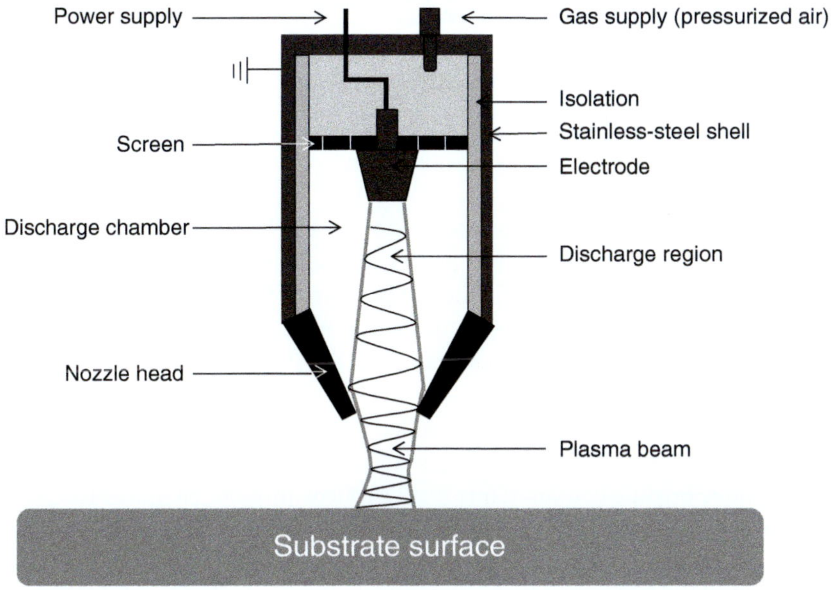

Figure 7.20 Representation of an atmospheric plasma system (schematic and not in scale).

Figure 7.21 Atmospheric plasma treatment of a metal surface. Source: Dr. Jürgen Klingen, Aboso-Consulting.

- high process speed,
- wide process window and robust processing,
- potential-free plasma source,
- inline capability, as no vacuum technology is required, and
- suitability for robotic applications.

There are several ways in which the plasma process can be used to improve material surfaces.

The plasma activation of plastic surfaces has already been described in detail above. In summary, the process offers a particularly efficient surface modification that provides significantly improved adhesive properties for components made of plastics, metals, and glasses. The process is environment friendly, easy to use, and particularly effective for low-energy materials such as PE and PP.

Plasma Cleaning This variant can also be applied to remove contaminants from surfaces before they are used in the adhesive-bonding process. The process is able to effectively remove organic contaminants from surfaces through chemical reaction with the plasma elements to form gaseous products, which are washed away by the continuous gas flow of the plasma device. The process is used for numerous materials, as well as for surfaces with complex geometries, and often replaces wet chemical-cleaning processes that may contain critical chemicals.

Plasma Polymerization This method of plasma processing enables the creation of highly crosslinked polymer layers on the surface, about 1 nm to 10 μm thick, formed from suitable gaseous monomers or other reactive substances. To achieve the desired properties of the corresponding functional layers, the reagents and process conditions are varied. A basic requirement for the success of this process is that the molecules of the gas from which the plasma is formed contain atoms capable of polymer formation.

Plasma polymerization is able to offer the following features:

- three-dimensional networks with high density,
- very good mechanical strength,
- high chemical resistance,
- adjustable surface energy,
- high barrier effect against gases and liquids, and
- high-temperature resistance.

Plasma polymerization, which can also have a cleaning effect, is used successfully in a wide range of industrial applications. Such coatings are capable of producing long-term stable adhesive bonds and in the case of aluminum-based substrates enable extremely high and long-term stable corrosion protection.

7.3.4.3.6 Laser Treatment

As a modern, safe, and environment-friendly method for cleaning and modifying surfaces to be bonded, laser treatment is increasingly being used in adhesive-bonding technology and represents a fast and easy-to-use alternative to conventional cleaning methods such as sandblasting, dry ice, or solvent cleaning. With this material-friendly method, surfaces can be cleaned and layers of defined thickness can be removed in sharply defined areas. The temperature increase of the substrate is moderate, since only a very thin layer of the base material is exposed to heat during laser treatment. The process is capable of removing troublesome organic or inorganic substances from the surface of all common materials ranging from metals as aluminum, magnesium, stainless steel, and chrome to various plastics, also including paints, greases, oils and flash rust, particles, adhesives, and oxide layers. For laser treatment, a powerful, focused laser beam is passed over the material to be cleaned, and the radiant energy of the laser beam is immediately converted into thermal energy. The resulting abrupt rise in temperature causes the dirt layer to vaporize, and non-vaporizable particles are mechanically blasted off the surface by the resulting plasma shock wave. By using high-intensity laser processes, it is also possible to change the structure and roughness of metal surfaces, such as aluminum. For example, very rapid melting of the uppermost 1–5 μm thick layer of aluminum surfaces followed by abrupt cooling can produce a sponge-like, amorphous-microcrystalline surface structure with very good mechanical strength. Due to its porous nature, it is capable of absorbing and anchoring adhesives, especially low-viscosity ones, resulting in excellent adhesive performance after curing. Such bonded aluminum composites prepared especially with epoxy adhesives show a significant improvement in long-term aging resistance, which can even surpass the performance of bonded components with substrates prepared by chemical surface treatments.

7.3.4.4 Summary: Physical Surface Treatment Methods

- Physical surface treatment methods (Table 7.8) are suitable for improving the strength and aging resistance of bonded components with metals, glasses, and plastics.
- In addition to increasing strength, they are often capable of significantly improving the aging resistance of bonded components based on metals such as aluminum.
- Especially in the case of low-energy plastics, surface treatment methods such as flame, corona, and plasma treatment as well as Pyrosil flame coating are very often used to enable the wettability of these surfaces and thus allow high-quality bonded components with these materials.

Table 7.8 Overview of physical surface treatment methods and their suitability for the treatment of metals, plastics, ceramics, and glass.

	Flame treatment	**Pyrosil®**	**Corona**	**Low-pressure plasma**	**Atmosphere plasma**	**Laser**
Materials	Plastics	Metals, glass, ceramics, some plastics	Plastics	Metals, plastics	Metals, plastics	Metals, plastics
Bond strength	++	++	++	++, plastics	++, plastics	++, metals
Aging resistance	++	+	++	++, plastics	++, plastics	++, metals
Cleaning	–	–	–	+, all materials	+, all materials	++, all materials
Activation	++, plastics	++, all materials	++, plastics	++, plastics	++, plastics	++, metals (microstructure)
Equipment cost	Low	Low	High	High	High	Very high
Safety effort	Very high	Low	Low	Low	Low	Low
Energy cost	Moderate	Low	Low	Low	High	High
Environmental impact	Low	Low	Low	Very low	Very low	Very low
Special suitability for adhesive bonding	Increase of initial strength and aging resistance of plastics	Good aging resistance of metal bonds with epoxy and acrylate adhesives	Increase of initial strength and aging resistance of plastics	Increase of initial strength and ageing resistance of plastics, cleaning of all surfaces	Increase of initial strength and aging resistance of plastics, cleaning of all surfaces	Effective cleaning of all materials, creation of suitable oxide layers on metal substrates

+, good effect; ++, very good effect; o, moderate effect; –, no effect.

References

1 Kollek, H. (1996). *Reinigen und Vorbehandeln.* Vincentz Verlag.

2 Jopp, H. (1995). Ein Beitrag zum besseren Verständnis der Wirkungsweise des Strahlens auf die Festigkeit und Beständigkeit von Metallverklebungen. PHD Thesis. Technische Universität Kaiserslautern.

8

Concept (Step 2b) – Adhesives, their Selection, and Definition of Concepts

8.1 Introduction

In Chapter 7, the reader has become familiar with the materials used as substrates in bonding, their surfaces, and the corresponding methods for cleaning and treatment to achieve safe and high-quality bonds. Thus, the reader already has a good picture of what to consider when selecting substrates and their advantages and disadvantages in bonding processes. He is also able to assess the available surface treatment methods in terms of their suitability for the job in hand.

A comprehensive knowledge of adhesives, their applications, and their selection is still lacking for the development of concepts for the proof of feasibility. In addition, an intensive look at what loads might act on the bonded component and what these have to do with the quality of the bond is necessary.

This is taught in Chapter 8 and thus, at the end of Chapter 8, the reader can propose 1–3 very well-thought-out concepts for the adhesive-bonding job. These include the substrates, the associated surface treatment, the adhesives selected, and the dimensions that these adhesive structures must have.

As an overview, the main tasks and corresponding deliverables of Step 2b are shown in Figure 8.1:

8.2 Adhesives for Industry and Craft

8.2.1 Introduction

Adhesives are materials that are applied to the parts to be joined in their existing physical state and then, after bonding, form a solid adhesive layer that is produced either by a chemical reaction or physical curing. If we now take into account the ways in which adhesives can be chemically structured, their performance in terms of strength or long-term behavior, and the corresponding variants for their processing, we arrive at thousands of different formulations that are in use in practice. For this reason, it is very difficult for the user to find the most suitable adhesive for the required material combination and processing possibilities.

Adhesive Bonding in Five Steps: Achieving Safe and High-Quality Bonds, First Edition. Jürgen Klingen.

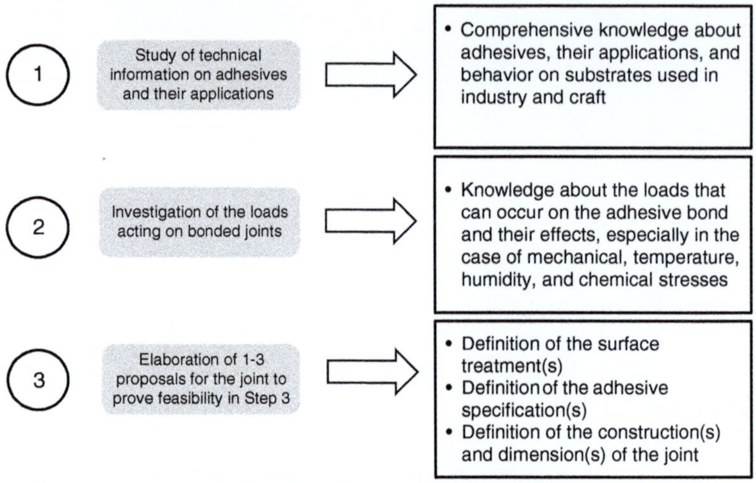

Figure 8.1 Major tasks and accomplishments of step 2b.

Looking only at the adhesive strength (overlap shear strength) of structural and elastic adhesives used in industry and trade, following short list can be applied:

- *Structural bonding in aerospace:* 30–40 MPa,
- *Structural bonding in industry and trade:* 15–25 MPa,
- *Elastic bonding* with PUs and silicones: <5 MPa and with acrylic foam tapes: <1 MPa.

The systematic classification of adhesives can be based on various criteria, although not all forms of presentation are helpful to the user and facilitate the selection. However, there are two classification principles, which might be the most suitable for the process developers, who need to select an adhesive for the required adhesive-bonding related challenges.

These are distinguishing the adhesives according to:

- *Curing mechanism* – divided into chemically and physically curing adhesives, and
- *Number of components* – divided into 1K and 2K adhesives.

8.2.1.1 Adhesive Classification by Curing Mechanism

When classifying adhesives according to curing mechanism, a distinction is made between chemically and physically curing systems (Figure 8.2). In chemically curing systems, the adhesive layer is formed by a chemical reaction of the reactive monomers or prepolymer molecules present in the adhesive.

8.2.1.1.1 Chemically Curing Adhesives

In their original form, chemically cured adhesive systems consist of low-molecular weight and thus flowable substances in which reactive chemical groups are present that can react with each other under certain conditions. After curing of

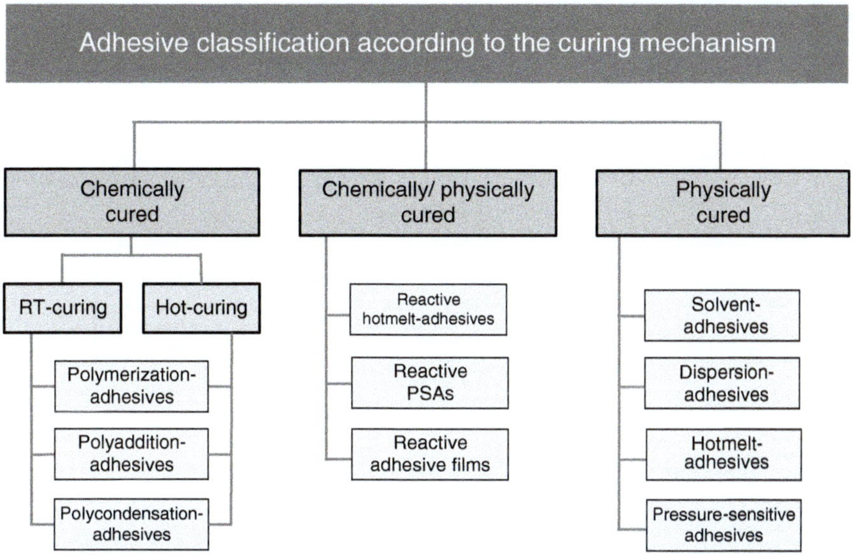

Figure 8.2 Adhesive classification according to the curing mechanism.

these components, polymeric substances with high-molecular weight and high mechanical strength are formed.

The curing of these systems takes place in different way, for example in the case of:

- *2K adhesives* – by two reactants that are mixed together before joining,
- *1K PU adhesives* – by the action of another component present in the environment (moisture in air), and
- *1K epoxy adhesives* – by an additional physical treatment (heat).

When using chemically curing adhesives, it is extremely important that the chemical reaction takes place only after the substrates have been joined. Otherwise – in the case of advanced conversion of the reactive components – the viscosity of the adhesive would be critical for proper wetting of the surface, resulting in inferior bonds. The chemical reaction for curing often takes place at room temperature, with the partners involved in the adhesive bond only gradually converting into a macromolecular substance. Since the adhesive cannot transmit any forces in the uncured state, the parts to be joined must be fixed, at least during the initial phase of curing. In most cases, the curing time can be shortened by adding heat.

8.2.1.1.2 Physically Curing Adhesives

When physically bonding systems are processed, no further chemical reaction takes place in the adhesive joint. These adhesives already contain the polymer compounds, which are brought into a wettable state by dissolving in appropriate solvent systems or by exposure to elevated temperatures. Joining usually takes place by pressurization, after which adhesion forces can build up to the surfaces of the parts to be joined. A characteristic feature of physically cured adhesive is that in the course of

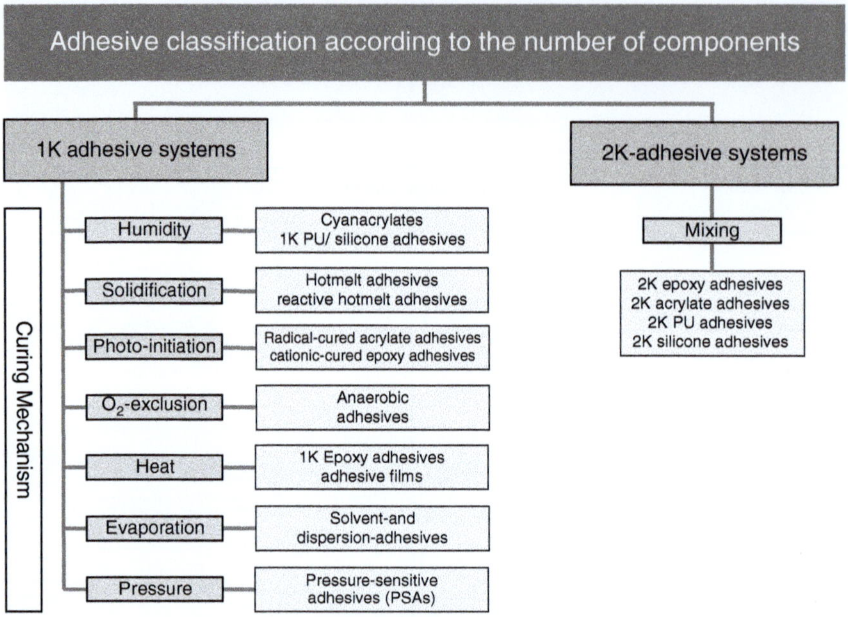

Figure 8.3 Adhesive classification according to the number of components.

the curing process, the polymer components in the adhesive are not changed in the chemical sense; they remain as they were in the original – non-cured – adhesive.

8.2.1.1.3 Chemically/Physically Curing Adhesives

Classification by curing mechanism also includes a type of hybrid system where a combination of both curing mechanisms is used. One example is reactive hotmelt adhesives. These adhesives require a combination of both curing mechanisms to fully cure. They first solidify physically so that sufficient handling strength of the bonded component is achieved, and then fully cure chemically to final strength. This occurs through an additional chemical reaction, usually with moisture from the environment.

8.2.1.2 Adhesive Classification by Number of Components

Classifying adhesives by number of components into one-component (1K) and two-component (2K) adhesives is much more useful to the users who are planning the process to produce an adhesive-bonded component (Figure 8.3). This is because it takes much more account of the interests of processors in industry and trade, especially with regard to user-friendliness and complexity of the adhesive curing process.

8.2.2 1K Adhesive Systems

Since 1K adhesives do not require the additional process step of mixing, they are easier to process than 2K adhesives and are used as delivered by the adhesive supplier. Therefore, these adhesives are very user-friendly and allow for low complexity in the curing process. However, once the parts have been joined, they require either another reactant – e.g. moisture in the case of moisture-curing systems – or an additional physical treatment – e.g. UV radiation in the case of photoinitiated systems – to cure to a solid adhesive layer. One exception is the increasingly used pressure sensitive adhesives (PSAs) in form of tapes, which do not require any step other than the application of pressure during the joining process.

In 1K systems, the adhesive layer is cured by:

- *influence of moisture* – as with cyanoacrylates, 1K PU adhesives, and RTV-1 silicones,
- *solidification of a molten polymer* – as in the case of hotmelt adhesives,
- *irradiation with visible or UV light* – as in the case of photoinitiated adhesives,
- *absence of oxygen (anaerobic)* – as in the case of anaerobic adhesives,
- *application of heat* – as in the case of heat-curing adhesives,
- *evaporation of solvents and water* – as in the case of solvent- and water-based adhesives, and
- *exclusive application of pressure* – as in the case of PSAs.

8.2.2.1 Cyanoacrylate Adhesives

8.2.2.1.1 Overview

Cyanoacrylate adhesives, also called *super glues*, were introduced by Eastman Kodak in the 1950s under the name *Eastman 910* and are now proven components in modern bonding processes. These adhesives are used primarily for bonding metals, plastics, and elastomers.

The most remarkable properties of these adhesives are:

- single-component and therefore easy to handle,
- solvent-free, and
- very fast curing.

Thus, these reactive adhesives meet today's manufacturing requirements and therefore have a strong position in industry and crafts with numerous applications. In addition, they are used for body tissues in medicine and as *super glues* for a wide variety of household and *do-it-yourself* applications.

8.2.2.1.2 Chemistry/Composition

The chemical structure of these adhesives is derived from α-cyano-acrylic acid, with the ester group containing various alkyl groups such as methyl, ethyl, butyl, or ethoxy groups. Because of its proximity to two electron-withdrawing groups (the cyano and ester groups), the double bond in the cyanoacrylate monomer is extremely sensitive to nucleophilic attack. So, there is always a tendency for negatively charged groups to attach, a feature of great importance for the curing of cyanoacrylate adhesives. Thus, weak nucleophilic substances such as OH ions – which are always present in water – very easily start the chemical reaction (polymerization) for curing. The required OH ions are supplied by the small amounts of water on the substrate surfaces or by the humidity in the surrounding air. In general, bases are suitable for initiating the polymerization of cyanoacrylates, whereas acids inhibit (retard) the polymerization process [1, 2].

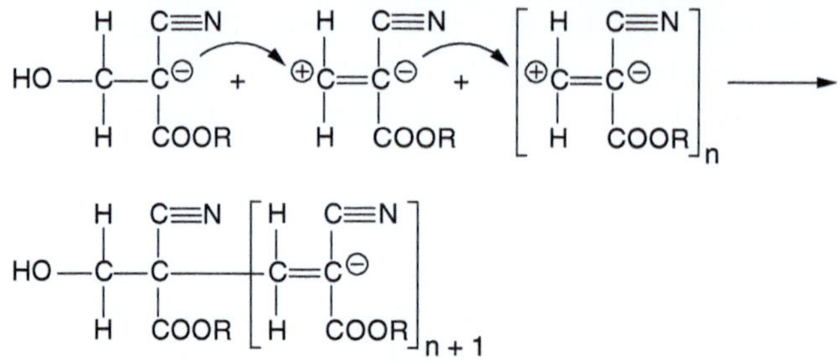

Anionic polymerization of an α-cyano-acrylic acid ester to form a cyanoacrylate adhesive (1).

Cyanoacrylate adhesives are often modified to obtain the following property changes:

- viscosity increase and adjustment of thixotropy by addition of, e.g. PMMA to enable gap bridging up to about 0.20 mm,
- flexibilization by elastomer particles,
- retardation of curing through the use of long-chain alkyl esters as well as the addition of acid-based reaction inhibitors, and
- stabilization by radical scavengers and Lewis acids.

8.2.2.1.3 Curing/Processing

Cyanoacrylates are easy to process, but unmodified do not allow gap bridging of more than 0.10 mm. Therefore, soluble high-molecular-weight acrylates and methacrylates are often added to the base monomers. After joining the substrates, curing (polymerization) of the adhesive takes place in a few seconds to minutes, triggered by the moisture present on the substrates or/and in the ambient air. Curing on less *adhesive-friendly* and acidic materials, such as wood and cork, can be enabled or accelerated by the use of activators.

8.2.2.1.4 Properties

A wide variety of different cyanoacrylate adhesives are now offered to the user in the form of colorless, clear liquids with a slightly pungent odor.

The adhesives differ in terms of

- the chemical base, in particular the ester groups, which determines the suitability of the adhesive for bonding certain materials such as elastomers, plastics, or metals, and
- the viscosity, which depends on the application method and the properties of the material to be bonded.

Influence of Ester Group Cyanoacrylate adhesives with various ester groups are offered, with ethyl-ester-based systems covering by far the largest range of applications. These adhesives show good strength and aging behavior, especially when bonding with plastics and elastomers. Other cyanoacrylate adhesives are based on methyl, butyl, or alkoxy esters, with adhesives based on methyl esters being the first in industrial use. Because of their unfavorable aging behavior on elastomers and plastic surfaces, this type of adhesive became increasingly less important. However, due to their relatively good temperature and chemical resistance, they are still considered for metal bonding. Due to their – compared to the methyl- and ethyl-based systems – longer-chain structure and thus lower curing speed, butyl-ester-based systems are preferred for medical applications. In this field, they are used for bonding human and animal skin, since hardly any heat is generated during polymerization of these monomers. This property of butyl-ester-based adhesives is also of great advantage for the stress-free bonding of heat-sensitive plastics. Due to their higher-molecular weight, alkoxy-ester-based cyanoacrylate adhesives have very low odor, which is very often an advantage and preferred by users. However, lower strengths and curing speeds must be expected with these systems compared to ethyl ester adhesives.

Cyanoacrylate adhesives are not crosslinked, are therefore thermoplastic in nature and exhibit relatively low flexibility. As a result, they exhibit high creep values, especially at elevated temperatures, which can have a negative effect on the quality of the adhesive bond. The average temperature load-bearing capacity of cyanoacrylate adhesives is around 70–80 °C under continuous load, and 100–110 °C is possible for short periods. Due to their chemical structure, these adhesives are not particularly resistant to the attack of moisture. However, to compensate for these disadvantages, the formulations can be upgraded by adding crosslinkers or thermally stable polymers. The mechanical properties of cyanoacrylate-based adhesive layers are adjusted by selecting the appropriate alkyl esters, with the elasticity of the adhesive layer increasing as the length of the alkyl esters increases. Cyanoacrylate adhesives are offered from brittle-hard, low-viscosity standard products to thixotropic and slow-curing versions. Also, specifically fine-tuned products with for example improved dynamic strength, temperature and humidity resistance, and good behavior in bonding materials with different coefficients of thermal expansion are available.

8.2.2.1.5 Applications

Because of their very fast cure, high yield, and ease of use, cyanoacrylate adhesives are used for many industrial and craft applications. After manual *drop-by-drop* or semi-automatic application of the liquid adhesive and subsequent joining of the substrates, which is done under light pressure, a thin adhesive film is formed that usually cures in a few seconds. Cyanoacrylate adhesives are ideal for bonding small and flat surfaces (Figure 8.4). Increased curing speeds and improved properties when bonding critical substrates are often realized by adding suitable activators.

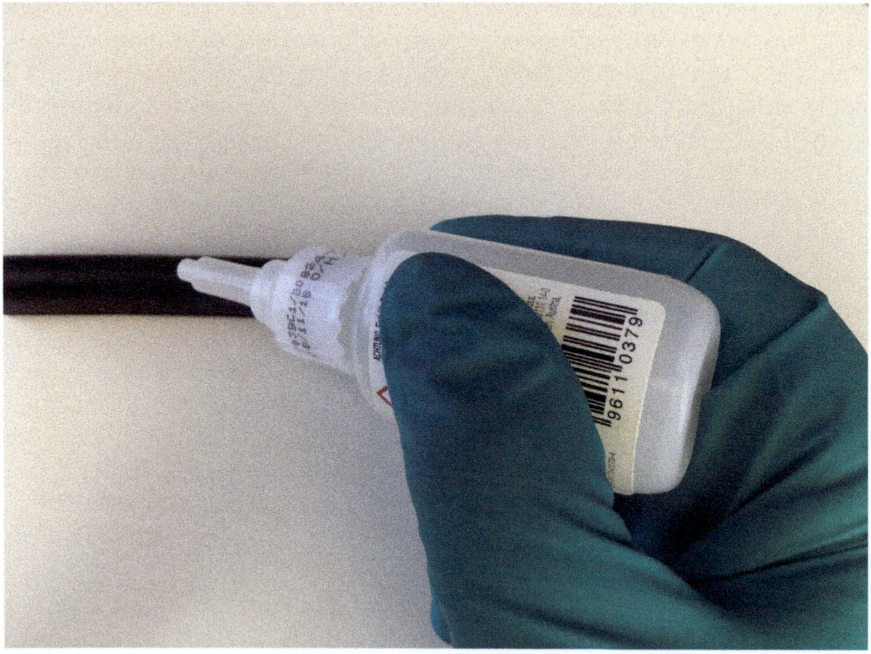

Figure 8.4 Dispensing a cyanoacrylate adhesive onto an EPDM seal used for car doors. Source: Dr. Jürgen Klingen, Aboso-Consulting.

The adhesives are successfully used in numerous bonding applications with metals, plastics, elastomers, glass, ceramics, wood, leather, and cork. However, the main applications are certainly the bonding of plastics and rubbers, where they are used in particular for bonding rubber profiles and elastomers that are difficult to bond, such as EPDM. Very good adhesive bonds can also be produced with PS, SBR, natural rubber (NR), PMMA, and PC, but special primers must be used for applications with non-polar plastics such as PE and PP. Here, the primers not only enable an increase in adhesive strength, but also lead to a considerable improvement in aging resistance. Due to the presence of many hydroxide groups on glass surfaces, cyanoacrylate adhesives cure extremely quickly on such substrates. This creates high stresses in the boundary layer, which leads to a dramatic deterioration of the high initial bond strength over time. This effect is further exacerbated by heat and moisture stresses, which is the reason why high-quality, durable adhesive bonds on glass surfaces cannot be achieved with cyanoacrylate adhesives.

Medical Applications In the medical field, special cyanoacrylate adhesives based on butyl and octyl esters are increasingly being used and are successfully replacing traditional sutures in plastic surgery and medical wound care. Special colored adhesives made of *n*-butyl-2-cyanoacrylate are applied here, which – in the presence of tissue moisture – polymerize within 60 seconds and enable very good adhesion to body tissue. These adhesives are colored, as the color intensity of the adhesive layer is used to assess its thickness. Special applications for cyanoacrylate adhesives in medicine include repair of middle-ear membranes, treatment of liver wounds, and bonding of tears in body tissue in groins.

Profile of Cyanoacrylate Adhesives

Short statement:

- Cyanoacrylate adhesives, also known as *super glues*, form thermoplastic adhesive layers and stand for easy handling, very fast curing, and high strengths. They are mainly used for small-area thin-film bonding with a focus on plastics and rubber-elastic materials.

Physical data:

- adhesive strength at room temperature: approximately 25 MPa for ethyl-ester-based systems and 15 MPa for butyl-ester-based systems
- adhesive strength at 100 °C: approximately 12 MPa for ethyl-ester-based systems and 8 MPa for butyl-ester-based systems
- elongation at break: very low
- service temperature: maximum 80 °C, short term 110 °C.

Special features:

- Cyanoacrylate adhesives cure extremely fast,
- show thermoplastic behavior with moderate resistance at elevated temperatures,
- show low moisture resistance,
- allow very good bonding on almost all elastomer surfaces,
- are not suitable for bonding large areas, and
- can usually only be used for very small gap bridges of approximately 0.10 mm.

8.2.2.2 1K PU Adhesives

8.2.2.2.1 Overview

PUs were discovered in Germany by Otto Bayer in 1937 and, since then, the material has been developed for a wide variety of applications. Although the basic patents have expired, new ones are constantly being added, proving the attractiveness of these materials. Today, we encounter PU products in everyday life, for example as a component of numerous coatings, foams, and dashboards. For some time now, PUs have been used in adhesives and sealants, with 1K PU adhesives playing a prominent role in elastic bonding with countless materials in industry and craft.

For bonding applications, PU materials are used as:

- *low-viscosity isocyanate hardeners* for 2K adhesive systems, which are added to the reactive hydroxyl-group-containing part and then cure to form a solid 2K PU adhesive, or
- *viscous prepolymers* for 1K adhesive systems, which are the reaction products of a previous reaction of poly-isocyanates with polyols still containing reactive isocyanate groups, which cure with water or moisture to form a rubber-elastic 1K PU adhesive.

8.2.2.2.2 Chemistry/Composition

PUs are the reaction products of isocyanate groups (R—N=C=O) with chemical compounds containing a mobile hydrogen atom, such as the hydrogen atom in an OH group of alcohols. Isocyanate groups are very reactive, which can be explained by the positively charged carbon atom located in the cumulative nitrogen–carbon–oxygen double-bond system.

According to the following mechanism, the reaction of an isocyanate group with an alcohol produces a urethane:

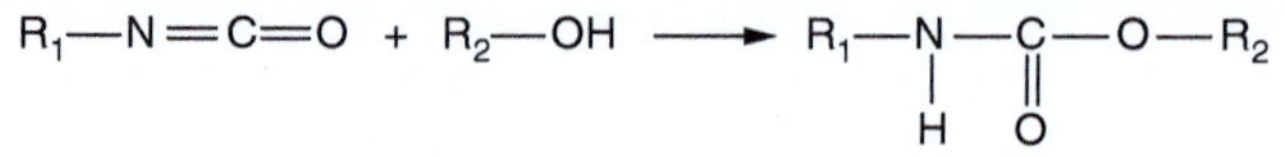

Formation of urethane from isocyanate and alcohol (2).

However, PU-based plastics, including adhesives, require polymers with high-molecular weight, which can only be obtained by using reactants with at least two functional groups. Thus, linear PUs are obtained by reacting bifunctional isocyanates (di-isocyanates) with bifunctional alcohols (diols):

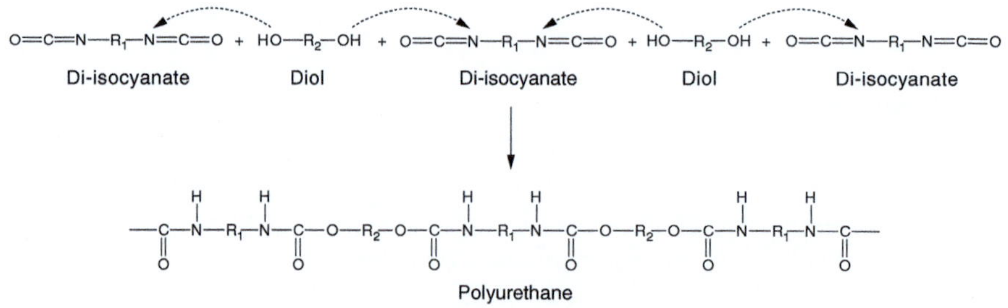

Formation of polyurethane from a di-isocyanate and a diol (3).

1K PU adhesives consist of prepolymers formed in upstream reactions of poly-isocyanates with polyols. In their original form, however, these adhesives still contain free reactive isocyanate groups. After the substrates have been bonded, these groups react with moisture, releasing carbon dioxide, to form urea according to the following mechanism:

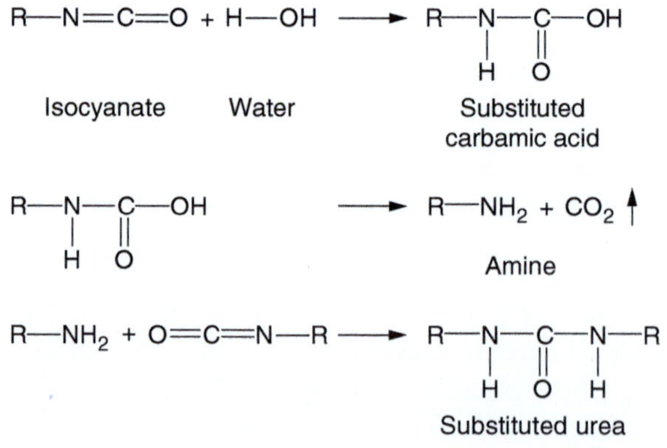

Formation of a substituted urea from isocyanate and water (4).

Even at room temperature, the urea groups just formed can react further with free isocyanate groups present in the adhesive. This results in the formation of biurets and a crosslinked polymer network. In the synthesis of 1K PU adhesives, the reaction mechanism proceeds essentially as described above. However, to obtain elastic adhesives with good strengths, prepolymers must be used that consist of higher-molecular-weight polyols with a stoichiometric excess of isocyanate. Thus, the adhesive contains both already-reacted urethane units and free reactive isocyanate groups, which allow the reaction with moisture described above.

8.2.2.2.3 Curing/Processing

Since the curing of 1K PU adhesives in the joint is accomplished almost exclusively by water from the air, penetrating the adhesive from the edge to the inside of the adhesive layer, a sufficient relative humidity of at least 40% is required in the processing rooms to allow reasonable curing times (Figure 8.5). Large-area bonding with metals or other moisture-impermeable materials is only possible with considerable restrictions in terms of curing speed.

It has already been mentioned that CO_2 is formed when the isocyanates in the adhesive react with water. This can lead to bubble formation in the adhesive layer and thus to a reduction in adhesive strength. This effect is favored by the use of porous substrates and can be reduced by a lower concentration of isocyanate groups in the formulation, a higher viscosity of the adhesive or/and longer curing times.

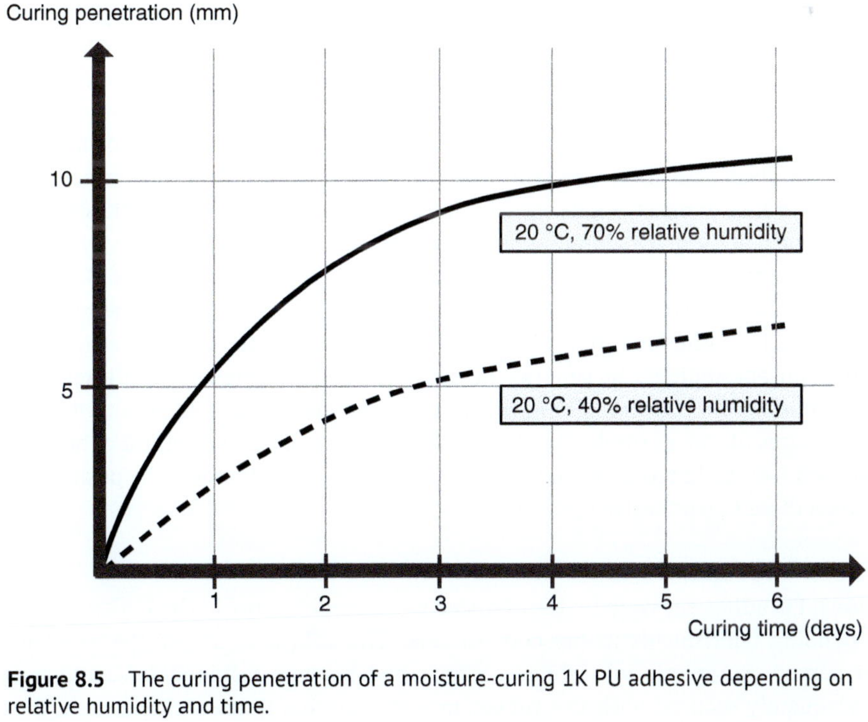

Figure 8.5 The curing penetration of a moisture-curing 1K PU adhesive depending on relative humidity and time.

Booster Systems With so-called *booster systems*, it is possible to accelerate the moisture-curing process of 1K PU adhesives. These systems contain pastes with hardeners that are matched to the corresponding adhesive formulation. With the aid of special application heads, these are injected into the adhesive to form thin layers. This allows short diffusion paths of the hardener into the base adhesive, and the production of a virtually isotropically cured adhesive layer in a short time. The use of this method enables 1K PU adhesives to be cured independent of the humidity present in the processing room. However, with this feature, there is a change from an easy-to-handle 1K to a more complex 2K adhesive system, which must be accepted by the user.

8.2.2.2.4 Properties

Due to the excellent availability of raw materials and many years of experience in research and development, numerous PU adhesives with a wide range of properties are available today. These elastic adhesives are compatible with a wide range of materials, including painted surfaces, ceramics, and some plastics. Primers, adhesion promoters, or activators are often used to improve the adhesion in the bonded joints. In the case of porous or absorbent substrates such as wood or cardboard, care must be taken to ensure that the adhesives, which are usually highly viscous, can establish the full-surface contact required for adhesion buildup. PU adhesives have good resistance to water and slightly acidic or alkaline liquids and only short-term resistance to fuels, mineral oils, and greases. They are not resistant to strong inorganic acids and bases and to the organic solvents commonly used. A characteristic property-determining feature of PU adhesives is the presence of hard and soft domains in the polymer structure. Thus, soft domains with low glass transition temperature are formed by the long-chain polyether or polyester groups, which contribute to the elasticity of the adhesive. Hard domains, formed by the reaction of di-isocyanates with short-chain diols or diamines, form units with high glass transition temperatures and sometimes even crystal structures. They contribute to the strength and high-temperature behavior of the adhesive [3]. The presence of these two domains explains why PU adhesives occur with two glass transition temperatures, one for the elastic domain and one for the hard domain. Due to the diversity of variants available on the market, it is not easy to accurately distinguish PU adhesives from other adhesive systems in terms of behavior and application possibilities. In general, PU adhesives exhibit higher glass transition temperatures than silicone adhesives, but come nowhere near the glass transition temperatures of, for example, highly crosslinked epoxy-based structural adhesives of around 100 °C. In terms of strengths, PU adhesives could therefore be positioned between these two adhesive systems.

8.2.2.2.5 Applications

In 1963, PU adhesives were first introduced by General Motors in the United States for bonding windshields in the body of cars. The adhesives solved the problem that curved panoramic windshields, which had become fashionable, could not be adequately secured with the rubber profiles commonly used. Since then, the use of these adhesives has expanded significantly to numerous applications in industry and craft, including elastic bonding of steel, aluminum, painted surfaces, and plastics. In the automotive sector, they are still used extensively for bonding windshields – as introduced by GM – and additionally for numerous plastic parts. Other applications can be found, for example, in container construction, truck bodies, or the assembly of caravans and mobile homes.

Today, almost all windshields in the body of a car are joined with 1K-PU adhesives and these adhesives are also used when replacing damaged windshields. First, the old cured adhesive layer is cut through from the edge, thus removing the damaged windshield. Then, the 1K PU adhesive is applied directly to the cut layer (Figures 8.6 and 8.7), and subsequently the new pane is positioned, fixed, and cured by the moisture in the air in a relatively short time.

The advantages of using PU adhesives instead of mechanical fastening of the panes are obvious:

- 30–40% increase in body stiffness,
- increased crash resistance in the event of an accident,
- reduction of vehicle mass,
- improved tightness of the windshield system by compensating for body tolerances,
- reduction of driving noise due to damping properties of the adhesive, and
- possibility of automated windshield installation by robotic systems.

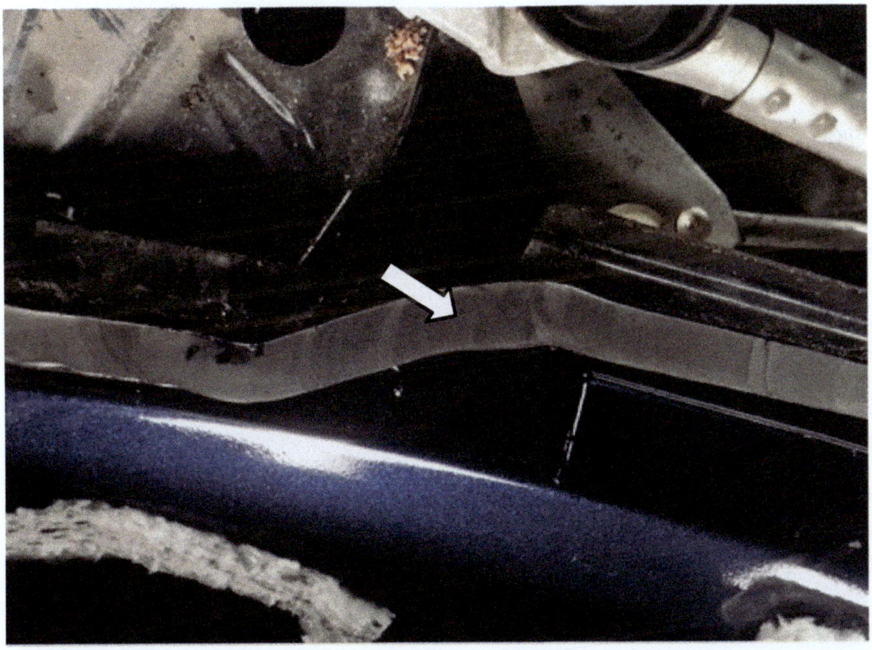

Figure 8.6 Old cut-open adhesive layer before re-application of 1K PU adhesive to repair the windshield of a car. Source: Dr. Jürgen Klingen, Aboso-Consulting.

Figure 8.7 Application of a 1K PU adhesive to an old cut-open adhesive layer to repair the windshield of a car. Source: Dr. Jürgen Klingen, Aboso-Consulting.

Profile of 1K PU Adhesives

***Short statement*:**

- Moisture-curing 1K PU adhesives are rubber-elastic adhesives with relatively low creep resistances that bridge large gap widths and are often used to bond materials with very different coefficients of thermal expansion.

***Physical data*:**

- tensile strength: 3–6 MPa
- elongation at break: 200–600%
- service temperature range: from −40 to 100 °C.

***Special features*:**

- 1K PU adhesives are sensitive to UV radiation,
- show low media resistance compared to epoxy adhesives,
- tend to form a foam structure due to CO_2 evolution during curing,
- often requires primers,

- allow the adhesive to be applied to the cut adhesive layer for repair, as with windshields, and
- the adhesives can be painted over.

8.2.2.3 1K Silicone Adhesives (RTV-1)

8.2.2.3.1 Overview

In the early twentieth century, silicones were recognized as a new class of materials and made available industrially. Today, they can be found in a wide variety of applications where their elasticity and/or heat resistance are valued, including high-temperature rubbers, materials for building protection, heat-resistant and elastic bakeware, body implants, as well as adhesives and sealants. 1K moisture-curing silicone adhesives (RTV-1) are rubber-elastic and are widely used for bonding glass and materials with very different coefficients of thermal expansion. Their special ability to withstand both extremely low and extremely high temperatures is highly valued in many demanding applications.

8.2.2.3.2 Chemistry

RTV-1 silicone adhesives originally consist of poly-dimethyl-siloxanes with OH groups at the ends, which are then modified with multifunctional silanes and thus initially blocked. This silane modification prevents both premature, undesirable crosslinking during storage and – after application of the adhesive and joining of the substrates – curing to the adhesive layer due to moisture.

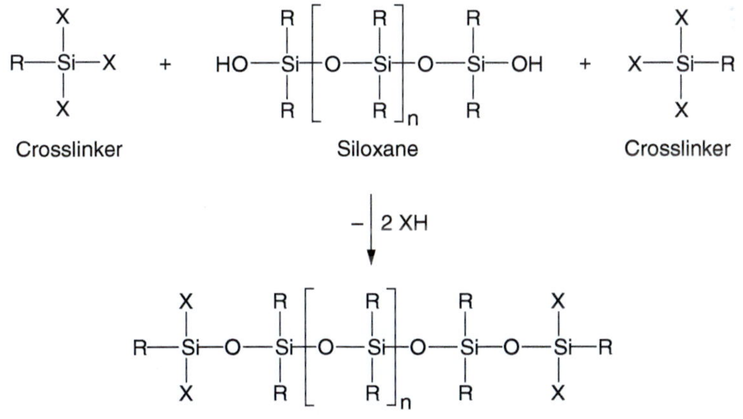

Silane modification of OH-terminated poly-siloxanes (5).

After joining the substrates, the hydrolysis-sensitive silane groups at the end of the poly-siloxane are decomposed by reaction with atmospheric moisture (hydrolysis) to form corresponding cleavage products, the nature of which depends on the chemical composition of the silane groups. Subsequently, the poly-siloxane chains are crosslinked via oxygen bridges. Depending on the type of products released during hydrolysis of the silane groups, 1K PU adhesives are classified as acetate, amine, and neutral systems, which release acetic acid, amines, oximes, and alcohols accordingly during curing.

8.2.2.3.3 Curing/Processing

During the curing of 1K silicone adhesives, very characteristic smelling cleavage products are often released, which – as already mentioned – depend on the chemical nature of the silane products used.

Acetate Systems These adhesives can be recognized by their strong odor of acetic acid, which serves as an indicator of the degree of curing. Therefore, the bonded components must not be stressed in the application as long as the odor of acetic acid is perceived, which indicates that the curing work has not yet been completed, and thus the bonded joint does not yet have the appropriate final strength. Acetate systems should not be used on acid-sensitive materials such as lead, iron, marble, and limestone. On copper and brass surfaces, these systems promote the formation of green rust.

Amine Systems They have a *fishy* odor during curing, were used for a long time but are used less frequently today. However, due to their very high physical performance and chemical stability, they are still preferred in industrial applications where high loads are applied to the bonded joints. Examples include applications in power plants, in air-conditioning system construction, and in the food industry for the construction of counters or superstructures in cold stores. A special feature of the amine systems is the autocatalytic effect of their corresponding cleavage products, which enables complete curing of bonds up to an overlap length of 5 cm. Another advantage is that they can be used to join corrosion-sensitive substrates, where acetate systems often fail.

Neutral Systems These alcohol- or oxime-releasing adhesives are most commonly used in industrial applications today, cure without strong odor, but do not offer the excellent aging resistance of acetate or amine systems. They are offered as *general-purpose* products exhibiting good adhesive performance on many substrates, including plastics, such as PC, PMMA, and PVC, as well as metals, glass, and minerals. Their neutral nature also makes them suitable for bonding corrosion-sensitive substrates.

Since RTV-1 silicone systems react with ambient humidity, the curing speed of such systems is largely determined by the diffusion of water molecules into the adhesive layer, depending on humidity, temperature, and layer thickness. At 23 °C and a relative humidity of 50%, the curing rate is about 1–2 mm per day, but correspondingly longer curing times are recorded if the adhesive layer is cured at lower relative humidity. The complete network of the silicone adhesive is usually formed after several days.

8.2.2.3.4 Properties

1K silicone adhesives are very elastic by nature and have some unusual properties compared to most organic adhesives due to their basic inorganic structure. One outstanding feature is their resilience of up to 90%. This refers to the ability of a material to return to its original shape after mechanical stress.

High-Temperature Performance Silicones also offer excellent temperature resistance of around 120–180 °C, which in the case of iron-oxide-modified acetate or oxime systems can be significantly improved up to 250 °C in the long term and 300 °C in the short term. The reason for the – compared to organic adhesives – significantly higher temperature resistance of these systems is in particular the high covalent-bonding energy of the Si—O bond (368 kJ/mol) compared to the C—C bond (348 kJ/mol).

Low-Temperature Performance Another exceptional property of silicones is their very high flexibility at low temperatures. The reason for this property is the high chain mobility of poly-siloxanes caused by the presence of different bond angles in the macromolecules, which loosen the polymer network. Thus, in the Si—O—Si group there is a bond angle of 143° and in the O—Si—O unit a bond angle of 110°. This fact also explains the extremely low values of the glass transition temperatures of the poly-siloxanes, which are as low as −123 °C.

Constant Modulus Over a Range of 120 °C Another special feature of 1K silicone adhesives is their constant modulus of elasticity in the temperature range from −40 to +80 °C, which allows hardly any embrittlement (at low temperatures) and no significant softening (at higher temperatures). Also, silicone adhesives show very good resistance to water, aqueous salt solutions, and dilute acids and bases. However, prolonged contact with organic solvents such as aliphatic or aromatic hydrocarbons, as well as esters and ketones, results in reversible softening, with no chemical changes in the adhesive.

8.2.2.3.5 Applications

The very elastic silicone adhesives allow very good adhesive bonds with materials that have very different coefficients of thermal expansion. Examples include the assembly of silicon solar cells or ceramic tiles with metallic materials, such as those used in aerospace applications. These adhesives are also able to establish excellent adhesion to silicate-containing materials such as ceramics, glass, or enamel, which can be explained by the chemical similarity of silicones to silicates. Due to their excellent UV and weathering resistance, silicone adhesives and sealants are very well suited for long-term outdoor use, especially on glass surfaces, which these adhesives have already impressively demonstrated over many decades. The most demanding application here is the manufacture of all-glass facades (structural glazing), in which silicone adhesives are used to bond insulating glass units to metal adapter frames, which are then mounted on the building to create a uniform, well-designed modern facade.

Profile of 1K Silicone Adhesives (RTV-1)

***Short statement*:**

- Moisture-curing 1K silicone adhesives are rubber-elastic and are widely used for bonding glass as well as materials with very different coefficients of thermal expansion.

***Physical data*:**

- tensile strength: around 1 MPa
- elongation at break: up to approximately 800%
- service temperature range: –120 to 280 °C, short term 300 °C.

***Properties*:**

- 1K silicone adhesives exhibit rubbery elastic behavior with low creep resistance (better than PUs),
- withstand both extremely low and extremely high temperatures,
- have excellent resistance to UV radiation and moisture,
- recover up to 90% to their original shape after loading,
- show a constant modulus of elasticity in a temperature range from –40 to 80 °C,
- show good electrical insulation properties,
- are susceptible to mold, and
- cannot be painted over.

8.2.2.4 Hotmelt Adhesives

8.2.2.4.1 Overview

Birch tar can be considered the first systematically produced hotmelt adhesive in human history. The black, tar-like distillate obtained from birch bark by dry distillation was a common all-purpose adhesive in the Stone Age, used primarily for shafting tools and weapons. Hotmelt adhesives are thermoplastic materials that are solid at room temperature and are applied to the substrate surface in a thin liquid state after melting (Figure 8.8). After cooling, a solid adhesive layer with good adhesion to the substrate is formed, which is predominantly in the amorphous or semi-crystalline state. Hotmelt adhesives are used in numerous applications in industry and the trades where ease of processing and rapid curing is required. The disadvantage of these systems is that they do not meet high-temperature requirements, which significantly limit their range of application.

8.2.2.4.2 Chemistry/Composition

Depending on the requirements of the application, various polymers and copolymers are used as raw materials for hotmelt adhesives, which essentially determine the properties of the adhesive layer, such as adhesion, tack, and temperature behavior.

The main polymers and copolymers used in hotmelt adhesive formulations are:

- PAs,
- saturated polyesters,
- ethylene vinyl acetate copolymers (EVA),

Figure 8.8 Application of a molten hotmelt adhesive onto a wooden surface. Source: Dr. Jürgen Klingen, Aboso-Consulting.

- polyolefins,
- block copolymers – styrene-butadiene-styrene (SBS) or styrene-isoprene-styrene (SIS), and
- polyimides.

Although the properties of the adhesive layer, such as adhesion, cohesion, and temperature behavior, are basically determined by the type of polymer used, chemical substances must be added to adjust special properties such as cohesive strength, viscosity, softening point, and solidification rate.

These include:

- tackifying resins such as hydrogenated rosin, terpenes, and hydrocarbon resins to increase tack, wetting properties, adhesion, and reduce viscosity during processing,
- plasticizers such as long-chain phthalic acid esters for use in polymers that lack flexibility,

- antioxidants such as aromatic amines or phenols as free-radical scavengers to reduce oxidative influences during processing of the melt under the influence of oxygen,
- fillers such as chalk or titanium dioxide as extenders and for cost reduction, and
- waxes to control open time, softening point, and viscosity.

8.2.2.4.3 Curing/Processing

To adequately wet the substrates, hotmelt adhesives are either melted for example by the use of manual hotmelt adhesive guns (Figure 8.9) prior to application and then applied to the material surface or introduced in the form of powders, films, or nets between the parts to be joined and then – under contact pressure and heat – converted to the molten state. The melting of the adhesives takes place in a temperature range of 120–240 °C, whereby the viscosity of the molten adhesives depends on the type of base polymer and its molecular weight. Spraying, extruding, or rolling processes are used to apply the melt to the substrate surface and subsequently, joining is carried out under pressure, after which the melt cools directly and thus cures. The solidification rate and the resulting open time of the adhesive depend on the difference between the processing and melting temperatures of the adhesive as well as the temperature and thermal conductivity of the parts to be joined.

A major advantage of these adhesives is that they can transfer forces immediately after cooling, thus allowing further processing of the just joined components directly after the bonding process. The temperature curve depending on time, which can be

Figure 8.9 Manual hotmelt adhesive gun and the corresponding adhesive in the form of dispenser sticks. Source: Dr. Jürgen Klingen, Aboso-Consulting.

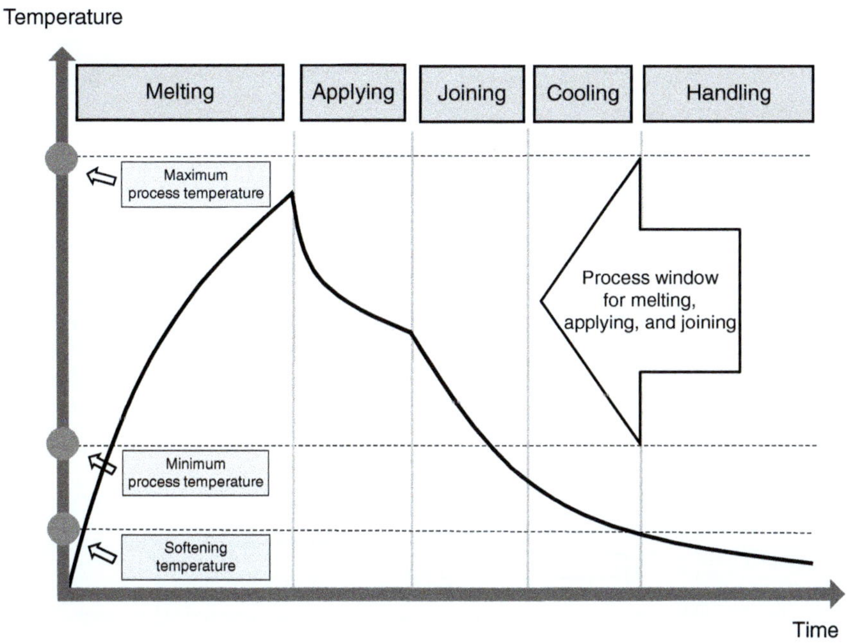

Figure 8.10 Temperature profile during the processing of hotmelt adhesives.

observed during the processing of hotmelt adhesives from melting to their handling strength, is shown in Figure 8.10.

A non-crosslinked hotmelt adhesive is by nature heat-resistant only up to its processing temperature, which very often makes it unsuitable for applications where high-temperature performance is required. A special feature of bonding with hotmelt adhesives is that they can be detached by heat and can be rejoined after this step without losing performance. When bonding metals, which naturally have very good thermal conductivity, the solidification of the molten adhesives progresses very rapidly and thus often prevents complete wetting of the substrate surfaces. Preheating of the parts to be joined is therefore recommended in such cases.

Process Temperature Control The development of viscosity as a function of temperature throughout the bonding process with hotmelt adhesives is shown in Figure 8.11. and, because of the considerable influence of viscosity on the properties of the adhesive layer, the temperature must be very well controlled. For this purpose, the temperature range during melting, applying, and joining the adhesive on the material surface can be divided into two ranges, one with very strong (Range 1) and one with less dependence of viscosity on temperature (Range 2). In Range 1, even small changes in temperature lead to an extraordinarily large increase in viscosity, and poor temperature control there can easily lead to insufficient wetting of the adhesive on the substrate surface and thus to poor bonding.

In Range 2 – which represents the proposed process window of the adhesive – temperature fluctuations do not have such a strong effect on the viscosity of the

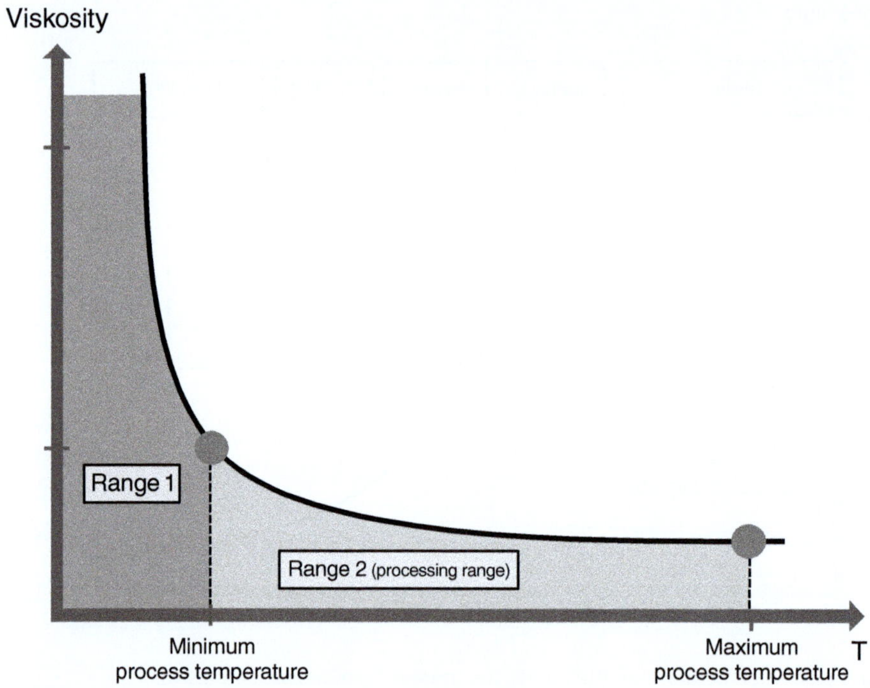

Figure 8.11 Temperature ranges when melting, applying, and joining hotmelt adhesives.

processed adhesive. Here, the viscosity remains relatively constant at a low level, which allows good wetting of the surface and constant adhesive layer thicknesses. Therefore, the processing temperature of hotmelt adhesives must be selected accordingly so that processing in Range 2 takes place above the minimum processing temperature. The processing of adhesive systems therefore takes place in the range of approximately 120–240 °C – in exceptional cases up to 260 °C – and depends on the chemical nature of the adhesive, in particular its base polymer and the respective molecular weight. Upwards, the processing temperature is limited by the temperature that still allows suitable melt stability. Since it determines the time between bonding and subsequent further processing of the bonded components, the difference between the melting and processing temperatures of hotmelt adhesives is important; the smaller the difference, the shorter the time to achieve good initial strengths for further processing.

Depending on the base polymers, hotmelt adhesives are processed at the following temperatures:

- *PAs*: 120–240 °C, mostly >200 °C,
- *EVA copolymers*: approximately 150 °C (maximum 200 °C),
- *polyolefins*: 120–200 °C,
- *amorphous poly-alpha-olefins*: 170 °C,
- *polyesters*: 120–200 °C,
- *PUs*: 130–180 °C, and
- *rubbers*: 130–180 °C.

8.2.2.4.4 Properties

Hotmelt adhesives cure *automatically*, so to speak, by solidification or crystallization of the melt and thus obtain their full adhesive strength without any further external action. In this respect, they are superior to other adhesive systems, which usually require much more processing and additional external action for curing. This is the reason why hotmelt adhesives are frequently used in fast-running automated manufacturing processes. The temperature resistance of these adhesives results from their viscosity, which is essentially determined by the molecular weight of the base polymer. Thus, the higher the viscosity at the processing temperature, the higher the strength of the adhesive at higher temperatures. In this context, the formation of a semi-crystalline structure after cooling of the polymer melt is advantageous for good strength and favored by slow cooling of the bonded components. If the bonded components are cooled too quickly for economic reasons, this is prevented. High-molecular weight hotmelt adhesives with correspondingly high viscosities enable high adhesive strengths typically of approximately 15–20 MPa at room temperature.

Compared to their fellow competitors, the chemically curing and solvent-based adhesives, hotmelt adhesives offer the following advantages:

- as they are free of solvents and volatile monomers, place no or only minor demands on environmental protection, occupational health and safety,
- offer greater processing flexibility, since no components need to be evaporated or mixed,
- enable very high production speeds, as solidification times are in the range of seconds, and
- create thermally removable and recoverable components.

However, the following disadvantages must be considered when using hotmelt adhesives

- lower temperature resistance and higher creep at elevated temperature and continuous static mechanical load,
- significantly higher processing viscosity compared to solvent-borne and monomer-based adhesives,
- higher swelling (mostly reversible) and higher solubility compared to crosslinked adhesives,
- considerable heat load on the materials due to the high temperature of the melt,
- necessary preheating of highly thermally conductive substrates, and
- the provision or purchase of melting equipment.

Curing Hotmelt Adhesives In cases where adhesive strength and/or long-term aging resistance need to be improved, curing hotmelt adhesives can be used. Such systems allow subsequent crosslinking of the adhesives after the entire bonding process has been completed. Thus, after cooling to room temperature, the joined parts already have sufficient handling strength for the subsequent manufacturing steps, and curing to final strength is subsequently affected by humidity or elevated temperatures.

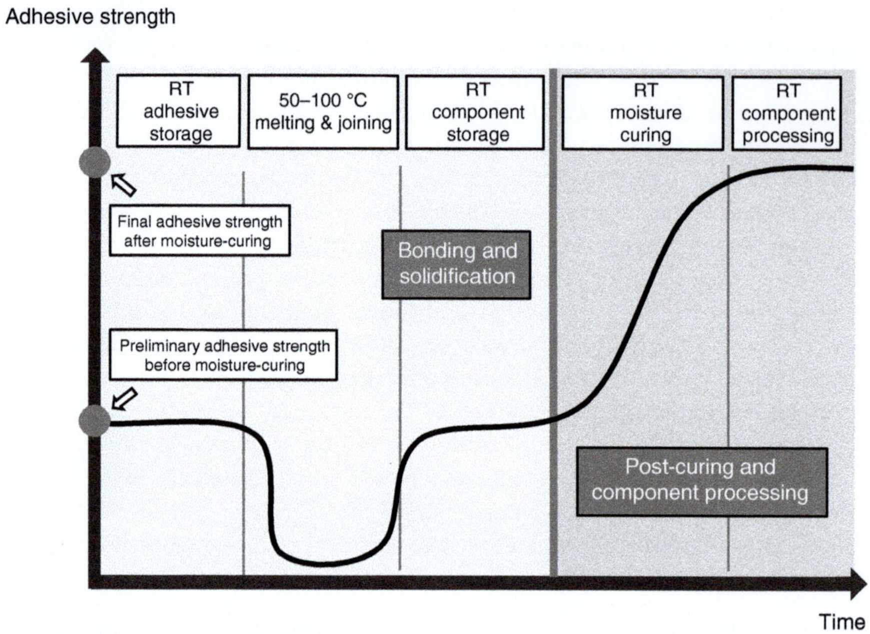

Figure 8.12 Development of adhesive strength of a moisture-curable hotmelt PU adhesive depending on time (not in scale).

Such adhesive systems combine the advantages of classic hotmelt adhesives with those of structural adhesives. Various reactive hotmelt adhesives are available today, which can be divided into moisture-curing PUs and heat-curing epoxy systems in terms of their crosslinking behavior. Figure 8.12 shows the development of bond strength during processing of a moisture-curing PU hotmelt adhesive from storage of the adhesive through melting and joining to curing at room temperature and, at the end of the process, to final strength of the bonded component.

PU hotmelt adhesives based on polyester polyols represent the most widely used post-curable hotmelt adhesive systems, in which the curing reaction occurs due to the moisture in the air diffusing into the adhesive layer. The result is adhesive coatings with strengths and heat resistances that far exceed those produced with uncured thermoplastic hotmelt adhesives. Typically, such adhesives are melt processed at relatively low temperatures of 50–100 °C and used in a temperature range of −40 to 100 °C, offering bond strengths of about 8 MPa with elongations at break of up to 600%. Epoxy-based curing hotmelt adhesives are typically melted at temperatures of about 60–80 °C and then applied to the substrates. These systems contain curing components that are in a non-reactive solid state at room temperature, but initiate curing at elevated temperatures. Dicyandiamide (DICY), which has a melting point of 200 °C and is insoluble in the epoxy resin up to about 100 °C, is often used for this purpose. By heating the adhesive system to above 100 °C, dissolution of the dicyandiamide in the epoxy resin takes place and the adhesive cures to its final strength. Such adhesive layers exhibit thermoset properties that allow bond strengths of up to 10 MPa (tensile strength) and elongations at break of about 20%.

Bonded components with these adhesives can be permanently exposed to temperatures up to about 100 °C.

8.2.2.4.5 Applications

Hotmelt adhesives are used for bonding applications with numerous materials in almost all industrial and craft sectors, as well as in household and *do-it-yourself* applications. Preference is given to these systems for applications where easy and fast application of the adhesives is required, as well as fast direct processing of the bonded components. However, applications are limited to those with relatively low-temperature resistance requirements. With the use of curable hotmelt adhesives, this limitation no longer exists.

General-purpose hotmelt adhesives are offered in various forms for applications in the following fields:

- *various industries* – for automatic labeling of containers in high volumes,
- *paper and packaging industry* – for sealing paper, cardboard, and tin packaging (cans) and for the production of cigarette paper and address labels,
- *shoe industry* – for various applications as a substitute for solvent-based contact adhesives,
- *woodworking industry* – for bonding wood in general and specifically for the manufacture of door and window frames,
- *automotive industry* – for bonding insulating and upholstery materials, headlight spreaders in the metal frames and hubcaps,
- *electrical engineering* – for bonding coil windings, heat-shrinkable tubing, and
- *manufacture of hygiene products* – for articles such as baby diapers.

High-performance hotmelt adhesives based on polyamides, polyesters, and polyimides are used, for example, in:

- *apparatus engineering* – for difficult bonding tasks with painted and coated surfaces,
- *automotive industry* – for the manufacture of seat heaters, parcel shelves, and roofs,
- *wood construction* – for applications with high demands on bond strengths,
- *construction industry* – for windows and acoustic ceilings, and
- *textile industry* – for various garment interlinings.

Labeling of Bottles and Containers Hotmelt adhesives have also proven their worth in the wrap-around labeling of glass bottles and containers made of metals or plastics, offering different adhesive properties depending on the container material. Due to their high initial tack and short setting time, they are recognized in this field as fast bonding adhesives with very good *take-on* characteristics. As a result, they enable the rapid application of paper and plastic labels via rolls or nozzles at speeds of up to 60 000 containers per hour. Water-soluble or dispersible hotmelt adhesives are needed for returnable and non-returnable bottles, meeting the industry requirement that the adhesives must be washable and not interfere with the subsequent cleaning or recycling process of the bottles. For this, various systems have been developed to ensure that, for example, the bottles can be reused without any problems or that the plastic material can be recycled cleanly according to type.

Heat-shrinkable Tubing In the electrical industry, hotmelt adhesives are used to produce heat-shrinkable tubing to protect cables from water penetration. For this purpose, PE hoses crosslinked chemically or by electron beam are coated with hotmelt adhesives, then expanded at high temperature and subsequently cooled. By applying heat again to the cold hose, the material shrinks until it reaches its original shape. In this application the PE is coated with a predominantly PA hotmelt adhesive, which allows simultaneous bonding of the shrink tubing to the corresponding substrate by applying heat.

Wood-bonding Applications Hotmelt adhesives based mainly on EVAs, PAs, and amorphous polyalpha olefins (APAOs) are used for various applications in the wood industry. However, their thermoplastic behavior and limited heat resistance, as well as their tendency to cold flow, make them unsuitable for load-bearing joints. Crosslinked hotmelt adhesives solve this problem in the wood industry by providing sufficient strength to perform stacking and subsequent processing of components. Then, during storage for hours and days, the adhesive strength is subsequently built up to its maximum.

In addition to improving materials, there has also been a strong focus in recent years on improving the handling of hotmelt adhesives. This has led to a wide range of dispensing forms offered today, including sticks for applicator guns, granules for automatic dispensing, or adhesives packaged in films. Such products are used in the manufacture of PSA tapes, which – including packaging – are dropped into the melt reservoir, then melted, and applied to a backing.

Profile of Hotmelt Adhesives

Short statement:
- Hotmelt adhesives are thermoplastic adhesives used for bonding applications with various materials in industry and the trades, where easy application of the adhesives as well as rapid further processing of the components are required and the high-temperature demands are relatively low.

Physical data:
- adhesive strength: up to 20 MPa (depending on base polymer, molecular weight, and degree of crystallization)
- elongation at break: approximately 5–150%, with post-crosslinking systems up to 600% (PU based) or 20% (epoxy based)
- service temperature range: depending on the softening ranges of the base polymers, for post-crosslinking PU systems −40 to 100 °C.

Special features:
- Hotmelt adhesives enable very fast bonding with cycle times in the second range,
- are thermally detachable and can also be bonded again,

- allow almost unlimited storage time,
- are relatively easy to swell and dissolve,
- show high creep behavior under continuous static load (especially with EVA, less with PP) and elevated temperatures, and
- their solidification rate or open time depends on the difference between melting and processing temperature.

8.2.2.5 Photoinitiated Adhesives

8.2.2.5.1 Overview

Photoinitiated adhesives, which are cured with the assistance of visible or ultraviolet light, are well established in today's world of adhesives (Figure 8.13). When these systems are exposed to light, radicals or cations are generated that cause polymerization of the base resins in the system and thus cure the adhesive. In this class of adhesives, a distinction is made between radical-curing systems, which require continuous irradiation to cure completely, and cationic-curing systems, which require only brief triggering with light to cure the adhesive layer completely. Photoinitiated adhesives are successfully used in many applications for precise and fast bonding of components with small surfaces.

8.2.2.5.2 Chemistry/Composition

The preferred materials for photoinitiated adhesives are based on acrylates (radical polymerizing) and epoxies (catatonic reacting). For curing, the adhesives require radiation with energy densities suitable to initiate polymerization of the base resins directly or indirectly. For this purpose, UV radiation or visible light is absorbed by the photoinitiators that are part of the adhesive formulation and, after transfer to a high-energy state, enables very rapid curing of the adhesive.

Radical-curing Systems These systems are characterized by the polymerization of monomers or oligomers with reactive carbon-carbon double bonds (C=C) initiated by radicals. Resins with vinyl functional units such as acrylate or methacrylate

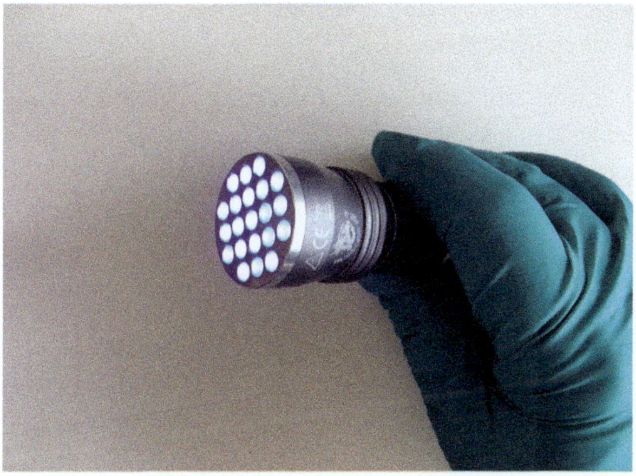

Figure 8.13 Radiation source for manual curing of photoinitiated adhesives. Source: Dr. Jürgen Klingen, Aboso-Consulting.

groups are used for this purpose, often compounded with reactive diluents to achieve the required processing viscosities. After completion of the curing reaction, these are covalently incorporated into the adhesive layer and are thus part of the adhesive network.

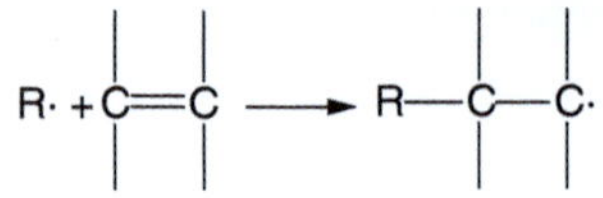

Initiation of the reactive system through the formation of monomer radicals (6/1).

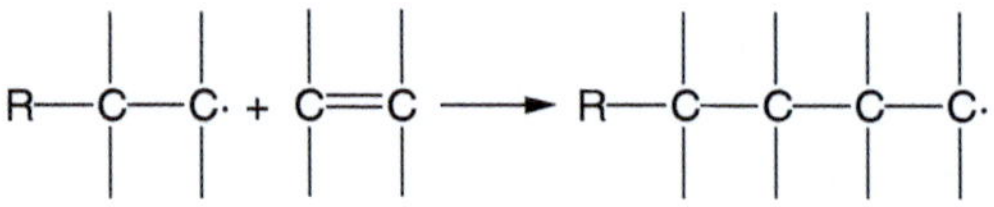

Formation of chain start molecules (6/2).

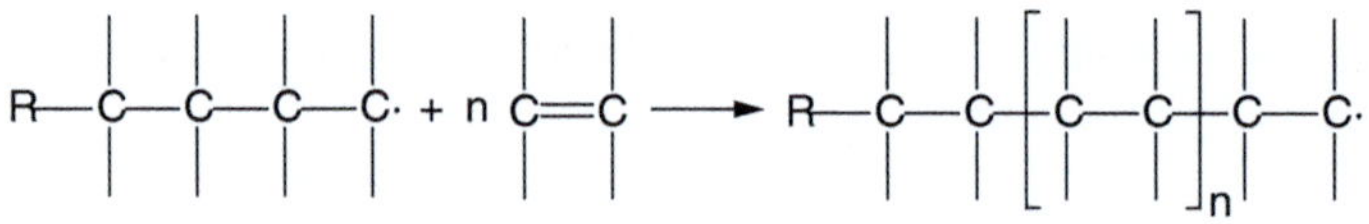

Chain growth to form the adhesive layer (6/3).

To enable monomers or oligomers that do not contain carbon-carbon double bonds (C=C), which are required for radiation curing, acrylate groups are incorporated into such molecules. Compared to the adhesive containing only acrylates, such systems can improve the properties of the adhesive layer, e.g. in terms of adhesion as well as temperature or aging resistance.

The following monomers are used for this purpose:

- epoxy acrylates,
- urethane acrylates,
- polyester acrylates,
- polyether acrylates, and
- silicone acrylates.

In the case of free-radical reacting acrylates, at least one substrate must be permeable to UV or visible light. Characteristically, these systems require continuous irradiation; switching off the light stops the reaction immediately and can lead to incomplete curing of the adhesive layer.

Cationic-curing Systems Monomers based on cycloaliphatic epoxides that cure by ring-opening polymerization are particularly suitable for such systems. In this reaction scheme, a cation attacks the epoxide ring, opens it, and then the reaction with another epoxide ring takes place. Subsequently, further ring openings are carried out and curing proceeds in the form of a chain reaction. In photoinitiated epoxy adhesives, the cation is generated by irradiating special onium salts with UV or visible light, which initiates the ring-opening polymerization and thus cures the adhesives. Due to the very fast rate of ring-opening polymerization, relatively little time is available for joining the substrates after initiation of the systems. However, this is only relevant in the case where the adhesive is irradiated before joining the substrates. If bifunctional epoxies are used, highly crosslinked adhesive layers with good mechanical properties and low shrinkage are formed.

Compared to radical-curing systems, the cationic-curing adhesives offer two major advantages:

- *First* – there is no oxygen inhibition during polymerization, which in the case of radical curing leads to chain breakage and thus to interruption of curing.
- *Second* – the course of the polymerization reaction continues after initiation without further irradiation of the adhesive layer, which is not the case with radical-curing systems.

8.2.2.5.3 Curing/Processing

Effective curing of photoinitiated adhesives requires a good match between the absorptivity of the initiator and the emitted radiation of the light source, and to this end the radiation source must be selected according to the chemistry of the adhesive system. Often, metallic surfaces reflect the radiation and accelerate the curing reaction as well as influence the crosslink density near the substrate. Transparent substrates are not necessarily transparent to the UV radiation used, such as PC and PMMA, which contain UV stabilizers that slow or even block UV curing. For successful bonding with photoinitiated adhesives, they must be understood as a *system* where only proper matching of the adhesive components and UV emitters will lead to success.

For this purpose, the following questions must be answered:

- *What wavelengths do the photoinitiators require?*
  The wavelengths required for curing photoinitiated adhesives are in the UVA or visible range from about 320 to 520 nm, using only a small section of the broad spectrum of a light source in which the absorption of the photoinitiator overlaps with the emitted radiation. Thus, for high efficiency, the emission characteristics of the light source and the absorption characteristics of the photoinitiator must be well matched.
- *Which light sources should be used?*
  Today, light sources with different lamp geometries and different wavelength ranges are available to the user. UV radiation is often generated by discharging gases, which are then bundled by special reflectors and thus converted into focused emission. However, a major disadvantage of gas-discharge lamps is the high proportion of IR radiation, which leads to energy losses and temperature increases. This can have a negative impact on the substrates and the adhesive

layers. For this reason, modern LED radiation sources with very narrow emission bands that do not have a disturbing UV or infrared component in the spectrum are increasingly being used today.

- *What UV radiation dose is necessary for curing?*
  The irradiation dose with the unit mJ/cm^2 is characterized by the light intensity multiplied by the irradiation time. Usually, the minimum UV dose required for complete curing of an adhesive layer is communicated in the corresponding adhesive data sheets. However, the light intensity (the unit is mW/cm^2) seen by the adhesive layer is not a fixed physical value and depends on several parameters, including the spectral ranges of the light sources, the distance of the source from the substrate, the type of substrate, the lamp reflector, and the age of the light source. Although minimum required irradiance doses are often communicated by vendors, the exact value depends on the application and therefore must be determined empirically by the user. For common photoinitiated adhesives, the minimum irradiation doses range between 100 and 500 mJ/cm^2, although the increased degree of contamination of the filters and reflectors during production, as well as the aging of the radiation sources, is associated with a decrease in intensity. Therefore – to be on the safe side – an oversizing of the light intensity up to about 50% is suggested.

A key distinguishing feature of photoinitiated systems compared to most of the other adhesive systems is their extremely fast cure speed, ranging from seconds to minutes. For this reason, these adhesives are often used for large-volume, small-area bonding of articles that must be manufactured with high precision and in good quality.

In terms of processing, photoinitiated adhesives are divided into three main systems:

- *Photoinitiated acrylic adhesives* – These adhesives are cured at room temperature by free-radical polymerization using UV or visible light, with at least one joining part transparent to the applied radiation. For joining, a material surface is coated with the liquid adhesive, and after the parts are joined, irradiation is applied through the transparent substrate, which allows complete curing of the adhesive layer in the range of 1 second to several minutes. When setting adhesive thicknesses >0.5 mm, it is recommended to check the conversion of monomers in the adhesive layer and thus the degree of cure. Photoinitiated curing acrylic adhesives allow a long open time before bonding, a very fast curing after irradiation, and very high strengths of the adhesive composites. However, due to the oxygen sensitivity of the adhesives, oxygen must be excluded during the curing reaction, either by using oxygen-impermeable substrates or by curing in an inert oxygen-free atmosphere.
- *Photoinitiated epoxy adhesives* – These adhesives are cured at room temperature by cationic polymerization, which occurs by UV or visible light. After the liquid adhesive is uniformly applied to a substrate, it is initiated with a sufficient dose of UV or visible light for 3–60 seconds. The parts are then joined and, without further irradiation, final strength is achieved in up to 24 hours, with higher temperatures

accelerating curing and leading to increased bond strength. The advantage of such a cationic-curing system is that the parts being joined do not have to be transparent to visible or UV light. However, the shorter open time after irradiation and the longer total cure time compared to the radical-curing systems can be considered as a disadvantage.

- *Dual-curing adhesives* – When radiation curing is not feasible due to unsuitable substrate geometries resulting in shadow zones, the use of light-curing adhesives with a second curing mechanism can be considered. For this purpose, radiation-curing adhesives are offered with three different additional curing methods, namely anaerobic, moisture, and heat-curing systems. The areas of the adhesive that cannot be reached by light are then cured by the anaerobic function (exclusion of air and metal contact), moisture diffusion into the adhesive, or exposure to heat. For this purpose, moisture-curing systems contain isocyanate or silanol groups and heat-curing adhesives contain temperature-sensitive peroxides.

8.2.2.5.4 Properties

Depending on the functionality of the monomers used, the free-radical-curing acrylic adhesives exhibit thermoplastic, and rarely thermosetting, properties. The network density can be adjusted by the raw materials used and determines the elasticity and deformability of the cured adhesive layers. Acrylate-functionalized epoxies, polyurethanes, polyesters, polyethers, or silicones are used for this purpose. The radically curing adhesives require at least one radiolucent joining part and usually cure after a few seconds to minutes. The result is relatively impact-resistant adhesive layers that can be applied permanently at temperatures up to approximately 120 °C, exhibit adhesive strengths of up to approximately 25 MPa and elongations at break of up to approximately 50%. The catatonic-curing epoxy adhesives also allow use in applications with non-transparent materials. Before or after joining, only a few seconds are needed to initiate the liquid adhesive, which then fully cures in up to 24 hours without further irradiation. The result is thermoplastic, rarely thermoset, adhesive layers based mostly on cycloaliphatic epoxies. The bonds, which are permanently resistant up to a temperature of 120 °C, have adhesive strengths of about 25 MPa and elongations at break of up to 20%. In addition, they offer good resistance to oils, solvents, and moisture. However, compared to free-radical-curing acrylics, these systems are much more sensitive to water and significantly more expensive.

8.2.2.5.5 Applications

Photoinitiated adhesive systems are increasingly used in industry and craft due to their precisely controlled, rapid curing, but mostly require at least one substrate that is transparent in the wavelength in which the adhesive is cured. The resulting adhesive layers not only exhibit excellent adhesive properties and high transparency, but also often offer processing advantages such as high production speeds, short cycle times, and favorable process costs. Key applications include glass bonding in the field of optics, such as lens bonding, bonding of glass fibers and safety glass, as well as applications in glass design. Also, in the furniture industry, photoinitiated

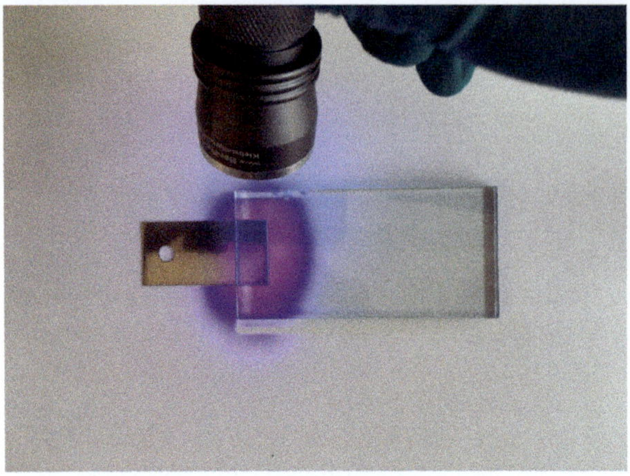

Figure 8.14 Photoinitiated curing of a glass–steel adhesive bond with an acrylic adhesive. Source: Dr. Jürgen Klingen, Aboso-Consulting.

adhesives are used for bonding glass to glass, glass to metal, or transparent plastics (Figure 8.14). In this area, they outperform cyanoacrylate adhesives, which do not offer the required aging resistance, and 2K epoxy adhesives, whose curing speed is too low. Other transparent materials bonded with these adhesives are plastic materials such as PMMA, PS, PC, ABS, or PA. There are also many examples of bonding with photoinitiated adhesives in electronics, where they are used to seal connectors and relays, for example, and to enable fast and precise assembly of smart cards. In the medical supply industry, these adhesives are used to bond needles in syringes. Other applications of the adhesives include bonding artificial fingernails and dental fillings in dentistry. Dual-curing systems are used very successfully in the automotive industry to align and bond rear-view cameras, which enables precise positioning of the cameras. First, the optics are precisely fixed in less than one second by irradiation of the adhesive, and then complete curing takes place thermally at 80 °C.

Profile of Photoinitiated Adhesives

Short statement:

- Photoinitiated adhesives, which usually form thermoplastic, rarely thermoset adhesive layers, are often used for small-area bonding with transparent materials where high production speed, precise positioning of the adhesive, and direct further processing of the bonded components are required.

Physical data:

- adhesive strength: up to 25 MPa
- elongation at break: approximately 50% for free-radical-curing acrylates, up to 300% for urethane acrylates, and 20% for cationic-curing epoxies
- service temperature range: −30 to 120 °C, up to 180 °C for short periods.

Special features:

- Photoinitiated adhesives enable very precise and fast bonding with cycle times in the second range,

- radical-curing acrylate systems require at least one light-transmissive substrate, steady irradiation during the entire curing process and are inhibited by oxygen,
- cationic-curing epoxy systems can be cured in combination with two opaque substrates, are sensitive to water, and more expensive than the radical-curing acrylates, and
- dual-curing products, which have two curing mechanisms, are used in applications with unsuitable substrate geometries causing shadow zones that are not accessible to radiation curing.

8.2.2.6 Anaerobic Adhesives

8.2.2.6.1 Overview

These special adhesives, which cure in the absence of oxygen, were invented in 1953 by Dr. Vernon Krieble of Trinity College in Hartford, Connecticut, USA, and simultaneously represented the start of the Loctite company. This company introduced the anaerobic adhesive in 1956, claiming that this unique joining material had finally solved the age-old problem of loose nuts and bolts in machinery and equipment. Anaerobic adhesives remain liquid in the presence of atmospheric oxygen. For curing, they require the exclusion of air and – in addition – metal ions on the substrate surfaces, which are naturally abundant when metallic substrates are used. After curing, these adhesives form thermoset layers, which are typically used for locking screws and for making shaft–hub connections and seals (Figure 8.15).

8.2.2.6.2 Chemistry/Composition

The anaerobic adhesives are also called *liquid adhesives* because they remain in a liquid state as long as they are in contact with the oxygen in the air. When – after the substrates are bonded – the oxygen is no longer present in the joint, the polymerization reaction begins in a very short time at room temperature, catalyzed by metal ions present on metal substrates. The base monomer used for anaerobic adhesives is tri- or tetra-ethylene glycol di-methyl acrylate (TEGMA), which, like photoinitiated

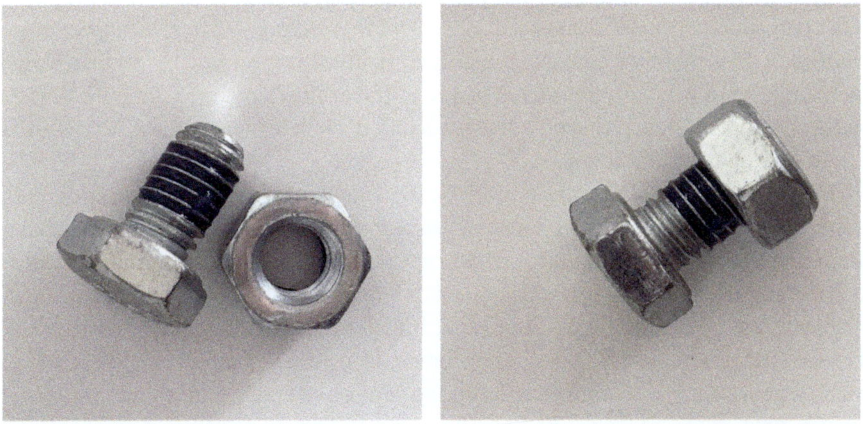

Figure 8.15 Screw locking performed by an anaerobic adhesive. Source: Dr. Jürgen Klingen, Aboso-Consulting.

or 2K acrylate adhesives for example, undergoes radical polymerization for curing. In addition, cumene hydroperoxide, saccharin, and *N*,*N*-dimethyl-*p*-toluidine are part of the formulation of such adhesives, with the cumene hydroperoxide acting as the radical initiator of polymerization.

Reaction Mechanism The complete chemical reaction mechanism upstream to enable free-radical polymerization to cure anaerobic adhesives is relatively complex. First, the residual oxygen in the adhesive joint is consumed. Then, metal ions of a lower oxidation state are formed at the material surface, which subsequently catalyze the peroxide to form the active radicals required for radical polymerization. As with the photoinitiated and 2K acrylic adhesives, the curing of the anaerobic adhesives ends up being a free-radical polymerization of acrylate monomers, in this case using TEGMA as the reactive acrylate monomer.

Here again in detail are the individual steps necessary for curing the anaerobic adhesives:

- First, the residual oxygen present in the adhesive joint is consumed by the reaction of saccharin with *N*,*N*-dimethyl-*p*-toluidine,
- then, an aminal is formed that releases metal ions from the surface of the bonded part and transfers them to a lower oxidation state,
- the metal ions in the lower oxidation state catalyze the decomposition of the peroxide starter into active radicals (possible only in the absence of oxygen), and
- finally, the starter radicals start the radical polymerization of the TEGMA.

The process as a cycle process guarantees the regeneration of the aminal and thus the high number of active radicals necessary for effective curing.

8.2.2.6.3 Curing/Processing

Anaerobic adhesives can only be used for bonding substrates that are able to avoid diffusion of oxygen into the adhesive layer and provide metal ions, especially iron and copper. This is the case, for example, in bonding applications for the production of shaft–hub connections and for screw locking. Before joining the substrates, the adhesive does not cure because the polymerization is inhibited by atmospheric oxygen and after joining, it polymerizes, due to the exclusion of oxygen, but only in the adhesive joint. Thus, uncured adhesive located at the edges remains liquid and can therefore be easily removed.

When bonding so-called *active* materials, the metal ions required for curing are present, whereas when joining so-called *passive* materials, such as plastics, additional metal-containing accelerators (metal sprays) must be used. Alternatively, the metal ions required for these *passive* materials can be provided with an appropriate copper- or brass-containing wire brush that transfers the metal ions to the material surfaces. *Active* surfaces include copper, brass, iron, steel, aluminum (containing copper), and bronze. *Passive* surfaces include stainless steel, chromium, zinc, nickel, tin, silver, gold, chromate, and metal oxide coatings, as well as plastics, elastomers, painted metals, glass, and ceramics. Because of oxygen diffusion from the adhesive edges, the maximum adhesive thickness for bonds with such adhesives is limited to about 0.3 mm. Final strength is achieved after about 3–24 hours. By applying heat for 30 minutes at 120 °C, curing of anaerobic adhesives is possible even without the presence of metal ions on the surface.

8.2.2.6.4 Properties

These relatively brittle and inflexible adhesives are available with adhesive strengths tailored to specific applications. They offer high adhesive strength as well as very good resistance to oils, solvents, and moisture and enable high-temperature applications up to approximately 150–175 °C. In addition to their adhesive function, these adhesives are also frequently used for sealing purposes.

To cover different gap sizes, anaerobic adhesives are offered in different viscosity ranges:

- *very low viscosity:* 10–20 mPas,
- *low viscosity:* 20–200 mPas,
- *medium viscosity:* 200–2000 mPas,
- *viscous:* 2000–20 000 mPas, and
- *pasty:* 20 000–100 000 mPas.

Modifications to anaerobic adhesives allow for performance enhancements such as:

- *absorption of oils and greases* from the surface without loss of performance,
- *high-temperature resistance* up to a continuous operating temperature of about 250 °C, and
- *dual curing* with, for example, UV radiation, where sufficient initial strength for further processing steps is achieved in seconds by UV irradiation, and then the entire adhesive layer is cured to final strength by the anaerobic reaction mechanism.

8.2.2.6.5 Applications

In the appropriate applications, these anaerobic adhesives can be viewed as an additional chemical interlocking, rather than an adhesive that produces full-surface adhesion for bonding something. Typical applications for anaerobic adhesives include screw locking, shaft–hub connections, and surface sealing. One advantage of this curing method is that adhesive that escapes from the joint remains liquid and can therefore be easily removed at any time.

For locking screws, 0.2–0.4 mm thick adhesive layers are used, which are characterized by compressive strengths of 5–40 MPa and high shock and vibration resistance. The systems are designed according to the requirements of the particular applications and present different torques for loosening the screws in case of repair. Anaerobic adhesives are also used in the shrinkage of shaft–hub connections intended for high-temperature applications. The shrinkage effect is achieved by cooling the hot-applied hub and filling the surface roughness with adhesive, allows a very effective interlocking between shaft and hub. In the automotive industry, the properties of anaerobic adhesives are highly valued in the assembly of engines or transmissions. Special versions with improved temperature resistance and the ability to absorb oil are used, for example, for surface seals in gearboxes. In this case, the adhesive layer has high cohesive force but low adhesion to the sealing surfaces, which is advantageous in this application because it allows better disassembly in case of repair.

Profile of Anaerobic Adhesives

Short statement:

- Anaerobic adhesives form relatively brittle thermoset adhesive layers with low flexibility, good strength as well as good temperature resistance and are mainly used for locking screws, making shaft–hub connections and sealing surfaces.

Physical data:

- shear strength: 5–40 MPa
- elongation at break: approximately 5%
- service temperature range: −60 to 150 °C, up to 180 °C for short periods.

Special features:

- Anaerobic adhesives allow easy adhesive application and are very suitable for automation,
- can only handle small gap bridges,
- for curing require the exclusion of oxygen and at the same time the presence of metal ions (preferably copper or iron ions),
- remain liquid outside the joining zone and the adhesive can be easily removed,
- are primarily metal adhesives with good cohesion, high resistance to dynamic loads, but relatively low adhesion properties,
- have good media and heat resistance due to the thermosetting character of the adhesive layer,
- require an additional supply of metal ions to the surfaces when bonding plastic substrates, and
- offer an additional sealing in addition to the screw-locking function.

8.2.2.7 1K Epoxy Adhesive

8.2.2.7.1 Overview

Epoxy resins, synthesized from *epichlorohydrin* and *diphenols*, were invented independently by the Swiss Pierre Castan and the German Paul Theodor Schlack in the late 1930s.

For structural metal bonding in industry and craft, adhesives based on epoxy resins are the most widely used systems, e.g. in metal construction, automotive and aerospace industries, due to their very good adhesive strength combined with high temperature and chemical resistance as well as excellent long-term stability.

8.2.2.7.2 Chemistry/Composition

The high-temperature curing epoxy adhesives consist of resin and hardener, which are already mixed in the required ratio and cure at temperatures above 80 °C after joining the substrates. The curing of these adhesives is achieved by chemical reaction of the very reactive epoxy groups with suitable monomers via a polyaddition mechanism [4, 5].

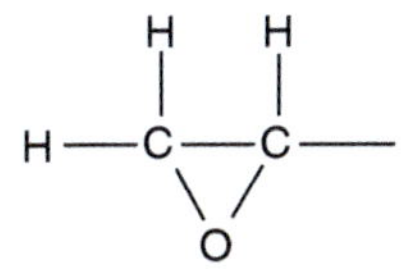

Epoxy group (7)

Depending on the molecular structure, there are different epoxy resins such as:

- *bisphenol A*-based resins,
- *cycloaliphatic di-epoxides,*
- *aliphatic epoxides,* and
- *epoxidized fatty acids.*

Many types of epoxy resins are available on the market for the formulation of adhesives, but by far the majority are based on *bisphenol A*. These resins are produced by reacting *bisphenol A* with *epichlorohydrin* and are preferably used in applications where very good strength is required, especially at elevated temperatures. Here, this property is achieved by the presence of the aromatic ring in the *bisphenol A* unit. Due to the special electron arrangement, the epoxy resins exhibit relatively high polarity and thus very good reactivity, which is another reason for the extensive use of these liquid epoxy resins for adhesive production.

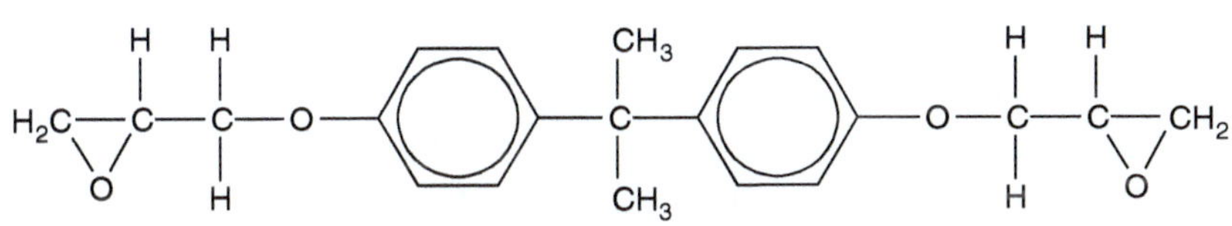

Di-glycidyl ether of *bisphenol A* (8)

Other epoxy resins commonly used in adhesives are *cycloaliphatic di-epoxides*, such as *dicyclopentadiene dioxide*, which form very dense crosslinked networks due to the shorter distance between the reactive groups, leading to high-temperature resistance of the corresponding adhesive layers.

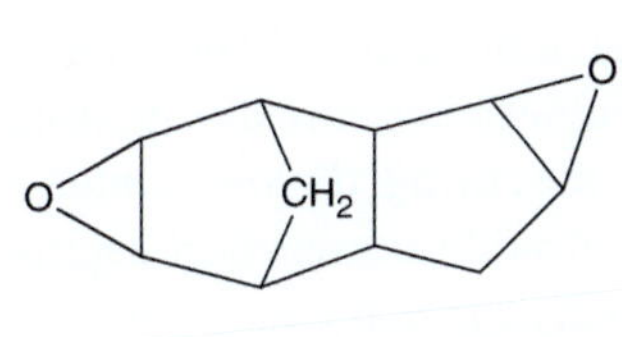

Di-cyclopentadiene dioxide (9)

Epoxy resins can react with polar molecules in two fundamentally different ways, with the reaction always involving an opening of the epoxy ring:

- *addition polymerization* – taking place in 1K and 2K epoxy adhesives, and
- *ring-opening polymerization* – taking place in cationic photoinitiated epoxy-based adhesives.

Addition Polymerization When 1K or 2K epoxy resins are used to form adhesive layers, addition polymerization is usually used, in which highly polar groups (hardeners) attack the epoxy ring and open it at the same time. This reaction preferably proceeds with amino groups, so that the reaction products – the cured adhesive layers – contain two groups with high polarity. These are –NH and –OH groups, both of which have a positive effect on the adhesion behavior of the corresponding adhesives.

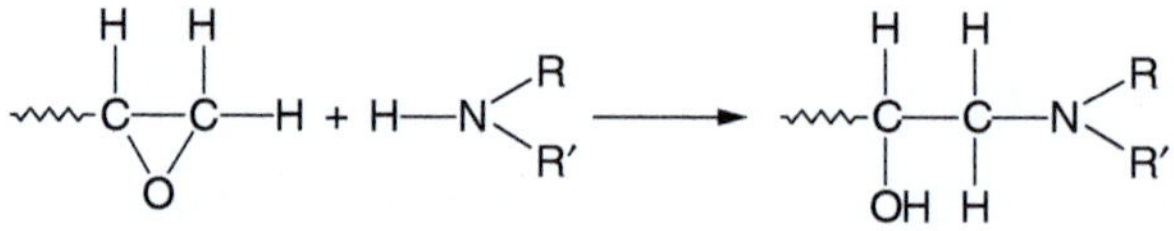

Reaction of epoxy resin with an amine (10).

In such systems, the crosslinking density of the adhesives can be very well controlled by the specific selection of multifunctional amines.

Typical representatives are:

- *diethylenetriamine (DETA)* – H_2N-CH_2-CH_2-NH-CH_2-CH_2-NH_2 and
- *triethylenetetramine (TETA)* – H_2N-CH_2-CH_2-NH-CH_2-CH_2-NH-CH_2-CH_2-NH_2.

In cases where the amine component is added only in a smaller amount than stoichiometrically required, reactions of the formed OH groups with excess epoxide groups are possible, resulting in polyether structures.

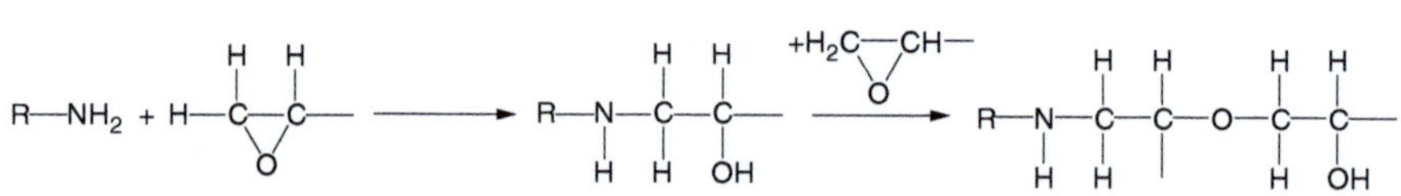

Reaction of the formed OH groups with excess epoxy groups in the adhesive (11).

For the formation of adhesive layers with 1K epoxy adhesives, *dicyandiamide* is used as hardener, which is insoluble in liquid epoxy resins at room temperature and therefore does not allow the adhesive to cure during storage [6]. When the temperature is raised to about 150 °C for curing, *dicyandiamide* is dissolved in the adhesive formulation, and then the polyaddition reaction begins. Such curing systems are used in the automotive industry for many applications, including hybrid bonding in body-in-white applications, where adhesives are combined with spot welding for joining. Another application is hem flange bonding of add-on parts, such as hoods, trunk lids, and doors. In these applications, the heat of around 180 °C generated during the drying of the e-coat is used to dissolve the *dicyandiamide* in the adhesive and enable subsequent curing.

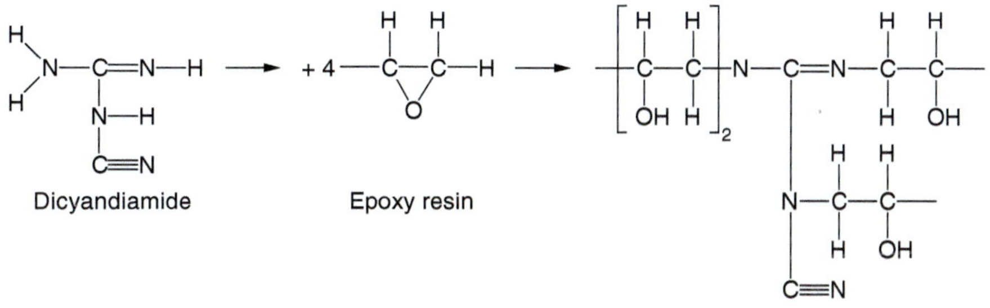

Epoxy curing with *dicyandiamide* (12)

Modification of Epoxy Resins Special curing agents such as aliphatic polyamines are used to improve the flexibility and peel strength of 1K epoxy systems. However, this modification comes at the cost of lowering the glass transition temperature and shear modulus of the corresponding adhesives. By adding elastomer particles or creating a finely dispersed rubber-elastic phase chemically bonded to the epoxy matrix, so called *toughened epoxy adhesives* with high impact and peel strength can be developed without affecting their glass transition temperature and shear modulus. In the automotive industry, specially modified, high-temperature-curing 1K epoxy systems with the ability to absorb oil residues from metal surfaces without affecting the adhesive strength are used very successfully to bond oily steel and aluminum sheets.

8.2.2.7.3 Curing/Processing

1K epoxy adhesives are stored at room temperature and can cure in a few minutes to hours at temperatures ranging from approximately 80 to 190 °C.

Application of these adhesives to substrates is accomplished by:

- *bead application* – a relative slow process,
- *swirl spraying* – fast and carried out with special heads, and
- *nozzle spraying* – very fast and applied for low viscosity adhesives.

Curing of these systems is usually by dicyandiamide, but high-melting aromatic amines or poly-aminoamides are also used. A major advantage of 1K over 2K epoxy adhesives is that both components are already mixed at the adhesive manufacturer. This provides the user with ready-to-use, easy-to-handle systems that do not allow any metering errors.

8.2.2.7.4 Properties

1K epoxy systems cure to relatively hard and tough thermoset adhesive layers and generally exhibit higher adhesive forces of up to 40 MPa and chemical resistances than 2K epoxy adhesives. They are also superior to these in temperature resistance, withstanding heat loads from −60 to 180 °C permanently or up to 220 °C for short periods. Improved flexibility and excellent crash behavior, as required for car doors, are achieved by modifying the adhesives through the addition of elastomer particles or chemically linked, phase-separated rubber-elastic polymers formed in the epoxy matrix.

1K epoxy adhesives offer the following advantages and disadvantages:

***Advantages*:**

- very high strengths,
- very good solvent resistance,
- very good heat resistance, and
- good flexibility in formulating the adhesive.

***Disadvantages*:**

- exerts thermal loads on the bonded components due to exothermic curing reaction,
- limited storage time due to the presence of the hardener in the adhesive,
- additional equipment (ovens) required for curing the adhesive, and
- high-energy consumption for curing.

8.2.2.7.5 Applications

1K epoxy adhesives are preferably used for structural bonding of steel and aluminum. This is done in numerous applications in the automotive industry, especially when very high mechanical and/or high impact strength are required. Since these adhesives are predestined for automated robotic applications, they have been used there for the construction of structural parts and add-on parts around the clock, so to speak, for years [7]. Typical applications there include hem flange bonding of doors, trunks, and hoods, as well as structural joining in the underbody, pillars, and roof. Special 1K epoxy adhesive systems are used in these applications that require no treatment of the oiled steel or aluminum surfaces and yet easily achieve the outstanding corrosion resistance required for the relevant applications. For optimum flange filling, glass beads are often added to the adhesive as spacers to ensure a constant thickness of the adhesive layer in the bond.

To meet the requirements for such applications, such as very good bond strength over a wide temperature range, stiffness, crash resistance, and excellent corrosion resistance, the adhesives must have the following properties:

- moduli of elasticity above 1000 MPa,
- glass transition temperatures above 80 °C,
- high impact strength in a temperature range from −40 to +80 °C,
- good oil compatibility of the adhesives for good performance on oiled surfaces,
- wash-out resistance to the cleaning or e-coat baths used, and

- only moderate decrease in bond strength after prolonged exposure to changing environmental stresses (maximum 20% decrease in bond strength compared to the unaged sample).

The use of 1K epoxy adhesives in car body construction enables – in addition to very good adhesion to automotive substrates – increased body stiffness, crash and corrosion resistance, and also reduces vibrations and noise during driving. Their compatibility with other joining processes, such as spot welding, explains, among other things, the great success of these adhesive systems in bonding steel and aluminum sheets in automotive construction, a success that has lasted for several decades.

Profile of 1K epoxy adhesives

***Short statement*:**
- 1K epoxy adhesives are available as thermosets with excellent adhesive strength and temperature resistance after curing and used for very high demanding structural bonding applications with steel and aluminum substrates.

***Physical data*:**
- adhesive strength: up to approximately 40 MPa
- elongation at break: up to approximately 10%
- service temperature range: −60 to 180 °C, up to 220 °C for short periods.

***Special features*:**
- 1K epoxy adhesives demonstrate very high adhesive strength and very good resistance to mechanical and chemical stresses,
- require a few minutes to hours for curing at temperatures of approximately 80 to 190 °C,
- after modification, tough-elastic adhesive composites with excellent crash resistance are obtained,
- special variants show excellent adhesion to oiled steel and aluminum surfaces without applying cleaning and surface treatment processes, and
- show low creep and thus good long-term resistance to static loads.

8.2.2.8 Solvent-Borne and Dispersion Adhesives

8.2.2.8.1 Overview

These *general-purpose* adhesives supplied in organic solvents or water already contain polymeric components of very high-molecular weight. However, the polymer chains are not chemically crosslinked, and thus there is no polymer network built in the adhesive. To enable their wetting on material surfaces by a low-viscosity state, they are dissolved in organic solvents or dispersed in water, the solvents used and the water serving only as processing aids. After final curing, these media are no longer part of the adhesive layer. In the case of impermeable substrates, such as metal and glass, they must be completely evaporated before the substrates are joined. With porous or absorbent materials, they may be partially present in the adhesive and leave the bonded component after the joining process. Solvent and dispersion adhesives have good wetting properties and can be easily applied by various process methods. Their dried adhesive layers provide good bond strength to a wide variety of materials, including metals, plastics, rubber, leather, cork, felt, wood, cardboard, textiles, glass, ceramics, and concrete. Because these adhesives can bond almost *anything*, they are often referred to as *general-purpose* adhesives.

In terms of how they are processed, solvent and dispersion adhesives can be further categorized into:

- *solvent-borne wet adhesives* – which in their liquid state are placed in contact with one or two substrates and then bonded,
- *water-based wet (dispersion) adhesives* – which in their liquid state are placed in contact with one or two substrates and then bonded,
- *solvent-borne contact adhesives* – which in their liquid state are placed in contact with both substrates and before joining are dried and then bonded, and
- *water-based contact (dispersion) adhesives* – which in their liquid state are placed in contact with both substrates and before joining are dried and then bonded.

8.2.2.8.2 Chemistry/Composition

Solvent-Borne Adhesives The base polymers used for solvent adhesives are PVAC, chloroprene rubber (CR), acrylonitrile butadiene rubber (ABR), PUs, acrylates, PE copolymers with polar groups, PVC derivatives, and synthetic and natural rubbers. The composition of the solvents is basically matched to their dissolving and processing properties and consists mainly of esters, e.g. ethyl acetate, and ketones, e.g. 2-butanone; alcohol is added if necessary. Solvents with different vapor pressures are used to prevent thread formation when the adhesive is applied to the surface and skin formation when the adhesive layer dries. In contact bonding of suitable thermoplastic polymers, e.g. PVC, solvents capable of swelling or dissolving the material surface are often added, which results in molecular diffusion from the adhesive into the substrate. However, to avoid subsequent microcracks and thus inferior adhesive bonds, care must be taken with substrates that are sensitive to stress cracking, such as PMMA, PC, or PS. Solvent contact adhesives are offered with a solids content of 10%–30% and often contain additives such as waxes or resins for tack improvement and antioxidants, fungicides, and UV stabilizers. Although solvent-based adhesives are very user-friendly and offer good properties for a wide range of applications, low-solvent or solvent-free adhesive systems are increasingly being used today due to growing environmental awareness.

Water-Based (Dispersion) Adhesives These environment-friendly adhesives, often based on PVAC, CR and acrylates, offer similar properties and cover similar application areas as solvent adhesives. In these systems, the adhesive is dispersed in water rather than dissolved in organic solvents, with the size of the dispersed thermoplastic or elastomeric polymer particles ranging from 0.05 to 1.50 μm and the solids content typically in the 40–70% range. When the adhesive layers dry, the polymer particles align and coagulate, and a non-reversible film formation takes place on the material surface.

The following polymers are typically used for dispersion adhesives:

- *PVAC* – which provides good adhesion to cellulosic substrates such as wood and paper,
- *PVAC copolymers* – which provide good adhesion to many polar plastics,
- *acrylates* – with good aging resistance for the production of PSA tapes, adhesives for tiles, floor, and wall coverings, and laminating adhesives,
- *SBR* – for laminating adhesives, e.g. for bonding aluminum foil to paper,
- *PUs* – with very good adhesion to various substrates and due to their partial crystallinity of high cohesive strength, which are used in the footwear industry, in laminating applications, and film bonding, and
- *CR* – with high elasticity, cohesive strength, and media resistance, which are used especially in the footwear industry.

Compared to solvent-based systems, dispersion adhesives have some interesting properties that are important for the user. For example, very high-molecular-weight polymers can be used in the adhesives system without an unacceptable increase in adhesive viscosity.

8.2.2.8.3 Curing/Processing

In the case of wet bonding, solvent and dispersion adhesives are usually applied to both material surfaces and then – after a minimum drying time has elapsed – joined wet under pressure. If at least one of the materials to be bonded is porous, the adhesive can also be applied to only one surface before the joining process. However, in the case of impermeable substrates, complete removal of residual solvent or water from the bonded joint is required, which is aided by applying pressure to the bonded component.

When applying solvent or dispersion adhesives, the following times must be observed:

- *Minimum drying time* – In all cases, substrates must not be joined until the minimum drying time has elapsed, even if solvent-permeable materials are used. During this time, most of the solvent or water present in the liquid adhesive escapes and the strength of the adhesive layer steadily increases. Premature joining of the substrates would lead to delayed cohesion build-up of the adhesive layer and thus to extended production times.
- *Open time* – Within the open time of the adhesive, bonding is still possible without reducing the final strength. This time is between the minimum and maximum drying time.
- *Maximum drying time* – This time must not be exceeded. If this were the case, the adhesive layer would be completely solidified and there would be no tack on the adhesive surface. Joining of the parts is then no longer possible.

For joining solvent and dispersion adhesives, the following processes can be applied:

- *wet processing* – where they are placed in contact with one or two substrates in their liquid state and then joined (applied are wet adhesives), or

- *dry processing* – where they are placed in contact with both substrates, dried before joining, and subsequently bonded (applied are contact adhesives).

Wet Processing In this process, the liquid adhesives are applied to one or both substrates and joining occurs between minimum and maximum drying times. After the minimum drying time has elapsed and the wet solvent or dispersion adhesive has been applied, which is usually done on both material surfaces, the substrates are joined under pressure. If production speed is not an issue, very permeable materials can also be joined immediately after adhesive application. In this case, however, the bonded components have very low initial strength, which builds up steadily as solvent or water evaporates through the permeable substrates. In all cases, it is important not to exceed the maximum drying time before joining, as good bonds cannot be made after this time. For this processing method, high-solids, non-flammable dispersion adhesives are very well suited. They exhibit good adhesion and cohesion built up by evaporation of water, and drying time can be shortened by applying heat in the form of infrared radiation or hot air. When the dispersion adhesive layer is completely dry, the color of the adhesive film changes from white to transparent.

Dry Processing (Contact Adhesive Processing) In this process, the liquid solvent or dispersion adhesives are applied to both substrates and then the material surfaces are dried. After the adhesive surface is no longer tacky, the two substrates are joined by pressing them together with a pressure of >0.5 bar, with the resulting adhesion determined by the amount of pressure and not by the time of pressure application; the higher the contact pressure, the better the adhesion properties. In this state, both joined parts are firmly bonded to each other, due to diffusion and partial crystallization of the molecules of the adhesive. Thus, very high initial adhesion is provided and immediate further processing of the bonded components is possible. Their final strength is achieved after about three days. In the process described, which leads to completely dried adhesive layers, repositioning of the substrates shortly after joining is normally no longer possible. However, this can be made possible, if at least one joining part is porous. In this case, contact adhesive processing can also be carried out with partially dried adhesive surfaces. Since the semi-dry adhesive then still has a low viscosity, repositioning of the joined parts can be performed in this case even after the materials have been bonded. This property of re-positionality is particularly appreciated in contact adhesive processing in textile and leather bonding.

8.2.2.8.4 Properties

Solvent and dispersion adhesives generally produce uncured thermoplastic or elastomeric adhesive layers that do not provide the heat and creep resistance achieved with chemically cured adhesives. To increase the strength of these adhesives, slow-acting curing agents can be added to these polymer systems, which do not affect processing but allow crosslinking in a few days. Contact adhesives offer very good aging resistance and can also be used for materials that are critical in terms of adhesion, such as stainless steel. The solvent-based systems are often based on CR, ABR, SBR (styrene-butadiene rubber), and the water-based contact adhesives are usually made with polyacrylates. The disadvantage of solvent-based contact adhesives is the very high solvent content of up to 80% by weight, which requires special protective measures in the processing areas. For this reason, these solvent-based contact adhesives are often replaced by water-based systems, but in this case drying takes considerably longer. Acceleration of this process step is possible by adding heat, but this requires additional energy. Other disadvantages of water-based contact adhesives are the insufficient adhesive strength during drying and the lower strength and resistance of the dried adhesive layers. The reason for this is the inhomogeneity of the polymer layer and the tendency of the adhesive layer to redisperse under the influence of moisture. In any case, dispersion contact adhesives do not allow repositioning of the substrates after joining, as is possible with solvent-based systems in cases where at least one joining section is porous.

8.2.2.8.5 Applications

Wet processed solvent adhesives are used in numerous applications for various surfaces, especially for bonding paper and cardboard, but also for diffusion bonding of soluble thermoplastics such as PVC. For ecological reasons, however, the trend is away from solvent adhesives toward solvent-free systems. Wet processed dispersion adhesives are suitable for similar surfaces covered by solvent-borne systems, such as wood, paper, cardboard, cork, leather, felt, textiles, rubber, plastics and foams, glass, ceramics, and coated metals.

Areas of application for solvent- and water-based contact adhesives include the bonding of trim and soundproofing elements in vehicle construction, the manufacture of composite panels in the furniture and insulation sectors and, last but not least, various applications in the textile and leather industries, such as shoe manufacturing. Highly flexible variants of these adhesives are used for bonding PVC and parquet flooring.

Profile of Solvent-Borne and Dispersion Adhesives

Short statement:

- After curing, solvent-borne and dispersion adhesives are generally present as thermoplastics, which are used in wet and contact bonding processes for joining materials such as paper, cardboard, wood, fabrics, leather, and plastics (less of metals) in the automotive, furniture, textile, and leather industries.

Physical data:

- adhesive strength: up to approximately 10 MPa
- elongation at break: 5–100% for wet adhesives, up to 400% for contact adhesives
- service temperature range: 80 °C for wet adhesives, 100 °C for contact adhesives.

Special features:

- Solvent and dispersion adhesives consist of dissolved high-molecular-weight polymers that are not crosslinked and cure without chemical reaction only by evaporation of the solvent or water,
- must be joined within a minimum and maximum drying time,
- are applied by wet processing (wet adhesives) on one or two sides and still contain significant amounts of solvent or water after joining the substrates; final strength is achieved only after complete drying of the bonded component,
- are applied by dry processing (contact adhesives) to both substrates and dried before joining; they obtain their final cohesive strength by diffusion and/or partial crystallization of the molecules after joining the materials under pressure, and
- tend to creep, especially at elevated temperatures, since the dried adhesive layers are thermoplastic in nature.

8.2.2.9 Pressure-Sensitive Adhesives (PSAs)

8.2.2.9.1 Overview

The first industrially used PSA tape was invented in 1925 by Richard Drew of the 3M Company in St. Paul, USA, who after persistent experimentation succeeded in producing a removable adhesive tape with good adhesion properties. This so-called masking tape enabled car paint shops to paint two-tone cars, which had just become fashionable, safely and in good quality. After further research in his laboratory in St. Paul, Richard presented the first transparent adhesive tape in 1928, which heralded the triumphant advance of PSA technology that continues unabated to this day. PSAs can be found everywhere in everyday life, for example as adhesive tapes, labels, stickers, and adhesive notes (Figure 8.16). They have the ability to transfer forces even after relatively light contact with the surface and can thus take over the function of bonding in the mechanical sense. As supplied, these adhesives are highly viscous polymer systems in which the properties of both a liquid and an elastomer are present. This property enables them to fully adapt to the roughness of the surfaces of the parts to be joined and thus to produce good bonded joints. Extensive research and development work in recent decades has led to the availability of high-performance adhesive tapes in addition to conventional self-adhesive products. Such systems are increasingly able to successfully replace liquid structural and elastic adhesives in a wide range of applications. Modern assembly and manufacturing processes benefit from the simple and fast application of these systems, and in use they score points for their good resistance to mechanical stresses and environmental loads.

PSAs exhibit a number of unique properties that include:

- no change of state during processing,
- a tricky balance of rheological properties,
- soft enough to behave like a liquid necessary to wet surfaces,
- stiff enough to resist peeling and shearing forces,
- sufficiently cohesive strength to hold together when removed, and
- enough cohesive strength to withstand stresses for long periods [8].

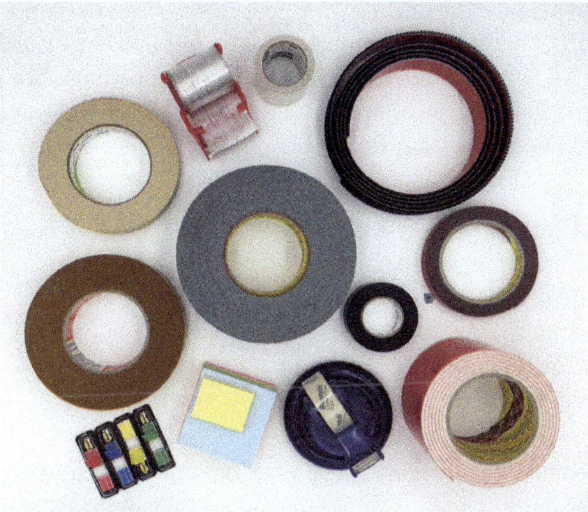

Figure 8.16 A selection of PSA products used in industry, craft, and household applications. Source: Dr. Jürgen Klingen, Aboso-Consulting.

PSAs are viscoelastic materials that exhibit properties that are highly dependent on the time period of a force application and temperature [9]. They are permanently tacky and adhere without requiring more than the pressure of a finger or hand. These products do not require any activation by water, solvents, or heat, and after bonding provide good adhesive properties, sufficient cohesive strength, and elasticity to be removed cleanly from the substrate [10].

8.2.2.9.2 Chemistry/Composition

For a long time, PSAs based on natural rubber dominated the relevant applications; however, to achieve the required tack and adhesion, the rubber material has to be compounded with tackifying resins and additives [11]. The shortage of natural rubber during World War II created the need to develop alternative materials for PSAs. Since 1929, it has been known that polymers based on acrylic acid esters provide the tackiness needed for PSAs [12]. These synthetic polymers are transparent and colorless and have much better aging resistance compared to rubber-based formulations. Today, after decades of successful continuous research and development, various types of acrylic copolymers are available, enabling the production of countless PSA tapes for a wide variety of applications, and in particular the manufacture of high-performance PSA foam tapes. For example, formulations based on acrylic acid and 2-ethylhexyl acrylate or iso-octyl acrylate are often used, which generally do not require tackifying resins to meet the requirements of a PSA [13]. For higher cohesion and thus better performance, PSAs are often chemically crosslinked. When block copolymers, which were introduced in the 1960s and are usually based on isoprene and styrene, are used, physical crosslinking is achieved by phase separation of the glassy styrene blocks in the elastomer matrix. However, due to their low-temperature resistance, these systems are not suitable for applications with corresponding high-performance requirements.

Nowadays, a variety of base polymers for PSAs are in use, which are often additionally compounded with, for example, resins, plasticizers, fillers, and stabilizers to achieve the desired PSA properties. Also, these polymers can be equipped with different functional groups. This results in a large number of adhesive formulations that, after coating on a wide variety of substrates, allow the design of PSA tapes in the form of, for example, adhesive tapes, self-adhesive films, and sticky notes that are used every day for a wide range of applications.

The following polymers are used for PSAs:

- natural, synthetic isoprene-based and butyl rubbers,
- acrylic copolymers,
- SIS or SBS block copolymers,
- polyvinyl ethers,
- polyesters, and
- polysiloxanes.

Natural Rubber PSAs When raw natural rubber is used in PSA formulations, it must be mechanically digested to reduce molecular weight allowing uniform consistency and good solubility in organic solvents. Alternatively, constant-viscosity (CV) grades are used, which already have a low-molecular weight in their as-supplied state. After the rubber is dissolved, tackifiers, plasticizers, and oils must be added to set the rheology required to be a PSA capable of wetting surfaces as supplied [14].

The Dahlquist criterion An important indicator for the basic requirement a PSA must meet is the Dahlquist criterion, which was established by Carl Dahlquist of the 3M company in the 1960s [15]. It defines the requirement for tack that a material must offer to be a PSA and thus be able to wet surfaces at room temperature. According to the criterion, it is required that the G' (storage modulus) value after deformation of the corresponding material for 1 second at room temperature must be less than $3 \cdot 10^5$ Pa. If this is the case, materials are able to wet surfaces and thus allow the required tack. For good peel behavior in the debonding phase, when the PSA is peeled from the substrate, G' must increase significantly, when going from 1 second to 0,01 second of deformation time. To provide good shear behavior, G' of such material must not drop too much, measured G' after very long times of deformation. PSAs based on natural rubber meet this criterion, but only after the addition of, for example, tackifiers, plasticizers, and oils. Thus, they obtain the rheology capable of wetting surfaces as supplied. These PSAs can also be crosslinked to improve cohesive forces, and stabilizers and fillers can be added to extend service life and improve overall performance.

The formulation of a natural rubber-based PSA consists of three main components, which are composed of as follows:

- 100 parts natural rubber,
- 50–120 parts tackifier – rosin or its derivatives, terpene or hydrocarbon resin, and
- a small amount of antioxidant – amines, phenols, or di-thiocarbamates.

Optional ingredients are plasticizers, pigments, or other elastomers, e.g. SBR. Natural rubber PSAs are often crosslinked by sulfur, sulfur-donating compounds (e.g. alkylphenol sulfide), or di-functional isocyanates. These PSAs, which are typically compounded with completely nonpolar hydrocarbons, allow only dispersive molecular interactions at the material surface, which supports or makes the often-requested removability possible in the first place. Due to this nonpolar composition of the PSA caused by the absence of polar groups, its performance is therefore controlled almost exclusively by the rheology. To this end, the blend of elastomer, tackifier, and plasticizer must be fine-tuned to achieve the optimum property profile required for the particular application, such as good tack, peel, and shear behavior. For this effort, it is very helpful to consider the storage modulus G' at the deformation times relevant to the properties of PSAs.

The required G' behavior of a good PSA is shown in Figure 8.17, where the dependence of G' on the deformation time at room temperature of a typical PSA is demonstrated:

- *0.01 second deformation time* (1) – where for good peel performance the G' value must be much higher than at 1 second deformation time,
- *1 second deformation time* (2) – where for good tack performance the G' value must be less than $3 \cdot 10^5$ Pa (Dahlquist Criterium), and
- *longer deformation times* (3) – where for good long-term creep performance the G' value must not decrease too much.

The addition of tackifiers to the natural rubber leads to an increase in the glass transition temperature, whereas the plateau modulus of the final blend is

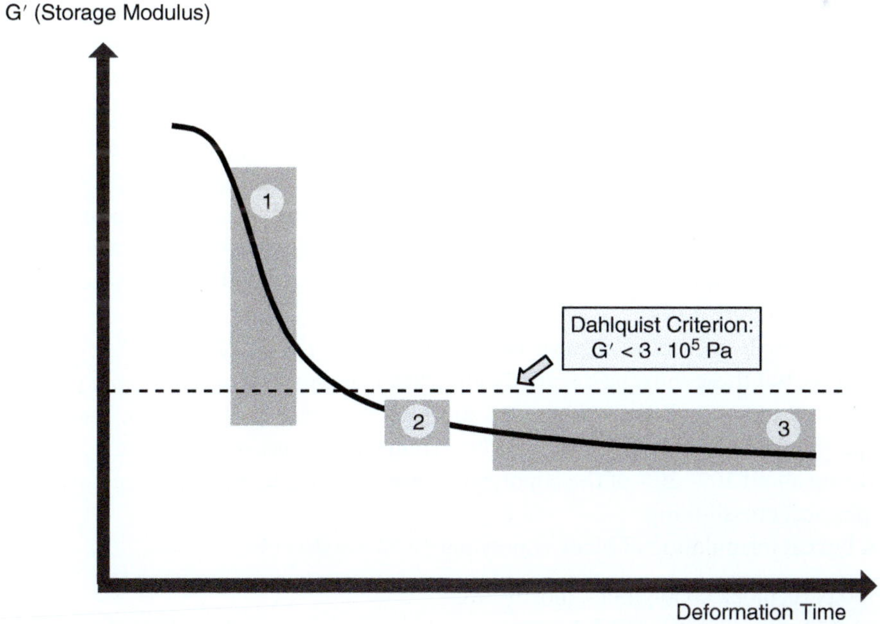

Figure 8.17 Areas of the storage modulus *G′* in a typical PSA curve relevant to the properties of PSAs.

reduced and thus, allowing the system to meet the Dahlquist criterion for tack and additionally the requirement for peel. The primary result of this modification is a significant increase in tack. However, peel properties are also greatly improved, caused by the significant increase of G' in the debonding time scale of 0.01 seconds. Like the tackifier, oils and liquid tackifiers also dilute the rubber and contribute to the reduction of the plateau modulus. This is good for tack but, unlike tackifier addition, has a negative effect on peel properties.

PSAs based on natural rubber offer the following advantages:

- low-cost raw materials are used,
- generally, show very good PSA performance,
- allow good adhesion to low-energy surfaces (base on its non-polar character), and
- enable good high-temperature resistance after additional chemical crosslinking.

There are also some disadvantages using natural rubber-based PSAs, which are:

- need for tackifier additions to meet the Dahlquist criterion for tack,
- relatively high variability of the rubber raw material,
- complex processing required to produce the PSAs,
- susceptible to UV and thermal degradation,
- high solvent consumption for the manufacture of the PSAs, and
- in some cases, the protein contained in the natural rubber can cause skin allergies.

Block Copolymer PSAs Block copolymers are thermoplastic synthetic rubbers with controlled molecular weight and architecture. Most commonly used for PSAs are linear SIS block copolymers consisting of polystyrene blocks (S) and polyisoprene blocks (I). In the PSA matrix, the styrene blocks are phase-separated with high glass transition temperature (hard segments) to provide physical crosslinking. The B blocks (soft segments) form the continuous, rubbery phase. Block copolymers also contain a very small amount of diblock, which cannot build bridges to other styrene domains and thus cannot participate in the physical crosslinking (Figure 8.18) [16].

Instead of linear SIS block copolymers, star-shaped structures are also used for PSAs, which typically have lower melt viscosity during adhesive preparation and allow the preparation of electron beam crosslinked PSAs with high-temperature resistance (Figure 8.19).

Block copolymers also require tackifiers to meet the Dahlquist criterion for tackiness, but these must be thermodynamically compatible with the rubber phase. They essentially *swell* the rubber and are thus essential for good tack and peel properties of the PSA. In contrast, the glassy, polystyrene-based hard phases – which make up about 10%–30% of the copolymer – are responsible for forming a network by physical crosslinking.

A typical formulation of block copolymer PSAs consists of:

- 100 parts block copolymers, mostly SIS,
- 80–120 parts tackifiers (solid and/or liquid),
- antioxidants and UV stabilizers.

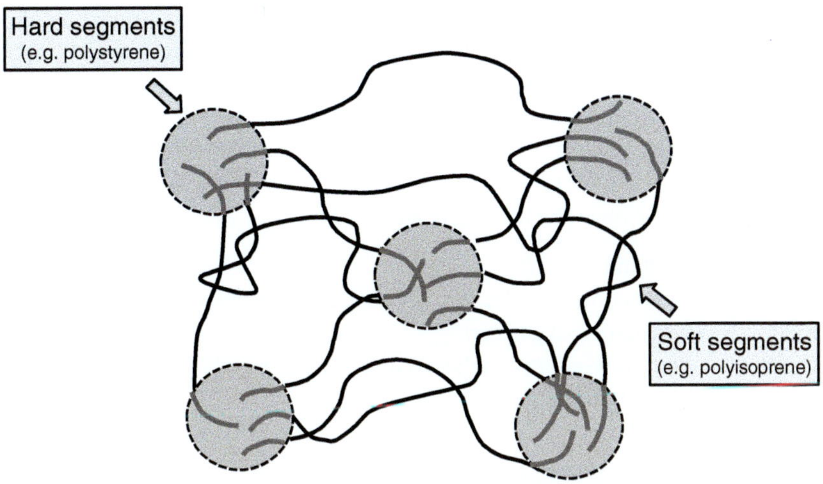

Figure 8.18 Organization principle of thermoplastic block-copolymers with polystyrene hard and polyisoprene soft segments.

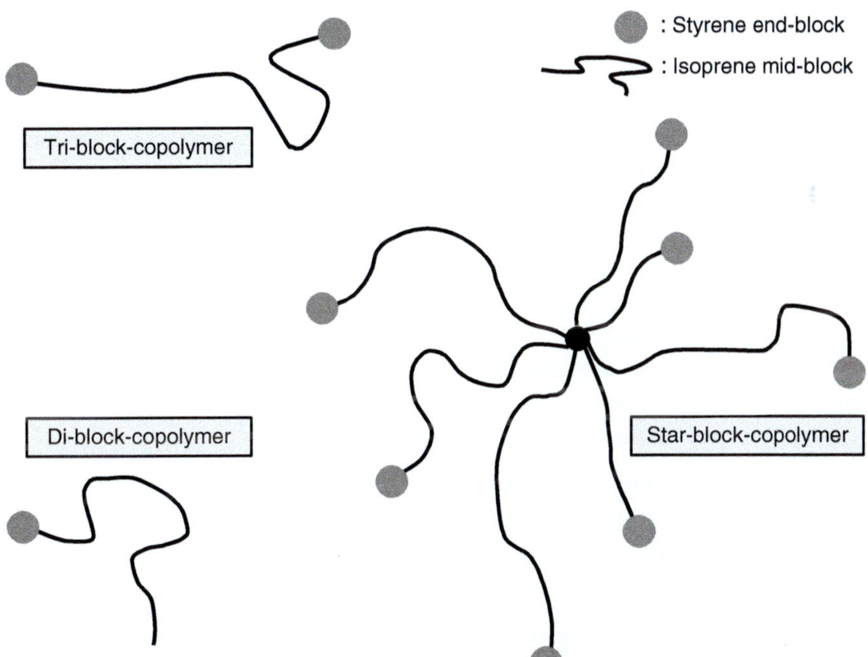

Figure 8.19 The three types of thermoplastic block copolymers found in PSAs.

Aliphatic hydrocarbons (C5), rosin esters, or poly-terpenes are used to tackify the isoprene phases of the block copolymer PSA. In addition, aromatic resins with high glass transition temperatures that are compatible with the end blocks can be added to increase the temperature resistance of the PSA. These accumulate almost exclusively in the styrene domains, increasing their glass transition temperatures and thus the resistance of the PSA to thermal stress.

Block copolymer PSAs have the following advantages:

- low-to-medium raw material costs,
- allow excellent quality control of the synthetic block copolymer,
- very good PSA performance in general,
- offer good adhesion to low-energy surfaces (due to its non-polar character),
- can be produced by hotmelt process, and
- do not require a crosslinker.

The disadvantages of block copolymer PSAs are:

- require tackifiers to meet the Dahlquist criterion for tack,
- susceptible to UV and thermal degradation, which does not apply to the hydrogenated versions,
- do not offer high-temperature performance.

Acrylic PSAs Unlike natural rubber or block copolymer adhesives, acrylic PSAs contain polar monomers that provide an additional variable to control performance. Polar acrylate monomers, including the commonly used acrylic acid, contribute to the rheological behavior of the adhesive and enable interfacial interactions between the PSA and substrate, as well as within the PSA. Thus, hydrogen bonds, acid/base, ionic, and dipole/dipole interactions take place, improving the adhesion and cohesion of the PSAs. This is of great advantage when polar substrates such as metal (and metal oxide), glass, and polar plastics, e.g. PVC, are bonded with these PSAs. Additionally, the polar acrylate monomers also allow good control of compatibility with additives, e.g. tackifiers and plasticizers [17].

Acrylic PSAs provide the following features:

- tacky already without tackifier addition,
- resistant to sunlight and oxidation,
- clear and non-yellowing, and
- compatible with human skin.

These adhesives are suitable for a wide variety of applications and, when properly formulated, provide good adhesion to low surface energy substrates. They exhibit very good plasticizer and gasoline resistance and, because of their breathability and moisture permeability, respectively, are ideally suited for medical adhesive tapes applied to human skin.

A typical acrylate-copolymer for PSAs consists of:

- 90–98 parts of higher alkyl acrylate esters, and
- 10–2 parts of polar monomers, often acrylic acid.

Optionally, 10–30 parts of lower alkyl acrylate replacing part of the higher alkyl acrylate esters, 0.01–1.0 parts of crosslinker, and 10–60 parts of tackifier can be added to improve cohesion or/and tack [18]. As described above, polar groups (such as —OH, $—NH_2$, —C=O, or —COOH) in the side chain increase adhesion to polar surfaces such as glass or metals. However, they also increase the glass transition temperature of the PSA, leading to an undesirable decrease in tack.

The alkyl acrylate esters are essential structural elements of acrylate PSAs, where the length of the side has following effects to the performance of the acrylate PSA:

- *Monomers with long side chains* (higher acrylate esters), such as 2-ethyl hexyl acrylate (glass transition temperature: −70 °C) lead to a reduction in the glass transition temperature of the PSA and thus increase the flexibility of the adhesive layer.
- *Monomers with short side chains* (lower acrylic acrylates), such as *t*-butyl acrylate (glass transition temperature: +35 °C), increase the glass transition temperature of the adhesive and thus also the hardness of the adhesive layer, but at the expense of adhesion.

The molecular weight also has a decisive influence on the quality of the acrylate PSA; the higher the molecular weight, the higher the cohesion of the resulting adhesive layer. If one wishes to further increase the tack of already-tacky acrylic PSAs or modify the viscoelastic properties, tackifiers can be added to act as diluents for the base polymer and improve compliance (softness) and wet-out. Low-molecular-weight tackifiers are used for this purpose and must be compatible with the acrylics. Typically, they are based on rosin esters, terpenes, and synthetic hydrocarbons, which have softening points in the range of 50–120 °C. To improve high-temperature resistance, acrylate PSAs can be crosslinked via the functional groups present in the polymer. Additionally, there are other advantages, such as improved creep resistance under mechanical loads, especially at higher temperatures, increase of cohesive strength, and better solvent resistance. On the negative side, crosslinking acrylic PSAs results in a possible slight decrease in adhesion and tack, increased likelihood of pop-off failure in shear, tack and peel tests, and poor stress relaxation behavior of the PSA.

Tackifiers To improve especially the tack and peel properties, the following tackifying resins are used for PSAs:

- *Rosin acids and esters derived* – are coming from wood extracts and are purified, esterified, or hydrogenated. Rosin acids are used for water-based PSAs, rosin esters for natural rubber, SIS block copolymer, and acrylic-based PSAs.
- Terpenes and modified terpenes (phenol or styrene) – are prepared by cationic polymerization of monomers derived from wood and citrus peels and optionally contain modifiers. Such resins are applied for natural rubber, SIS block copolymers, and in some cases acrylate-based PSAs.
- *Hydrocarbon tackifiers* – are produced by polymerization of monomers obtained from petroleum cracking, which can be hydrogenated to improve stability and

color. The result is aliphatic (C5), aromatic (C9), or mixed (C5 and C9 combined) resins. Aliphatic hydrocarbons are used for natural rubber and SIS block copolymers (mid-block). Aromatic hydrocarbons are applied for better temperature resistance of polystyrene (end-block) of SIS copolymers and increased tack of acrylate-based PSAs.

PSA Backings Tape properties such as tensile strength, elongation, and elasticity are largely determined by the materials used as backing for the PSA.

For PSA tapes, the following carrier materials are used:

- *polymer films* – made of PE, PVC, PET, and PA,
- *paper* – as non-woven or creped,
- *foams* – made of PE, PP, PU, or polyacrylates,
- *glass fabrics* – as nonwovens, and
- *metal foils* – e.g. made of aluminum.

Semi-structural PSAs In the 1990s, acrylic/epoxy hybrid foam tapes were developed to provide users with the benefits of PSA technology while at the same time being able to produce thermally resistant adhesive bonds with semi-structural properties. These unique systems are processed by the user in the same way as standard double-sided PSA foam tapes, but are thermally cured after joining typically at a temperature of about 140 °C for about 20 minutes. They are able to combine the special properties and ease of handling of PSA foam tapes with structural strength usually provided by liquid structural adhesives. This is made possible by the addition of a bisphenol A–based and dicyanamide-cured epoxy resin to the acrylic formulation of the PSA. Applications of these tapes include the attachment of mirror frames to automobile windshields and metal mounts for kitchen stove–viewing windows.

8.2.2.9.3 Processing

PSAs are materials that exhibit permanent tack at room temperature and allow various surfaces to be firmly bonded by low pressure, performed by the action of a finger or hand. To establish a firm bond to surfaces such as paper, plastic, glass, wood, or metal, they do not require additional activation by, for example, radiation, moisture, or heat, as is necessary with other 1K adhesive systems. Due to their viscoelastic behavior after joining, they flow into the irregularities of the material surfaces and are able to establish good adhesion through molecular interactions. Thus, in contrast to liquid adhesives, which must be processed in a low viscosity to paste-like form for good wetting, PSAs are capable of wetting substrate surfaces solely by being pushed against the surface of the material to be bonded. In this process step, the application of sufficient contact pressure has a major influence on the quality of the bond.

Things to Consider when Processing PSAs These products are dry and solvent-free and are quick and easy to process, preferably at a temperature between 15 and 35 °C.

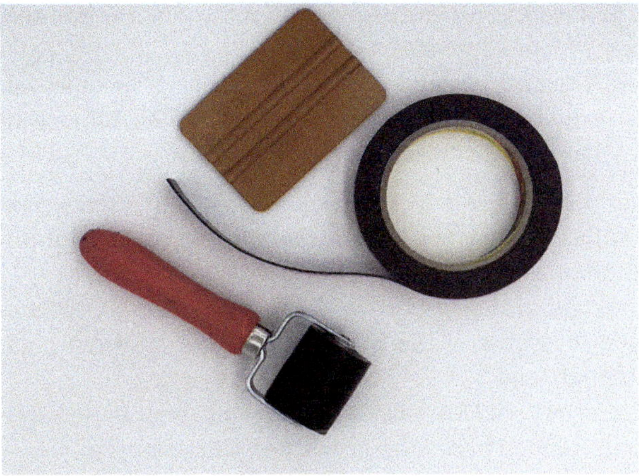

Figure 8.20 Tools for the application of PSA tapes. Source: Dr. Jürgen Klingen, Aboso-Consulting.

The bonded components have a high initial strength and can be mechanically loaded immediately after the joining process; however, final strength is usually achieved only after about 72 hours at room temperature. Due to the easily possible and often invisible contamination of the surfaces with fingerprints, gloves should always be worn when processing PSA tapes. For successful bonding, professional cleaning of the substrates is mandatory, and in a few cases surface treatment is necessary before the PSA is applied to the substrate. To avoid air entrapment, the joining process needs to be performed from one side under light pressure using elastic squeegees or rubber-coated rollers (Figure 8.20). After joining, the surfaces of the parts to be joined are wetted by applying a uniform pressure of approximately 0.20 MPa for a short time. Precise positioning of the PSA tape is required, as the bond already has a relatively high strength after the pressure has been applied. Immediately after joining, the bonded component should not be fully loaded, since the bonded joint usually undergoes a continuous time-dependent strength increase of 50%–100% of the initial strength. Thus, the final strength of composites with PSAs is often reached only after about 72 hours at room temperature.

8.2.2.9.4 Properties

PSAs represent viscoelastic materials with very high viscosity that, unlike radiation-, moisture-, or heat-activated 1K adhesive systems, form good bonds to material surfaces without altering their chemical or physical state. However, to produce such high-quality bonds, they must have a specific and characteristic set of rheological properties, which include

- *deformability* – to conform to rough surfaces,
- *good wettability* – at the appropriate surfaces to ensure molecular interactions, and
- *internal strength* – to withstand greater loads over a long period of time.

Thus, in the initial phase of the bonding process, PSAs must be able to quickly and completely wet the surfaces to be bonded, but the quality of wetting depends not only on the viscoelastic behavior of the adhesive, but also on the surface tensions of the adhesive and the surface energy of the substrates.

One important focus in PSA-related research and development is to tune the appropriate formulations to exhibit the appropriate behavior in the important phases (bonding and release) that a PSA will face in application:

- *during bonding on the substrate surface* – a PSA usually has about 1 second to deform and must exhibit sufficient flow behavior during this period to conform to the surface structure of the substrate,
- *during detachment from the substrate surface* – a PSA has about 0.01 second to deform and must relieve the stress in the bond immediately allowing high dynamic peel and shear strength, and
- *during long-term mechanical stress* – a PSA must be able to withstand static loads over a very long period of time and, if necessary, at elevated temperatures.

Compared to chemically and physically curing systems, PSAs with glass transition temperatures of about −20 to −70 °C offer good tack and peel performance, but relatively low shear strengths and resistance to creep. However, this drawback can be easily compensated for by appropriate sizing or applying the hybrid bonding method. The following should also be noted: The tack of a PSA is not necessarily related to its peel strength. Thus, systems with relatively low tack can have very high peel strength and vice versa. Therefore, as is often mistakenly assumed, a very tacky PSA is not necessarily a very good PSA.

When selecting PSAs for the appropriate applications, the following parameters must be considered:

- spontaneous wetting capacity of the PSA, indicated by good tackiness,
- specific interactions with the surface to produce good adhesion,
- cohesive strength of the adhesive layer, and
- thermal, chemical, and mechanical load-bearing capacity of the bonded components.

Acrylic PSAs Compared to other adhesives, acrylic-based PSAs usually offer exceptionally good aging resistance, even on surfaces that are difficult to bond. The so-called *centipede theory* attempts to explain this behavior, according to which the adhesive forms *thousands* of bonding points on the surface after joining [19]. However, after destruction of some of them by, for example, environmental influences, they are able to rearrange themselves and again form new bonding points on the substrate surface. This is only possible due to the special rheology of these systems and could explain the very good aging behavior of acrylate-based PSAs. Today, acrylate-based PSAs are offered for a wide range of applications where high demands are made on strength and aging resistance. Some product examples include transparent, medical, office, and foam tapes. Repositionable PSAs such as Post-it™ Notes and masking tapes feature peel forces of <10 N/25 mm, while permanent PSAs are positioned above this performance level. High-polarity pure acrylic PSAs exhibit temperature resistance up to 120 °C and represent the highest performance of all PSAs in terms of adhesion and cohesion force.

8.2.2.9.5 Applications

Today, PSAs are used in numerous industrial applications, as well as in the home, office, and do-it-yourself sectors. They generally make life easier and solve a variety of problems on a daily basis. In this context, PSAs are almost exclusively used as coated products, which are offered in roll form, as sheets or as die-cuts by countless manufacturers, distributors, and converters.

PSA products are available as:

- *single-sided adhesive tapes* – for fastening, masking, protecting, and packaging,
- *double-sided adhesive tapes* – for joining, fixing, mounting, and splicing,
- *double-sided adhesive foam tapes* – for high-demanding applications,
- *self-adhesive labels* – for marking diverse articles,
- *removable, so-called sticky notes* – for labeling with information (Figure 8.21),
- *self-adhesive damping products* – for vibration damping and noise reduction, and
- *electrically and thermally conductive tapes.*

Natural rubber and block-copolymer based PSA are used for the following products:

- packaging tapes,
- masking tapes for automotive and painting applications,
- duct tapes (Figure 8.22),
- electrical adhesive tapes,
- carpet tapes,
- splicing tapes,
- filament tapes,
- labels, and
- self-adhesive floor tiles.

Figure 8.21 A pad of removable sticky notes for labeling with information. Source: Dr. Jürgen Klingen, Aboso-Consulting.

Figure 8.22 Duct tape made of a PE-coated adaptable fabric with natural-rubber-based PSA, as used in numerous applications. Source: Dr. Jürgen Klingen, Aboso-Consulting.

Acrylic-based PSAs flow into the following products:

- office and household adhesive tapes,
- decorative and reflective self-adhesive films,
- medical adhesive tapes and ECG electrodes,
- Post-It™ Notes and flags,
- water-dispersible tapes for splicing and graphics,
- transfer tapes and labels,
- electrical adhesive tapes,
- transdermal patches for drug delivery,
- optically clear adhesive tapes,
- double-coated adhesive tapes, and
- high-performance foam tapes.

Applications of High-performance Foam Tapes In recent years, conventional mechanical fasteners such as screws, rivets, and spot welds, as well as liquid adhesive systems, have been replaced to a greater extent by high-performance foam tapes. Such systems, based on acrylic PSAs, feature good adhesive strength and high resistance to heat, UV radiation, and chemical attack. The tapes are constructed from a closed-cell acrylic core coated on both sides with functional acrylic-based PSA layers. The resulting non-separable, continuous viscoelastic unit enables the production of high-quality, stress-free bonded components. These unique adhesive systems have proven their worth in countless applications since their introduction more than 30 years ago and continuous further development up to the present day. Due to their viscoelastic behavior, they are able both to withstand dynamic forces and vibrations much better than conventional mechanical joining methods and to bond a wide range of materials with different surface properties. Here, these PSA systems demonstrate excellent bonding performance with peel forces on steel between 120 and 450 N/100 mm and adhesive strengths between 42 and 69 N/cm^2. A very unique and important application-related property of these high-performance foam tapes is that they are able to successfully compensate for differences in the linear expansion of substrates up to three times their thickness. Thus, the commonly used 1 mm thick foam tapes would be able to compensate for even 3 mm difference in length, demonstrating the excellent performance of these high-performance foam tapes in applications where materials with different coefficients of thermal expansion are used.

For this purpose, the following example, which represents a typical application of these foam PSAs and expresses their unique performance capabilities, which are particularly effective when materials with widely varying thermal expansions are bonded:

- According to the relationship $\Delta L = L_0 \cdot \alpha \cdot \Delta T$ (where L_0 = original length of the material, α = coefficient of thermal expansion, and ΔT = temperature difference), a temperature rise of 50 °C in metal substrates 1 m long causes an elongation of $\Delta L = 1.15$ mm in the case of aluminum with $\alpha = 23 \cdot 10^{-6}\ \mathrm{K}^{-1}$ and of $\Delta L = 0.80$ mm in the case of stainless steel with $\alpha = 16 \cdot 10^{-6}\ \mathrm{K}^{-1}$. Thus, a temperature increase of 50 °C causes a difference in length between 1 m long substrates of aluminum and stainless steel of 0.35 mm. Now, if these two 1-meter-long substrates are bonded and used in an environment where the temperature increases by 50 °C – e.g. from 20 to 70 °C – this difference of 0.35 mm can be easily absorbed by such foam tapes, because they are able to linearly expand three times of their thickness. Theoretically, a 120 μm thick version would be able to perform the task of length compensation, but a thicker version should be used to deal with possible unevenness of the surfaces involved.

Figure 8.23 illustrates the maximum expansion capability of acrylic foam tapes. As already mentioned, these tapes are capable of expanding up to three times their thickness resulting in a tan γ value of 1.5, which allows them to compensate very well for different thermal expansions of the substrates used, such as steel and aluminum. So, with a strip thickness of 1.2 mm, a difference in the linear expansion of such substrates caused by temperature increases would be compensated for up to 3.6 mm.

Good adhesive properties, the possibility of rapid further processing, tension-free bonding with a wide range of materials, high-temperature resistance, and the possibility of using application-optimized die-cuts make these modern high-performance PSA products an excellent alternative to conventional mechanical fastening systems and liquid adhesives. The products are offered in transparent, black, white, or gray versions with thicknesses between 0.05 and 3 mm and widths between 6 and 25 mm.

Applications for these acrylic foam tapes include:

- self-adhesive door seals in car doors,
- glass elements in metal frames for *structural glazing* applications,
- roof structures of truck trailers,
- container, bus, and wagon elements, and
- profiles for a wide variety of applications including stiffening.

Applications where the transparency of special designs is particularly appreciated are found in new modern manufacturing processes where glass-containing components are assembled, as in solar elements. Since in such applications the

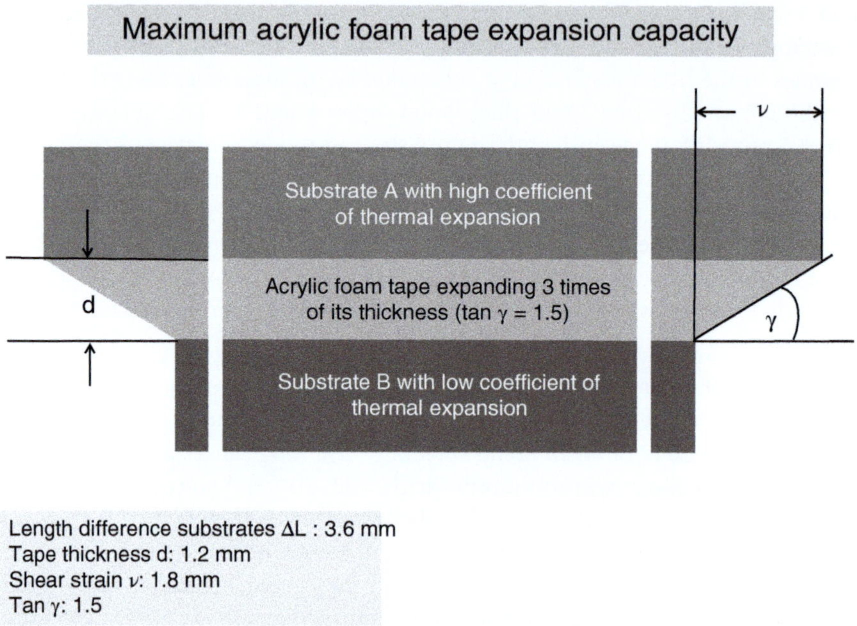

Figure 8.23 Example demonstrating the maximum expansion capacity of a typical acrylic foam tape.

transparency of products after environmental exposure is important, appropriate evaluations were performed using accelerated aging test procedures. For this purpose, bonded components were exposed to high temperatures and intense UV radiation for 3000 hours. After exposure to these stresses, transparency of such a tape decreased by only about one percent, from the original 88.2 to 87.3%.

8.2.2.9.6 New Developments

Much R&D work is performed on creating additional functionality such as thermal, electrical, or optical conductivity in the PSA, complementary to the adhesion and cohesion properties that PSAs typically provide. Another focus in research and development is the creation of so-called hybrid systems, in which chemical curing mechanisms performed by, for example, heat [20], radiation, and moisture are used to increase the strength of the PSAs. When such tapes are applied to the substrate, they initially exhibit the generally appreciated user-friendly property of PSAs and, after curing, provide the strength property that semi-structural and structural adhesives typically offer. Another focus of PSA-related research and development is on high-performance systems that allow good bonding to low-energy, often textured plastics such as PE or PP, but without additional treatment of the substrate surfaces.

Profile of PSAs

***Short statement*:**

- PSAs, which can be found everywhere in everyday life in the form of, e.g. adhesive tapes, labels, stickers, and sticky notes, are viscoelastic materials that have the ability to stick after light pressure and then fully conform to substrate surfaces to subsequently establish good adhesion.

Physical data (high-performance PSA foam tapes):

- adhesive force: about 0.40–0.70 MPa
- peel force: between 120 and 450 N/100 mm (on steel)
- service temperature range: −40 to 120 °C.

Special features:

- PSAs are very user- and environment-friendly products,
- adhere to the surface already after light pressure,
- show viscoelastic behavior,
- can be bonded to a very wide range of materials,
- as high-performance PSA foam tapes they allow excellent bonding with materials that have very different thermal expansions,
- exhibit high initial adhesion after joining and therefore rapid further processing of bonded components, and
- usually reach their final strength after 72 hours after joining.

8.2.3 2K Adhesive Systems

8.2.3.1 Introduction

2K adhesives cure at room temperature and, unlike 1K adhesives, require an additional process step before they are applied to the substrate surface (Figure 8.24). For this purpose, the individual components of the adhesive system must be mixed very well in the specified mixing ratio and applied to the substrate within the pot life. Particularly in the case of polyaddition curing, as used in 2K epoxy resins and 2K PU adhesives, a deviation from the specified mixing ratio leads to a change in the properties in the resulting adhesive layers.

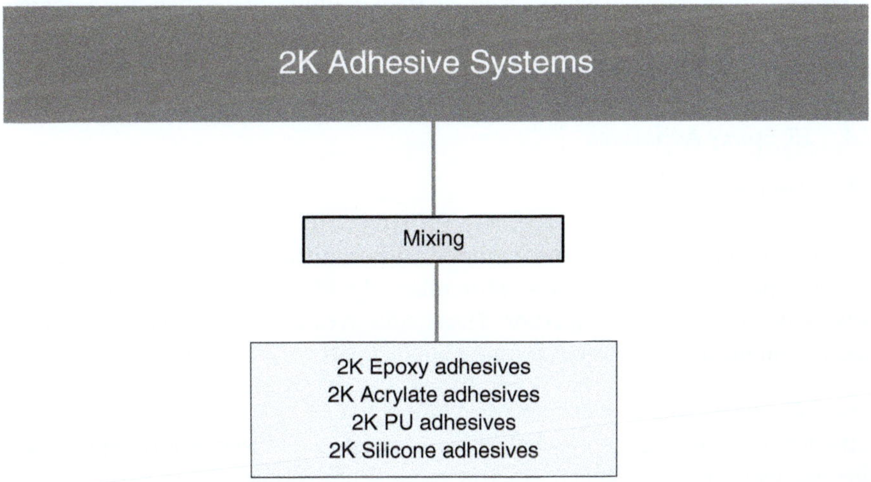

Figure 8.24 2K adhesive systems.

8.2.3.2 Curing and Processing of 2K Adhesives

2K adhesives are often dispensed and mixed manually from tubes and cans, and applied with user-friendly on-demand mixing and dispensing systems using dual cartridges, where the dual cartridges isolate the two resin components until they are needed. Mixing occurs as the materials are dispensed through the motionless mixing nozzle (static mixer) and is accomplished by hand tools or compressed air guns. Such systems allow dispensing, mixing, and application of the adhesive (spot or bead) to the substrate surface in just one step. After joining the adhesives are then cured at room temperature, although the curing time can be shortened by increasing the temperature.

When processing 2K adhesives, the following guidelines must be observed:

- Precise adjustment of the specified mixing ratio is crucial for good adhesive quality,
- immediately after mixing the two components, the curing reaction starts and the processing time (pot life) runs,
- within the pot life, which is specified by the supplier, the adhesive must be fully processed and the substrates joined,
- the pot life is dependent on ambient temperature and batch size,
- temperature increases reduce the pot life and as a rule of thumb, a 10 °C increase halves the pot life,
- when handling strength is reached, component movement is allowed, but larger loads cannot be applied, and
- once the final strength is reached, the bond can be fully loaded.

8.2.3.3 Applications of 2K Adhesives

When structural performance of bonded components and relatively simple and user-friendly adhesive processing are required, 2K adhesives are often the adhesives of choice. These systems cure at room temperature, but curing can be accelerated by heat. Epoxy as well as acrylic and PU-based 2K adhesives offer structural properties and provide good bonds to metals such as steel, aluminum, copper, brass, many plastics such as PC, PVC, GFRP, ABS, PMMA, as well as glass, ceramics, and wood. Depending on the application, these adhesives are flexible, hard, or tough-elastic.

8.2.4 2K Epoxy Adhesives

8.2.4.1 Overview

2K epoxy adhesives are used especially for structural metal bonding with high strength and aging resistance requirements. After curing, the 2K epoxy adhesive layers are present as thermosets with relatively high cohesive strengths and relatively brittle-to-elastic properties. These adhesives exhibit very good adhesive properties on most of the commonly used bonded parts.

8.2.4.2 Chemistry and Composition

For the formulation of 2K epoxy adhesives, in most cases bisphenol A–type epoxy resins are used (part a), which react at room temperature with a curing agent (part b), usually based on an amine. For this purpose, liquid, low-molecular-weight polyamides with amino groups at both ends of the molecules are preferred. These poly-amidoamines have low odor and react under low heat, but somewhat slower than the short-chain amines. A big advantage using these curing agents is that they allow easy adjustment of the optimum epoxy/amine (part a to part b) mixing ratio to 1 : 1, enabling easy processing of the adhesives with user-friendly on-demand mixing and dispensing systems using two cartridges. 2K epoxy adhesives are also offered as both electrically and thermally conductive variants, where electrically conductive systems are filled with up to approximately 60–80 wt% of metal particles (gold, silver, or copper), while thermally conductive adhesives are given an addition of up to 60–75 wt% of aluminum oxide, boron nitride, or metals such as aluminum or silver.

8.2.4.3 Curing/Processing

To ensure a high-quality bond, the individual components of the 2K epoxy adhesive system must be mixed correctly in a specific ratio and from mixing to the completion of the joining process the pot life must be strictly observed. Any deviation from the specified mixing ratio will result in proportions of low-molecular substances in the adhesive that have not participated in the curing reaction. These remain in the adhesive and thus lead to a reduced quality of the bonded component (Figure 8.25).

Although a deviation from the specified ratios of approximately ±5% does not usually have a significant effect on the properties of the adhesive layer, every effort should be made to ensure that the specified mixing ratio specification is met.

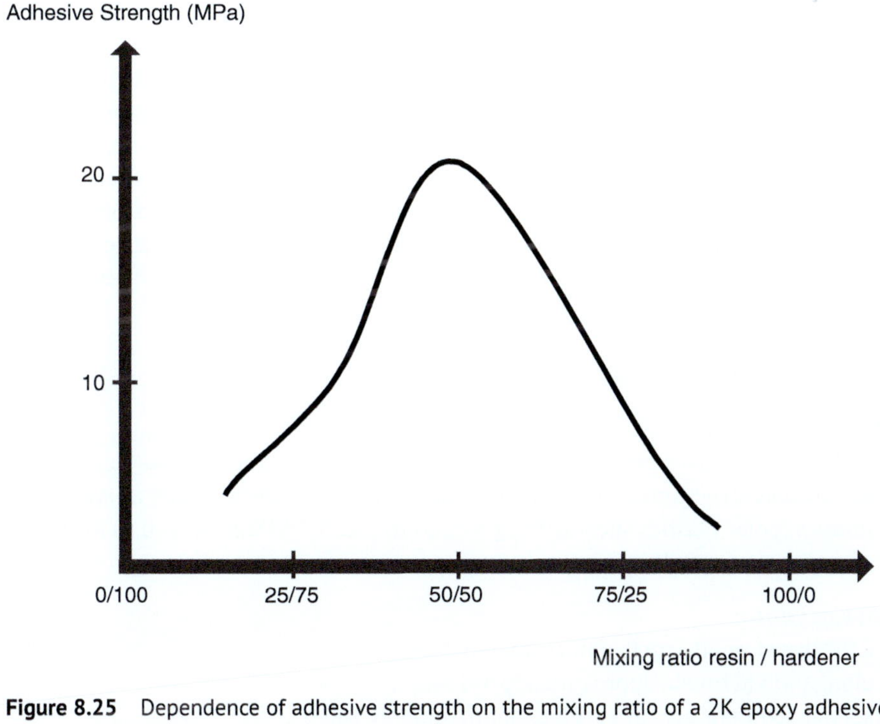

Figure 8.25 Dependence of adhesive strength on the mixing ratio of a 2K epoxy adhesive.

8.2.4.4 Properties

After curing, 2K epoxy adhesives present themselves as relatively brittle thermosets with high crosslink densities, offering high shear moduli and high glass transition temperatures, as well as very good adhesion properties to many materials. Therefore, these adhesives are highly desirable for the preparation of structural adhesive composites with high strength and long-term durability. 2K epoxy adhesives are often made tough-elastic by suitable long-chain reactive components, such as aliphatic polyamines and poly-amidoamines, to improve their conformability, impact strength, and stress relief in the adhesive layer. However, this reduces the glass transition temperature and thus the strength of the bonded components, especially at higher temperatures. This can be avoided. Thus, by adding elastomer particles or creating a finely dispersed rubber-elastic phase chemically bonded to the epoxy matrix, *tough-elastic* or often called *toughened* epoxy adhesives with high impact and peel strength can be constructed without affecting their glass transition temperature and shear modulus. This method provides an excellent compromise between high strength, good conformability, impact resistance, and stress relief in the adhesive layer. Incorporation of the incompatible rubber-elastic phase is achieved by adding about 5–20 wt% of reactive liquid rubber components, which, due to their incompatibility with the epoxy phase, separate to form a discontinuous (disperse) phase in the form of *rubber particles* that are chemically bonded to the epoxy resins. These rubber-elastic particles are able to reduce stress concentrations in the cured adhesive layer caused by the attack of dynamic mechanical loads and thus help to withstand strong impacts such as those encountered in a vehicle crash.

8.2.4.5 Applications

2K epoxy adhesives cover a very wide range of applications and are used in particular as structural adhesives for metal bonding with high requirements for strength and resistance to environmental stresses. Due to their relatively easy handling, which is ensured by modern application systems, they are very popular among users. These adhesives show very good adhesion to aluminum and steel as well as to many polar plastics. In modified form, they allow special applications, e.g. as electrically conductive adhesives in microelectronics or as thermally conductive adhesives in areas where heat must be kept away from sensitive electronic components.

8.2.4.5.1 Profile of 2K Epoxy Adhesives

***Short statement*:**

- 2K epoxy adhesives are presented as creep-resistant thermosets after curing and are used as structural adhesives for bonding various materials such as steel, aluminum, polar plastics such as PC, PVC, GFRP, ABS, PMMA, as well as for glass, ceramics, and wood.

***Physical data*:**

- adhesive strength: up to approximately 35 MPa
- elongation at break: approximately 5–15%
- service temperature range: up to approximately 100 °C.

Special features:

- 2K epoxy adhesives are cured at room temperature and represent dense crosslinked polymer networks that exhibit high long-term aging resistance and low creep behavior,
- after modification with poly-amidoamines, elastomer particles, or reactive liquid rubbers, tough-elastic (toughened) adhesive composites with excellent impact strength values are obtained,
- exhibit high resistance to physical and chemical stresses,
- can be used for a wide range of surfaces, including metals, polar plastics, glass, ceramics, and wood,
- are easy and fast to process with user-friendly on-demand mixing and dispensing systems, and
- require exact adherence to specified mixing ratios of resin (part a) and hardener (part b) for optimum strength.

8.2.5 2K Acrylate Adhesives

8.2.5.1 Overview

When elastic adhesive bonds with high strengths and fast further processing options are required, 2K acrylic adhesives are often the material of choice. Their very good compatibility with oily or greasy surfaces is also often highly valued, especially in the automotive industry. In many cases, the characteristic performance spectrum of 2K acrylic adhesives gives them a competitive advantage over high-strength but relatively brittle epoxy and very elastic but low-strength PU adhesives in demanding bonding applications.

8.2.5.2 Chemistry/Composition

These adhesives are cured by free-radical polymerization of methyl methacrylate and other esters of methacrylic acid, such as ethyl methacrylate. Di- and trifunctional esters of methacrylic acid are added for crosslinking. The toughness of the adhesive is usually achieved with the aid of chemically bonded elastomer particles, which are a phase-separated part of the acrylate matrix. Improved adhesion to plastic surfaces is achieved by incorporating polymerizable isocyanate components such as isocyanato-ethyl methacrylate into the polymer chain.

8.2.5.3 Processing

Compared to other common 2K adhesive systems, which require exact adherence to the specified mixing ratio of the components, these adhesives have the advantage of allowing much greater deviations in this respect. This is due to their radical curing mechanism as opposed to the polyaddition mechanism of, for example, 2K epoxy and PU adhesives, and allows somewhat easier and safer handling. Another important feature of acrylic adhesives is their very short curing times, which usually allow hand strength after a period of 5–120 minutes. This feature is of particular interest to the cost-conscious user from the point of view of automation and economy.

2K acrylic adhesives can be processed according to three process types:

- *"A-B" process* – For this process, two stable components are provided, one containing the accelerator and part of the acrylate monomers, and the other containing the hardener and the other part of the acrylate monomers. The application can be carried out either by dosing the previously mixed components onto only one material surface (variant 1) or by applying component A to one and component B to the other material surface (variant 2), with the mixing then being carried out by applying pressure to the parts to be joined. This is followed by curing, which takes only a few minutes. Frequently, adhesive application according to option 1 is carried out by very user-friendly double-cartridge on-demand mixing and metering systems.
- *"No-mix" process* – The hardener dissolved in a volatile organic solvent is applied to only one of the two substrates and then dried. Further processing takes place either directly afterwards or, if desired, at a later stage. For this purpose, the second adhesive component, which consists of a mixture of the acrylate monomers and the accelerator, is applied to the second substrate. Curing is then initiated by bringing the two substrates together, which leads to solidification of the adhesive layer after a short time.
- *"Mix" process* – Immediately before processing, the hardener is added to the monomer/accelerator mixture in the form of paste or powder. After addition, the bonding process must be carried out immediately within the specified pot life of the respective adhesive system.

8.2.5.4 Properties

The advantage of 2K acrylic adhesives over other 2K systems is the combination of high adhesive strength values of up to 25 N/mm^2 and, at the same time, high elasticity of up to 120% elongation. In a very short time, these adhesives enable hand and also relatively fast final strength of the bonded components. Their adhesion to plastic surfaces is better than that of the other competing 2K systems. Additionally, they are more suitable for automation and for special applications, their relatively high tolerance to oily or greasy surfaces is very popular.

8.2.5.5 Applications

Because of their unique combination of properties, 2K acrylic adhesives are often the adhesives of choice for appropriately demanding applications, outperforming high-strength but relatively brittle epoxy and elastic but low-strength PU adhesives. For structural bonding of low surface energy plastics, such as PE and PP, special types of acrylic adhesives have been developed in recent years that provide excellent adhesion without surface treatment otherwise required. Due to their non-polar nature, these materials, which are increasingly used in industrial applications, are very difficult to bond to each other and to other materials. They could be welded, but bonding these materials always required additional surface treatment of the substrate surfaces using physical methods such as corona and plasma treatment. Thanks to appropriate new developments, it is now possible to bond low-energy materials such as PE and PP without this time-consuming treatment of the parts to be joined.

8.2.5.5.1 *Profile of 2K Acrylates*

***Short statement*:**

- Acrylate adhesives are fast-curing, combine high strength with high elasticity and show good adhesion to plastics as well as compatibility with oily or greasy surfaces.

***Physical data*:**

- adhesive strength: up to approximately 25 MPa
- elongation at break: up to approximately 150%
- service temperature range: up to approximately 100 °C, short-term up to 180 °C.

***Characteristics/special features*:**

- Acrylic adhesives cure very quickly, usually in a few minutes at room temperature, to thermoplastic or thermoset adhesive layers,
- are very suitable for many applications, especially in metal bonding,
- combine high strength with good elasticity,
- show better adhesion to plastics compared to other 2K adhesives,
- show good compatibility with oily or greasy metal surfaces,
- allow application and mixing by very user-friendly double-cartridge on-demand mixing and dispensing systems (according to A-B method),
- temperature increase causes an acceleration of the curing speed, but also a decrease of the average molecular weight and thus a reduction of the strength,
- unlike other 2K adhesives, allow greater deviations from the specified mixing ratios,
- often require unusual and difficult-to-handle mixing ratios (e.g. 1 : 10) and hardener additions of a few percentages,
- often have a very strong odor,
- only allow relatively small areas to be bonded due to the very fast curing speed, and
- must be stored at approximately 8 °C.

8.2.6 2K PU Adhesives

8.2.6.1 Overview

2K PU adhesives consist of low-molecular-weight isocyanates or prepolymers (part a) and low-molecular-weight polyols or polyamines (part b). When bifunctional monomers are used, essentially linear, thermoplastic products are obtained, whereas tri- and higher-functional monomers produce highly crosslinked adhesive layers with a thermosetting character.

8.2.6.2 Curing/Processing

To achieve optimum strengths, exact compliance with the specified mixing ratio of the components must be ensured. The curing of the adhesives, which takes place at room temperature over a period of several hours to days, can be accelerated by applying heat. If necessary, the pot life can be varied by selecting the appropriate adhesive components and the curing times can be shortened by adding accelerators.

8.2.6.3 Properties/Applications

Depending on the selection of monomers, 2K PU adhesives appear after curing as hard to rubbery thermoplastic or thermoset adhesive layers with low creep resistance. Compared to 1K PU adhesives, an even greater variety of properties such as strength, elasticity, and resistance to chemical stress can be adjusted by selecting specific monomers. Applications of 2K PU adhesives can be found in the automotive industry, where these adhesives are used to bond metals to plastics and foams to textiles, and in the furniture industry, where they are used to bond PVC films to wood.

8.2.6.3.1 *Profile of 2K PU Adhesives*

Brief description:

- Depending on the choice of monomers, 2K PU adhesives produce hard to rubber-elastic thermoplastic or thermoset adhesive layers with low creep resistance, which are often used for larger area bonding applications in automotive and furniture industry.

Physical data:

- adhesive strength: up to approximately 25 MPa for structural, up to approximately 6 MPa for rubber-elastic adhesive types
- elongation at break: approximately 50% for structural adhesive types, 600% for rubber-elastic adhesive types
- service temperature range: from −40 to approximately 110 °C.

Special features:

- 2K PU adhesives show thermoplastic or thermoset adhesive layers with low creep resistance,
- offer lower UV and media resistance compared to epoxy adhesives,
- often require a primer to produce high quality bonds, and
- for repair bonding, allow the adhesive applied to the cut open old adhesive bead.

8.3 Adhesive Bonding on Metal, Plastic, Glass, and Wood Surfaces

8.3.1 Adhesive Bonding on Metals

Adhesive bonds with metals represent a significant proportion of all bonded components in industrial applications. Even before planning the bonding process, it is important to know the characteristic properties and adhesive behavior of the metal in question. With knowledge of the adhesive properties, the design principles, and the stress criteria, it is quite possible to transfer existing experience with a metallic material to the new tasks. When bonding metals, it must be taken into account that the metal surfaces to be bonded generally do not have the properties of the pure material.

Rather, it is usually a complex multi-layered structure consisting of

- the material itself,

- a boundary layer resulting from deformations,
- a reaction layer resulting from chemical reactions with the environment,
- an adsorption layer covered with non-species molecules, and
- a layer with possible impurities.

Appropriate surface treatment is essential for the successful bonding of metals. For this purpose, pickling solutions are still used to produce long-term aging-resistant bonds, which are particularly necessary in aircraft construction. For reasons of occupational safety and environmental protection, however, their use is very limited. In principle, care must always be taken to ensure that the surfaces to be bonded are absolutely free of dirt, grease, and oil residues. By subsequent grinding, brushing, or blasting, if necessary, supported by sealing of the adhesive edges (to protect against corrosion), adhesive components with sufficient resistance to environmental stresses can be produced for most applications. Especially when grinding, it should be noted that the application of excessive pressure can lead to surface tarnishing and distortion, which occasionally occurs in metals with low thermal conductivity. For this reason, high-grinding speeds must be avoided during machining. Glass beads, stainless-steel grit, or quartz sand are used for blasting, which must be iron-free to prevent surface rusting. The starting point of industrial metal bonding with adhesives is considered to be the process for bonding aluminum components in aircraft construction, which has been very well established since the middle of the last century. Since then, a large number of applications and adhesive systems have been developed by users and adhesive suppliers, so that metal bonding is now very well established in industry and craft. Almost all known chemically or physically curing adhesive systems are suitable for bonding metals, and high-performance acrylate-based PSA tapes are playing an increasingly important role due to their ease of handling and special properties.

8.3.1.1 Adhesive Bonding on Steel Surfaces

In industrial manufacturing and assembly, steel is one of the most commonly used materials that can be bonded well with adhesives, which is the reason why this material is joined with adhesives in a wide variety of applications there. In the automotive industry in particular, components for the body structure and add-on parts such as doors, hoods, trunk lids, and covers are bonded to a large extent using modern adhesive systems. In steel construction, too, the highest requirements are met with the appropriate adhesives. The great attractiveness of bonding steel is explained by the lower labor input and the great advantage that processing takes place completely without heat generation, which is not the case with welding. Particularly when processing sheet steel, the use of adhesives prevents negative influences on material properties, cooling stresses, tarnishing of the surface, or warpage of the components.

8.3.1.2 Bonding on Steel Sheet

Among the countless types of steel products, cold-rolled sheet plays a prominent role for adhesive applications. In numerous joining applications with this material in mechanical engineering and especially in car body construction, adhesive bonding is used exclusively or combined with other mechanical joining processes, e.g. spot welding [21]. It should be noted that sheet steel is often provided with special coatings on the surface to adjust special properties. These can consist of metals such as zinc, aluminum, tin and lead, organic coatings or polymers, and inorganic substances such as phosphates and chromates. The applied adhesive therefore does not wet the base material, but the applied coating. Therefore, the properties of these surfaces must be taken into account when selecting the surface treatment method and the adhesive system used.

8.3.1.3 Bonding on Oily Steel Sheet

One of the biggest challenges for bonding steel sheets is in what is probably the largest application of bonding in the automotive industry, where oil-coated steel sheets are used in the assembly of body structures and add-on parts [22]. The oil coating, which is not removed before bonding, serves to protect the sheets from mechanical damage during stamping operations and to provide temporary corrosion protection. The adhesive must therefore not only adhere well to the sheet surface, but also be able to absorb the oil present on the surface. In addition, the adhesive bond must meet the stringent requirements of the automotive industry in terms of long-term aging and corrosion resistance. The fact that it has been possible to develop adhesive systems and associated processes for this purpose can be regarded as a milestone in adhesive-bonding technology.

Using adhesives for joining steel provides the following features:

- invisible joints,
- lighter constructions and thus less weight of the component,
- safe bonding of different materials,
- no contact corrosion due to the insulating effect of the adhesive,
- no stress peaks due to uniform stress distribution in the bond,
- visually attractive designs,
- sealing of the adhesive joint,
- simplifying the manufacturing process,
- reduction of process costs,
- compensation of tolerances,
- good damping properties, and
- better impact resistance and crash behavior in vehicle construction.

The following points must be taken into account when using adhesive technology for metal bonding:

- cleaning of the joining parts is always required (exception is bonding of oily steel sheets in automotive applications),
- contamination of the surface must be avoided before the application of the adhesive,
- surface treatment is often required after cleaning,
- the design of the component needs to be adhesive bonding friendly,
- compared to mechanical joining methods, adhesive-bonded components show a limited strength at high temperatures, and
- non-destructive quality control of adhesive bonds is usually not possible.

8.3.1.4 Bonding on Galvanized Steel

To provide the steel with good corrosion protection, most of the metal-coated steel is coated with zinc before industrial use (Figure 8.26). Responsible for the anticorrosive effect of zinc is its much less noble character compared to iron, which is represented by a value of −0.76 V in the electrochemical series (iron has a value of −0.04 V). When an electrolyte attacks the metal surface, zinc acts as a *sacrificial anode* and goes into solution before the iron dissolves. This initially forms zinc hydroxide, which combines with carbon dioxide in the air to form a stable zinc hydroxide/zinc carbonate corrosion protection layer. This layer is well bonded to the zinc base material and therefore does not flake off. When bonding to galvanized steel, the behavior of the adhesive on the zinc hydroxide/zinc carbonate layer, as well as the mechanical properties of the elemental zinc layer and its adhesion behavior to the steel surface, must be taken into account. It should be noted that fresh zinc surfaces are still chemically unstable and therefore – in this state – behave unfavorably for bonding.

The coating of steel materials with zinc is mainly carried out by hot-dip (piece or strip) and electrolytic galvanizing. Due to the fine-grained nature of the resulting zinc coating, electrolytically deposited coatings exhibit significantly better adhesion behavior than hot-dip-galvanized coatings. This advantage is due to the fact that with this layer on the steel surface, there is a more homogeneous stress distribution in the bonded joint under mechanical stress. In the case of hot-dip-galvanized coatings (in pieces), relatively large crystallites form on the surface, which often have a

Figure 8.26 Hot-dip-galvanized steel railing at the bridge over the Neckar in Tübingen/Germany. Source: Dr. Jürgen Klingen, Aboso-Consulting.

negative effect on the adhesive strength. This can be explained by the fact that, due to their low flexibility and the resulting heterogeneous stress distribution in the bond, these layers cause a low adhesive performance already in the early phase of elastic deformation of the adhesive layer, which thus often leads to failure of the joined component even at low loads. This is different with strip galvanizing, where smaller crystallites are formed, which in practice leads to correspondingly good adhesive strengths. For this reason, strip-galvanized steel has already been bonded very successfully for many years for a wide range of automotive applications, including hem flange bonding of doors, engine hoods, trunk lids, and covers.

8.3.1.5 Surface Treatment of Galvanized Steel

The white rust that forms on the applied elemental zinc layer in an air atmosphere consists of zinc hydroxide, little zinc oxide, and very little zinc. This zinc rust is undesirable on the surface because it reduces the corrosion protection and thus the quality of the bonded joints. This is due to the fact that under the influence of moisture the white rust promotes moisture diffusion into the adhesive layer, which naturally has a particularly negative effect on the long-term stability of the bonded joint. Therefore, it must be completely removed from the elemental zinc surface before starting bonding processes with galvanized steel. To further improve the moisture resistance of bonds on zinc coatings, in addition to the removal of white rust adhesives with a high crosslinking density should be selected. These reduce water diffusion into the boundary layer and thus contribute to good moisture resistance of the bonded component. Mechanical surface treatment of the galvanized steel must be carried out carefully because of the possible damage to the relatively thin elemental zinc layer. Thus, after rough cleaning of the surfaces, steel brushes should not be used to remove white rust. Much more suitable is careful sanding of the layer with stainless-steel wool or fine sandpaper. To be on the safe side, however, the grinding process should be carried out with brushes having nylon bristles. After this careful mechanical treatment, the substrate must be cleaned with, e.g. isopropanol, acetone, or a universal cleaner before joining. For successful bonding of galvanized steel, in most cases careful grinding or brushing off of the white rust in conjunction with prior, and subsequent degreasing of the zinc surface is the completely sufficient procedure for preparing the surfaces [23].

8.3.1.6 Bonding on Stainless Steel

When high weathering and corrosion resistance for components are required, stainless steel is often the material of choice. Especially when the trend toward lightweight construction and modern design has to be reconciled with increasing cost pressure, modern joining methods such as adhesive bonding with easy-to-use adhesives or adhesive tape systems are increasingly replacing classic fastening methods such as welding, riveting, or screwing. In this context, bonding stainless steel results in less weight, an unchanged material structure, invisible joints, and the possibility of using a wide variety of material combinations. A wide range of adhesives and adhesive tapes is available for this purpose, which can be very easily integrated into standard manufacturing processes in industrial production.

Bonded stainless-steel joints are of great importance in architecture, for example, in facades, stainless-steel sheets are joined flush with corresponding substructures by structural adhesives. Bonded all-glass facades without visible fastening elements are also indispensable in modern architecture. Here, large-area glass elements are bonded to adapter frames made of stainless steel, which are then mechanically connected to a substructure. Adhesive-bonding technology is also used very successfully in interior design to join stainless steel to various materials, for example in the cladding of escalators, elevators, or ventilation systems. In medical technology, stainless steel is bonded for the manufacture of diagnostic and therapeutic devices as well as prostheses or implants. For example, stainless-steel cannulas are bonded in PP syringes or small lenses in stainless-steel objects for endoscopes used in minimally invasive surgery. In these applications, adhesive-bonding technology meets the requirements for both cost-effective microproduction and very demanding operating conditions.

8.3.1.7 Surface Treatment of Stainless Steel

As stainless steels contain at least 10.5% chromium, they have significantly improved corrosion resistance compared with unalloyed and low-alloyed steels. As a result, they are resistant to numerous aggressive media and require no further surface protection. The good corrosion resistance of stainless steels is based on the formation of a passive layer of chromium oxide only a few nanometers thick on the material surface. In the event of mechanical damage, this protective layer is very quickly rebuilt on the stainless-steel surface due to the excellent migration capacity of chromium. This passive surface of stainless steels is characterized by very low reactivity with the environment, which leads to their excellent corrosion resistance. However, this property poses a challenge for bonding these materials, as it severely restricts the formation of intermolecular bonds, which are necessary for good adhesive properties. Since stainless steels are mainly used for long-term resistant components, the corresponding bonded joints must also be designed for long-term stability. For this purpose, extensive studies on chemical and electrochemical surface treatments have been carried out and published, but these methods are only used to a very limited extent due to their complexity and aspects of environmental protection and occupational safety. Also, for a large number of bonding applications with stainless steel, it is not necessary to use these chemical and electrochemical treatment methods and, in most cases, mechanical treatment by grinding or blasting in combination with prior and subsequent cleaning of the surface results in very good bond strengths. Here, for the surface treatment of stainless steel by blasting, glass beads, broken glass, stainless-steel grain, or iron-free quartz sand are very suitable. When bonding stainless steel with adhesives, some special features must be taken into account. In the case of austenitic stainless steels, mechanical surface treatment by grinding or brushing must not be carried out at excessively high contact pressures and, in the case of machining, not at excessively high grinding speeds. Due to the relatively low thermal conductivity of these materials, this could lead to tarnishing and distortion of the surfaces. To avoid extraneous rust, the materials used for grinding and blasting must be iron-free and also, materials that have already been used for the treatment of unalloyed steels must not be used for the treatment of stainless steel.

If stainless-steel surfaces are treated by grinding, brushing, or blasting before joining, very well-bonded components with high initial strengths are obtained. However, the application of such bonded joints in humid atmospheres and, in particular, the action of salts in this atmosphere can lead to a significant reduction in bond strength, which must be taken into account when using the materials in the appropriate environment [24]. To solve the problem, adhesive bonds with higher aging resistance are produced with highly crosslinked thermoset adhesives such as 1K or 2K epoxy systems. Further improvement is promised by the application of the SACO process for surface treatment of the stainless steel, whereby the surface is also chemically treated with silanes in addition to the mechanical treatment.

8.3.1.8 Bonding on Aluminum Surfaces

Because it is light, corrosion-resistant, relatively soft, ductile, and easy to process, aluminum finds the most diverse applications as a construction material in many areas of industry and craft, especially in vehicle and aircraft construction. As a metal, aluminum has a high surface energy of 1200 mJ/m^2 and is therefore very well suited for bonding, but its tendency to oxidize in air must be taken into account. This natural process of passivation, in which a layer of oxide up to 50 nm thick is formed on the surface by atmospheric oxygen, results in the good corrosion properties of aluminum. Thus, the applied adhesives usually do not find pure aluminum on the surface, but the resulting oxidation products such as aluminum oxide, its hydroxide, and carbonate. These layers are morphologically complex and do not always have a positive effect on the quality of the bond due to their brittleness and porosity, which is of particular importance when highly durable adhesive-bonded joints are required.

8.3.1.9 Surface Treatment of Aluminum

When joining aluminum substrates with adhesives, the following surface preparation should be able to create composites with sufficient strength and aging resistance, suitable for many applications in industry and craft:

- Removal of oils, greases, and dirt particles with acetone, MEK, isopropanol, or *n*-heptane. After cleaning of the surface, the substrates should not be touched with bare hands,
- mechanical surface treatment by grinding (grain P 120 up to P 600), brushes, or sandblasting [25], which results in a fresh and thus chemically active layer with a larger effective surface, and
- cleaning with acetone, MEK, isopropanol, or n-heptane.

To avoid the uncontrolled build-up of the atmospheric oxide layers, which are critical for adhesive bonding, subsequently the joining process must be carried out within 10 minutes after grinding. Other surface treatment methods such as anodizing, titanium zirconium thick-film passivation, phosphating, or chromizing – often already applied by the suppliers – result in good aging-resistant surfaces, which can be bonded very well directly with structural epoxy and elastic PU adhesives as well as viscoelastic high-performance adhesive tapes. Higher corrosion resistance of aluminum substrates is achieved by the very frequently used so-called *anodic treatment process*, in which 5–25 μm thick oxygen-containing protective layers are produced by converting the uppermost metal layer into oxides or hydroxides. These layers – already applied by the suppliers – protect the bonded component from corrosion and also provide a good basis for direct bonding.

8.3.1.10 Surface Treatment for Long-term Aging-resistant Bonds

Due to the very early and extensive use of aluminum and its diverse alloys in aerospace industry, this metal is the most studied in adhesive-bonding technology and, because the aluminum alloy $AlCuMg_2$ was prescribed in the German standard DIN 53283 for the determination of adhesive strength, there is a wide range of adhesive-bonding investigations with this alloy. Hereby, very often the focus was on long-term aging resistance of aluminum compounds and the related analysis of the complex chemical and morphological properties of this ignoble material that are relevant to its corrosion behavior [26]. In contact to air, all aluminum alloys spontaneously form oxygen-containing protective layers, which vary in the morphological states and thicknesses. These oxides, hydroxide or carbonate-containing layers, which generally are in the nanometer range, exhibit different behaviors toward aggressive media. For this reason, the focus of many research projects in the past was on the development of surface treatment processes to produce the most stable aluminum oxide and hydroxide structures possible on the surface of the elementary aluminum.

The etching solutions proposed for aging-resistant aluminum compounds contain inorganic acids and chromium salts with oxidizing properties. For bonded joints with high requirements for strength and long-term aging resistance, the so-called *pickling process* is proposed for surface treatment. This etching process, which is critical from the point of view of occupational medicine and the environment, uses a mixture of 27.5% concentrated sulfuric acid, 7.5% sodium dichromate ($Na_2Cr_2O_7 \cdot 2H_2O$), and 65% water (all proportions by weight). Due to the toxicity of the chromizing baths, such processes are no longer up to date and are now only used for limited applications, such as in aircraft construction, but under strict environmental and occupational health and safety conditions. Therefore, sustainable chromate-free alternatives for industrial bonding applications with aluminum substrates have been developed in recent years, resulting in coatings based on, e.g. polyacrylates, silicon dioxide, or silica. Such treatments are easy to apply and are harmless from an environmental and occupational safety point of view.

For particularly aging-resistant adhesive bonds, Rider and Arnott describe a particularly effective process, which is not critical from the point of view of occupational medicine and the environment and which ensures very good adhesive strength and aging resistance of the resulting aluminum components [27]:

- After mechanical surface treatment of the material by blasting (cleaning/degreasing before and after blasting), the substrates are immersed in boiling water for 15 minutes and then coated with a silane-based adhesion promoter of 3-glycidoxy-propyl-trimethoxy silane in a 1% aqueous solution.

For the production of long-term resistant adhesive bonds with aluminum, a good balance must be achieved between the adhesive polarity, the chemical surface properties, and the morphological fine structure of the oxide layer. There is a clear relationship between the morphology of the surface and the corrosive damage in the bonded joint. A prerequisite for long-term resistant aluminum bonding is therefore a surface treatment capable of producing a stable nanoscale structured surface morphology. For this purpose, the surface treatment of aluminum is recommended in exceptional cases by well controlled conventional wet chemical processes and by new chromate-free alternatives as well as laser processes. In this context, the laser treatment process represents a fast, robust, and clean contemporary method with good compatibility with environmental, occupational health, and safety requirements. This method is capable of producing excellent long-term aging resistant-bonded components that are subjected to mechanical and environmental stresses common in the aerospace industry [28, 29].

It should be noted, however, that most industrial and craft applications do not require chemical, electrochemical, or laser treatment methods. Mechanical treatment, such as grinding and blasting in combination with prior and subsequent cleaning of the treated surfaces, in most cases results in adhesive-bonded components with good strengths and aging resistance.

8.3.2 Adhesive Bonding on Plastic Materials

Due to their special structure, plastics have some material- and process-specific properties that are relevant when designing bonded components with such substrates. For example, the tensile strength of plastics is about 10 times lower than that of metals, e.g. only about 35 MPa for PE. In contrast, the very common structural steel has a tensile strength of about 430 MPa. When bonding plastics with adhesives, similar strength values can be assumed due to the chemical similarity of the substrates to be bonded and the adhesives used. In contrast to metals, so-called *butt bonding* is possible with plastics, which should be avoided when bonding metal parts. Plastics exhibit significantly different deformation behavior under load than metals, which is decisive for the strength of a bonded plastic component. The characteristic measure of the stiffness of a material is the modulus of elasticity, which is about 210 GPa for mild steel, but only 1.4 GPa for polyamide 6. The stiffness of a bonded component is thus several orders of magnitude lower than that of metal composites. While the modulus of elasticity of metals is a material constant, this measure varies for plastics and depends on their molecular weight, degree of crystallinity, and morphological structure. Thus, the strength of a bonded plastic component is influenced both by the relatively low modulus of the plastic and, additionally, by its high-temperature dependence, which is not present in metals. Even the application of low stresses to the bond causes relatively strong elongations. This often leads to plastic deformation, which results in stress peaks in the bonded component and ultimately in failure of the bond. To avoid such critical stress concentrations, adhesives with sufficient deformation capacity such as PU, rubber-based adhesives, and self-adhesive foam tapes are preferred for bonding plastic materials.

On metal surfaces, the adhesives are able to establish intermolecular interactions, which contribute to the very good wetting on these substrates. In addition, the metals behave completely inertly toward the adhesives, so that neither dissolution nor diffusion processes are observed on contact.

8.3.2.1 Role of Diffusion and Solubility Characteristics

Due to their chemical similarity with the adhesives, which is reflected in comparable surface energies and strengths, plastic materials behave quite differently. Thus, during the bonding of these materials, in addition to the intermolecular interactions, dissolution or diffusion processes also occur, which are caused by the solvents used. In the application of chemically reacting adhesives, however, the liquid monomers used also play a major role here. The success of bonding thus depends both on the characteristic surface property responsible for the formation of adhesion forces, in particular polarity, and on the diffusion and solubility behavior of the solvents and monomers that are part of the adhesive formulation. The degree of adhesiveness of the individual types of plastic materials can thus be determined by their polarity and solubility, with the crosslinked thermosets exhibiting significantly lower solubility than the thermoplastics. Even within these groups, there are differences in the solubility of the materials. For example, PVC and PMMA are highly soluble in many solvents, while PE and PP are very sparingly soluble.

According to Lucke [30], the following evaluation is available for thermoplastics, which takes into account both the polarity and the solubility of the substrates (Table 8.1):

- A plastic material that is non-polar and insoluble – such as PTFE – cannot be bonded without treatment and is very difficult to bond with treatment.
- A plastic material that is non-polar but soluble – such as PS – is easy to bond.
- A plastic material that is polar and soluble – such as PVC and PMMA – is easy to bond.

As a material-specific parameter, the solubility of a plastic cannot be changed to a large extent. However, to improve the adhesive properties of plastics, it is possible to increase the polarity of non-polar plastics to a sufficient extent by generating polar functional groups on the surface. This is achieved with the aid of suitable surface treatment processes and is often the only way to enable the interactions on the surface that are necessary for good adhesion.

8.3.2.2 Bonding on Recycled Plastic

For some years now, there has been a social and economic trend for the increased use of recycled plastic materials. PE in particular is one of the main sources of plastic waste due to its widespread use as film in the packaging industry and is therefore particularly suitable for substitution with recycled material. But PP, the most widely sold plastic, also plays a major role in the material flow, especially in the industrial

Table 8.1 Adhesive-bonding properties of plastics as a function of their polarity and solubility in organic solvents.

Material	Surface tension (mJ/m^{-2})	Polarity	Solubility	Adhesive bonding capability (without surface treatment)
PTFE	18	Non-polar	Non-soluble	Not possible
PP	32	Non-polar	Hardly soluble	Not possible
PE-LD	31	Non-polar	Hardly soluble	Not possible
PS	35	Non-polar	Soluble	Good
PMMA	40	Polar	Soluble	Good
PVC	45	Polar	Soluble	Good
PET	43	Polar	Non-soluble	Limited
PA	43	Polar	Hardly soluble	Limited

sector. Even though efforts are currently being made to enable pure recycling in the long term, impurities such as color and printing ink or adhesive and foreign material residues can never be completely ruled out. Previous work has shown a clear influence of the preceding life cycle and the recycling process on the property profile and processability of the plastic. In studies conducted by research institutes, it was possible to observe also an influence on the quality of PE bonds with different recycled content, although the surface energy hardly differs. For example, while the virgin plastic bonded with a two-component acrylate adhesive achieves a bond strength of over 6 MPa and fails proportionally cohesively (CF), the recycled plastic has significantly poorer strength of approximately 3.3 MPa leading to an adhesion failure (AF).

8.3.2.3 Long-term Resistance of Plastic Bonds

Despite good initial adhesion, plasticizers diffused into the adhesive layer of plasticized plastics can lead to a drop in strength after some time and thus to failure of the bond. This effect occurs, for example, with plasticized PVC or EPDM and cannot be completely avoided by surface treatments. Therefore, when selecting materials, care should be taken to ensure that the parts to be joined do not contain plasticizers or that the plasticizers are crosslinked and thus chemically anchored in the polymer. Otherwise, the long-term resistance of plastic bonding is generally not critical if the substrates originally already contain polar groups or, as in the case of nonpolar PE or PP, polar groups are generated in the surface layer by suitable surface treatment processes [31]. In plastic bonding, the boundary layer – the area where adhesion takes place – usually proves to be very stable. This is because, in addition to the intermolecular interactions as the cause of adhesion, the molecular migration of low-molecular-weight substances of the adhesive such as solvent molecules and monomers into the joined part also ensures good uniform cohesion of the boundary layer.

8.3.3 Adhesive Bonding on Glass

There is no doubt that glass, due to its transparency and diverse design possibilities, is an important and often indispensable material that is increasingly being used by designers and architects in the course of modern design. This is because in cases where different materials, such as glass with plastics or glass with metals, are to be joined together, there is very often no alternative joining technique besides adhesive bonding to join these materials. Thus, glass elements, windows, or glass facades are joined quickly, securely, and permanently with the help of adhesives. Here, the right choice of adhesive system can compensate for the different thermal expansions of the joined materials – as they occur in glass–metal joints – and the vibrations as well as dynamic loads occurring in the application can be absorbed without any problems.

With a modulus of elasticity of 65–90 GPa and a tensile strength of 30–80 MPa, glass is one of the brittle materials. Plastic joining part deformations as well as bending at the overlapping ends of the bonded joints are therefore very limited. Thus, in glass bonding, the resulting stress peaks – as known from metal bonding – practically do not occur. However, stress concentrations caused by inhomogeneities in the glass are very critical for the quality of the bond. Another critical physical property of the glass to be considered for the adhesive-bonding process is the very low coefficient of thermal expansion α. For example, bonding window glass ($\alpha = 8 \cdot 10^{-6}\ K^{-1}$) and steel ($\alpha = 13 \cdot 10^{-6}\ K^{-1}$) with a 2K epoxy adhesive ($\alpha = 100 \cdot 10^{-6}\ K^{-1}$) results in a glass/adhesive layer/metal expansion ratio of 8 : 100 : 13, which leads to a considerable stress concentration in the adhesive-bonded glass–metal components.

Modern architecture demands thin frame and window profiles, and their easy integration into aesthetically and elegantly looking facades is often only possible with the appropriate adhesives and adhesive tapes. Thus, with the help of adhesive-bonding technology, it is possible to produce filigree, jointless, and frameless all-glass facades with maximum stability, in which the windows can be a stiffening element. In the design of bonded joints between glass and metal, the dimensions and temperature range of the application play a significant role. Here, the significantly lower thermal expansion of glass compared to metal during temperature changes leads to a not inconsiderable shear stress on the adhesive, which must be absorbed in form of deformation in the adhesive layer.

8.3.3.1 Selection of Adhesive for Glass Bonding

For this reason, flexible adhesives, for example based on PU, silicones, or acrylate as well as adhesive foam tapes, should generally be used for glass-to-metal bonding in larger components. Structural or relatively stiff bonds produced with UV adhesives, for example, are only useful if the bonding areas are relatively small. Especially in structural glass engineering, it is important to properly match the adhesive system requirements to the objective of the intended application. If the bonding is mainly responsible for transferring loads over small areas, an adhesive with high strength and stiffness is required. If, on the other hand, no significant forces need to be transmitted through the adhesive layer, elastic adhesives may be the materials of choice. This also allows the different thermal expansions of the joining partners to be compensated for. Adhesives that are very suitable for this purpose include PUs or silicone adhesives and increasingly used high-performance adhesive foam tapes. Most silicone adhesives do not require any surface treatment, while the PU adhesives and PSA foam tapes always need a silane-containing adhesion promoter. PSA foam tapes in particular are regarded as a fast, simple, and proven alternative to mechanical fastening methods when it comes to glass bonding. For example, high-performance viscoelastic double-sided acrylate-based foam adhesive tapes are capable of permanently and securely bonding glass to metal, plastic, or wood. The use of these products for the construction of glass facades results in considerable time savings, as the prefabricated glass panels can be easily and quickly bonded to the metal frame structure in the workshop and then immediately processed for final assembly in the relevant buildings. This is possible because these strips have already reached up to 80% of their maximum strength immediately after bonding.

8.3.3.2 Glass Bonds with Long-term Resistance

Due to the favorable surface energy of the glasses of around 300 mJ/m, very good wetting of the glass surface and thus good adhesive bonding should be easily possible. However, this is not always the case and good bonding to glass is often only possible if special agents are used to promote adhesion. The reason for this is the so-called *glass corrosion*, in which oxides of the elements contained in the glass, such as sodium, potassium, calcium, barium, or boron, are dissolved by water [32]. This leads to an increased concentration of OH ions and thus to an increase in the pH value. As a result, the physical properties of the material change in the contaminated areas, and roughening occurs as well as the formation of fine cracks on the surface. The consequence for transparent objects is that the glass materials become cloudy. This glass corrosion caused by absorbed water has a considerable influence on the formation of the adhesive forces required for successful bonding to the glass surface. Particularly in the case of glass with a high alkali content, the surface energy is significantly reduced compared to uncontaminated glass. Thus, when bonding the glass surface, the applied adhesive does not come always into contact with an ideal – oxide-rich – surface, but touches a glass surface with reduced surface tension, which is provisionally covered with OH groups generated by hydrolysis.

Influencing parameters of glass corrosion are:

- composition of the glass,
- pH value of the dissolving water on the surface,
- ion concentration of the water on the surface, and
- temperature of the environment.

For example, relatively soft water in dishwashers due to low calcium-ion concentration and acid rain due to low pH cause increased dissolution of minerals from the glass surface, resulting in more severe glass corrosion.

8.3.3.3 Surface Treatment of Glass

Prior to bonding, treatment of the glass surface is mandatory and under moderate application conditions, aqueous or alcohol-based cleaning and thorough drying

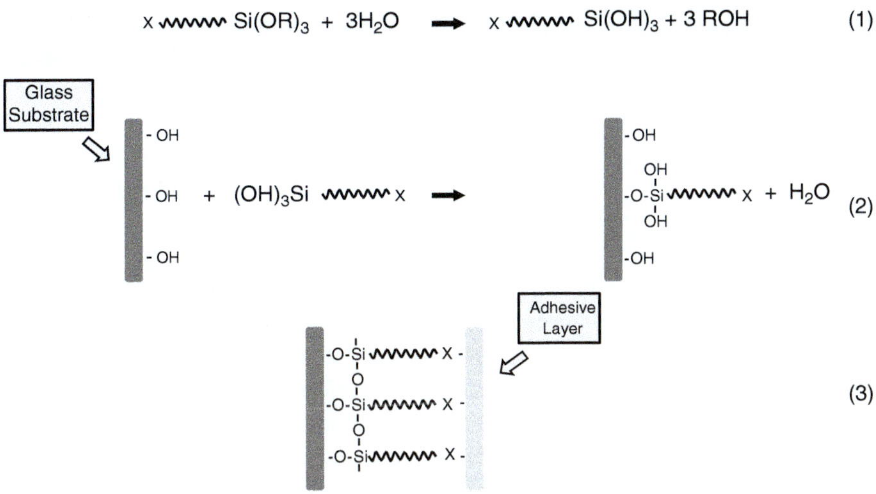

Figure 8.27 Reaction mechanisms of coupling agents with the glass surface and functional groups in the adhesive polymer.

of the glass surface at 40–45 °C are usually sufficient. However, if long-term stable glass composites are required, additional treatment with suitable silane-based adhesion promoters is necessary. With these so-called *coupling agents*, which are either applied separately or added to the adhesive, very long-term stable adhesive bonds with glass substrates are possible [33]. These adhesion promoters, form so called *chemical bridges*, and are thus capable of significantly improving the strength of glass adhesives and, above all, their aging behavior in the presence of a humid atmosphere. They have two chemical functionalities to form the bridge, one reacting with the glass surface and the other with the adhesive polymer (Figure 8.27).

The general structure of this *silane-coupling agent* coupling is $(RO)_3$–Si–R'–X, where RO is a hydrolyzable group such as an alkoxy group, X is an organofunctional group. The organofunctional group and the alkoxy group are capable of forming strong bonds with the adhesive and the glass surface, respectively. Thus, they are able *to couple* the glass surface with the adhesive through covalent bonds. In this process, the alkoxy groups of the coupling agent are first hydrolyzed to form alcohols and silanol groups (1). Then, the formed silanol groups and the glass surface undergo a condensation reaction that leads to the release of water and simultaneously establishes a covalent bond to the glass surface (2). The next and final step is the reaction of functional groups in the adhesive with the organofunctional group of the coupling agent to form the *bridge* between the two bonding partners, glass surface and adhesive layer (3). The organofunctional groups of the adhesion promoters include vinyl, allyl, amino, methacryloxy, and epoxy groups, which react with the appropriate functional groups in the adhesive layer.

As an example, the following surface treatment is suggested for bonding plastic bars and painted/unpainted metal items with acrylic foam tapes to window glass as well as glass and ceramic surfaces found in kitchen and bathrooms:

- After cleaning and drying the glass surface, a *silane adhesion promoter solution* (0.5% in isopropanol) is applied at room temperature using an absolutely clean and lint-free soft cleaning cloth. The adhesion promoter is applied evenly in one direction as a very thin layer. Immediately after applying the adhesion promoter, it is wiped off with a dry cloth. Subsequent application of the acrylic foam tape should be carried out within five minutes after drying (wiping off) of the adhesion promoter.

8.3.3.4 Adhesives for Glass Bonding

Due to the different thermal expansion behavior of the bonding partners in glass bonding, room-temperature curing adhesives and adhesive tapes should be preferred for this purpose. Since they are capable of deforming sufficiently to relieve any stress concentrations occurring in the bonded components, elastic adhesive systems based on silicones, PUs, and PSA foam tapes are very well suited for this purpose. The possible applications of 2K epoxy adhesives depend to a large extent on their constituents and the crosslinking density. Moisture-curing 1K silicones or fast-curing 2K silicones are very suitable for bonding glass without the use of a silane-based adhesion promoter, where the low-temperature flexibility and high-temperature stability of the silicones in the corresponding applications should be particularly emphasized. For example, in furnace construction, glass composites with 1K and 2K acetate silicone adhesives can be made heat resistant and used at temperatures up to 250 °C. In other glass applications, the excellent low-temperature flexibility of silicones is highly valued, allowing their use at very low temperatures down to −60 °C. Since glass is largely transparent in the UV range, UV-curable adhesives are also an option for glass bonding. It should be noted, however, that standard glass absorbs wavelengths below 300 nm in many cases. Also, when UV adhesives are used, only relatively small adhesive film thicknesses of <0.3 mm can be cured. However, bonding metals to glass with UV radiation only makes sense if the bonding areas are relatively small. The reason is the significantly lower coefficient of thermal expansion of glass compared to metal, which leads to a not inconsiderable shear stress in the adhesive layer during temperature changes, which cannot always be absorbed by the deformation of the relatively stiff UV adhesive.

8.3.4 Adhesive Bonding on Wood

For the user, wood presents itself as a *benign* substrate that allows mechanical anchoring of the adhesive film in the structure due to the inherent pore structure created by the evaporation of capillary water during the drying process. In addition, the chemical structures of the main components of the wood matrix – cellulose and lignin – provide multiple opportunities for the formation of hydrogen bonds due to abundant hydroxyl and other polar functional groups, which are needed for adhesion formation on the surface. Thus, even covalent chemical bonds with the adhesive are possible. Evidence of the use of adhesives to join wood in the history of mankind goes back a long way, as shown by archaeological finds of wooden objects and artifacts. Glues of natural origin, for example those derived from plant or animal organs, were used for the consistently non-load-bearing compounds. The production of glutin-based adhesives, which are glues made from skin or bone, represented a significant advance. Works of art and utensils made with such glues were considered particularly valuable and were found, for example, in the burial treasures of the Egyptian pharaohs. Plinius the Elder (23–79 after Christ) was the first to deal scientifically and systematically with the gluing of wood in his 37-volume work *Historia Naturalis*. He recognized that certain hardwoods are far more difficult to glue than softer varieties, a fact that is still current even with modern adhesives.

8.3.4.1 Requirements for Wood Bonding

In the first half of the 20th century, wood and wood adhesives became key materials, and the post-war years saw a veritable flood of new test methods and standards to increase the safety and reliability of wood structures. In the modern context of wood bonding, it is useful to consider two dimensions, distinguishing between *load-bearing – non-load-bearing*, and *industrial – craft*. In the craft segment, gluing is usually too costly or uneconomical due to small series sizes, which is why mechanical fastening techniques such as dowels, screws, or nails are used where loaded composites occur, for example on shelves in furniture. Also, when assembling load-bearing elements on the construction site, the mostly missing possibility to apply pressing pressure is a major obstacle for wood-bonding technology. These restrictions do not apply in the industrial segment. Here, cost-effectiveness and quality consistency are the dominant motives, and large quantities justify investments in complex plant systems and the development of sophisticated process sequences. Wood bonding is usually carried out under pressure, which can be generated in a variety of ways: from screw clamps in the craft sector to fully automated stacking and forming presses for industrial applications. The choice of a suitable adhesive system is based on the chemical and physical properties of the wood-based material, whereby a wide range of raw materials and formulation options are available to the user.

When bonding to wood, the following boundary conditions must be taken into account for the adhesives used:

- *moisture compatibility* – since wood always has a residual moisture content of 6%–15%, the adhesive must have a certain compatibility with water and water vapor, both during application and in continuous use,
- *stress relief properties* – wood is subject to moisture-related and thus climate-related dimensional fluctuations, which can result in high shear stresses at the joint interfaces that must either be absorbed by the elasticity of the adhesive or held in place by sufficiently high adhesion and cohesion forces, and
- *gap-filling properties* – since wood surfaces are never absolutely flat, even when planed or sanded, wood adhesives must be able to fill the gaps thus present.

Furthermore, in the case of wood bonding, it should be noted that certain types of wood are rich in ingredients that can diffuse back to the surface within a short time after mechanical surface treatment. These are low-molecular resins and oils, but also aqueous solutions of sugars or peptides. Such substances can weaken the bond, either by acting as a release agent or by softening the wood structure. For this purpose, it is recommended to increase the resistance of wood bonding by using adhesion promoters. It has been shown, for example, that the delamination behavior of glulam in the water test can be significantly improved even with difficult woods by using different adhesive systems after suitable primer treatment. However, this requires an additional manufacturing step, which is only justified where quality has the highest priority over quantity.

The following adhesive systems are usually used for bonding wood:

- PVAC dispersions (*white glue*),
- hotmelt adhesives,
- formaldehyde resin dispersions,
- post-curing PVA dispersions, and
- moisture-curing 1K PU adhesives.

8.3.4.2 PVAC Dispersions (White Glue)

The undisputed workhorse in wood processing is the so-called *white glue*. These are adhesive dispersions based on PVAC as a homopolymer or its copolymers in combination with acrylates or other functional monomers. Film formation takes place physically by drying away the water, whereby minimum temperatures below which filming no longer takes place must be maintained. This is typically between 2 and 15 °C. If the drying of the joint takes place below this temperature, no compact adhesive film is formed, but a chalky or crumbly layer without useful strength is observed. The curing process can be accelerated by applying higher temperatures using heated presses. However, this principle reaches its limits with thicker wood layers, because wood as a good heat insulator hinders the flow of heat from the press jaws into the adhesive joint. Therefore, the application of high-frequency fields is particularly suitable for drying, since these are absorbed much more strongly by the aqueous adhesive than the surrounding wood. This is caused by the high dipole moment of the water and allows, that even thicker components can be penetrated and cured.

8.3.4.3 Hotmelt Adhesives

These adhesives are characterized by the fact that they build up their adhesive strength through solidification or crystallization. In this respect, they are superior to physically and chemically bonding systems and are therefore the preferred choice for fast-running machine production processes. The base polymers used in wood adhesives are predominantly based on EVA, PA, and APAO. The adhesive layers of these hotmelt adhesives are thermoplastic, which limits their heat resistance. In addition, like all thermoplastics, they exhibit cold flow, and are therefore not suitable for load-bearing joints. Reactive hotmelts, which post-crosslink by means of moisture after application, combine the advantages of classic hotmelt adhesives with those of thermoset adhesives. Here, the thermoplastic bond is sufficiently strong to allow stacking and machining of the workpieces, and the increased strength values due to isocyanate crosslinking build up over hours and days during subsequent storage.

8.3.4.4 Formaldehyde Resin Dispersions

In contrast to the systems already mentioned, these adhesives belong to the class of chemically curing systems. Here, a polymerization reaction takes place in the adhesive joint during the pressing process, followed by a crosslinking reaction between the macromolecules formed in this way. As a result, the strength values increase sharply and the adhesive layer is converted from a thermoplastic to a thermoset state, the basic requirement for joints that are under constant load in the application. Thus, the choice of the appropriate adhesive system depends on the type of substrates to be bonded as well as on the future stress on the joint. Formaldehyde condensation resins are probably the most extensive class of chemically curing systems for wood bonding. Reactants of formaldehyde are ureas, phenols, or melamines, as well as all conceivable combinations thereof. The reaction partners are processed as two-component systems in aqueous solution and, in the case of industrial application, mixed immediately before the adhesive is applied to the wood surface. In the craft environment, on the other hand, ready-to-use mixtures in powder form are commonly used, which are then mixed with water into a reactive and applicable form.

8.3.4.5 Post-curing PVAC Dispersions

The performance of these adhesives can be positioned between the highly crosslinked reactive formaldehyde-based resin systems and the aqueous thermoplastic PVAC adhesives. Here, a bifunctional isocyanate component is added to the PVAC emulsion shortly before application, which reacts with the hydroxyl groups of the polymers to achieve crosslinking. The result is both improved water and heat resistance of the bonded components, with some of these systems even qualifying for certain load-bearing joints.

8.3.4.6 Moisture-curing 1K PU Adhesives

Solvent-free moisture-curing 1K PU adhesives have recently gained considerable importance in wood bonding, especially in engineered wood construction, with the following advantages over the classic formaldehyde condensation resins [34]:

- smaller quantities of adhesive are used, as they are solvent-free,
- less waste because of almost lossless application,
- can be used as supplied by the manufacturer without mixing,
- no reaction water is produced,
- good flexibility in selecting curing time, and
- inconspicuous adhesive joint with wood-like aging behavior.

8.3.4.7 Bonding Wood to Other Materials

All known reactive systems are suitable for bonding wood to other materials, whereby the requirement profiles of the planned applications are decisive. These can require high moduli, such as when bonding carbon fiber lamellae in the tension zone of load-bearing beam elements, which is realized with 2K epoxy adhesives, for example. At the other extreme, reactive systems often based on PU are used for bonding parquet to mineral substrates. The viscoelastic behavior of these joints dampens impact sound and is also capable of absorbing the high expansion and shrinkage stresses that occur in hardwood parquet when the climate changes.

8.4 Loads on Adhesive Bonds

8.4.1 Introduction

The loads on a bonded component caused by mechanical and environmental influences almost always occur simultaneously and are complex in nature. They can act on the bonded component both in the short or long term and statically or dynamically. In any case, they cause the bonded component to age, e.g. due to temperature, humidity, UV radiation, and chemical influences, which usually leads to a weakening of the bonded joint (Figure 8.28).

For the aging processes that take place in the adhesive bond, it plays a major role whether the environmental influences act on mechanically stressed or unstressed adhesive layers. For example, water molecules can diffuse into a deformed adhesive layer more quickly than into an unstressed one and thus more easily cause accelerated aging. It is a core task of development in adhesive-bonding technology to design the adhesive composite in such a way that it can withstand these external influences while retaining the original strength values over the entire service life.

8.4.2 Long-term Behavior of Adhesive Bonds

The durability and long-term behavior of bonded joints are particularly important in areas where the failure of a component results in the destruction of the overall structure and potentially significant consequential damage.

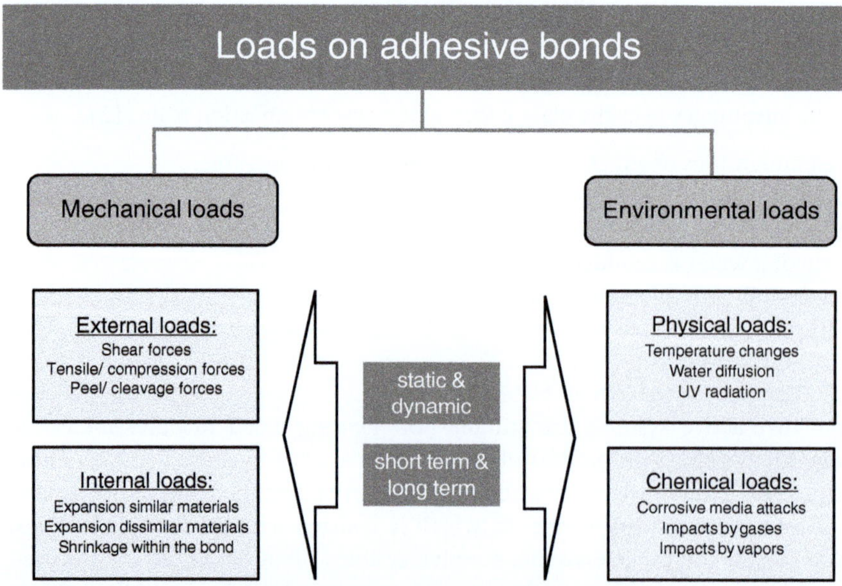

Figure 8.28 Loads on adhesive bonds.

The following factors determine the failure of a component:

- the aging behavior of the adhesive,
- the time-dependent mechanical behavior in the form of creep under load, and
- the stability of adhesion under the influence of the expected environmental conditions.

For this reason, it is not the strength behavior of an adhesive bond in the static short-term test that is decisive, but the strength values that result after time-dependent stress application under the influence of the loads relevant to the application. Since aging processes are generally very slow, the results of evaluations after accelerated aging in the laboratory are used for the development of an adhesive bond. By increasing the temperature or applying other more stringent aging conditions, a *time-lapse effect* is achieved, thus enabling the long-term behavior of an adhesive bond to be determined in a relatively short time using laboratory samples. The results of these tests, in the form of percentage strength reductions after a specified time, form the basis for determining the reduction ratio to be used when dimensioning an adhesive bond. Although it is quite common to use a generally valid reduction ratio for the corresponding influences, based on the values obtained after a specified time, this approach does not provide any information on the time dependence of the strength reduction of the component. Thus, it is possible that a large change in strength has occurred after a relatively short time, the further course of which then asymptotically approaches a limit value. For this reason, it is always advisable in practice to evaluate the bond strength over time in addition to the results after the specified total time and to include the related observations in the assessment for dimensioning bonded components.

8.4.3 Mechanical Loads

8.4.3.1 External Loads

Forces acting on a bonded joint can be caused externally or internally and always represent a significant load on the bonded component. One of the first questions the designer asks when planning a bonded joint is which external mechanical forces act on the bonded layer and how they are aligned. Are we talking about shear, tensile/compressive, or even peel or cleavage forces? Furthermore, are they static or dynamic, and if they are dynamic, at what frequency do they affect the bond? All these questions have to be answered and the results of the relevant considerations are an essential part of the planning and dimensioning of the component to be bonded.

8.4.3.2 Internal Loads

In addition to the stresses caused by the external mechanical loads acting on the bonded joint, residual stresses are present in the components, which are a source of danger for the reduction of the bond strength. They can be caused by differences between

- adhesive and similar substrates,
- adhesive and dissimilar substrates, and
- shrinkage of the adhesive layer.

8.4.3.2.1 Loads Caused by Similar Substrates

Since the coefficient of thermal expansion of adhesives is many times higher than that of metals, stresses can occur with temperature changes that amount up to 50% of the adhesive force. This effect is caused by different length changes of adhesive and substrate in the bonded component. Figure 8.29 shows the behavior described for the case where, for example, two aluminum sheets of the same quality (substrate A) and dimensions are bonded with an epoxy adhesive. Then, after joining and curing, the temperature is increased by 100 °C.

The coefficient of thermal expansion α of an epoxy structural adhesive used in industry and trade, for example, has a value of about $90 \cdot 10^{-6}\ K^{-1}$, while for aluminum it is about $23 \cdot 10^{-6}\ K^{-1}$. Young's modulus E_K of such structural epoxy adhesives can be assumed to be about 2000 MPa.

The residual stress σ occurring in such an adhesive joint can be simply calculated by

$$\sigma = E_K\,(\alpha_K - \alpha_M)\,\Delta T$$

with the subscript M for metal and K for adhesive. ΔT is the temperature difference in practical application. Such a calculation assumes elastic behavior of the adhesive layer and does not take into account the temperature dependence of the coefficients of thermal expansion. Another assumption is that the modulus of elasticity of the adhesive layer does not change in the range of the temperature difference ΔT, which is given when the application takes place below the glass transition temperature.

According to this calculation, the bonding of two identical aluminum components ($\alpha = 23 \cdot 10^{-6}\ K^{-1}$) with a structural epoxy adhesive ($\alpha = 90 \cdot 10^{-6}\ K^{-1}$) as demonstrated in Figure 8.29 builds up a stress of 13.4 MPa when the temperature is increased by 100°. However, this value is not achieved in practice because the residual stresses are relieved by the elastic–plastic behavior of the adhesive layer. This calculation assumes that the component is at a temperature below the glass transition temperature of the adhesive. Above the glass transition temperature, no permanent

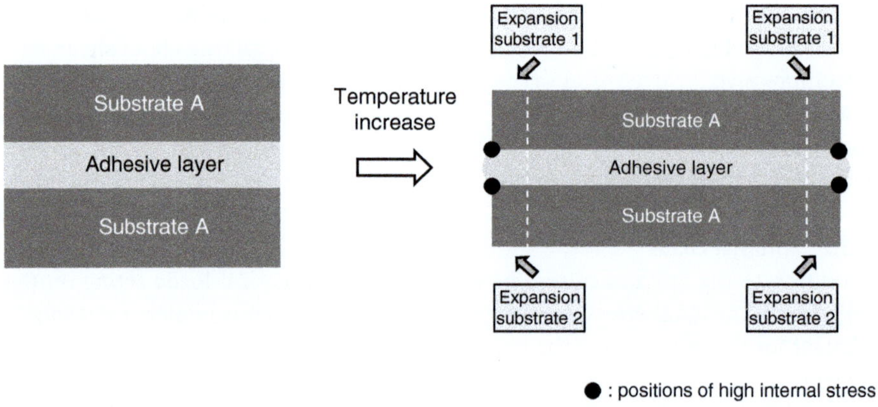

Figure 8.29 Internal loads due to thermal expansion of similar or identical substrates (schematic and not to scale).

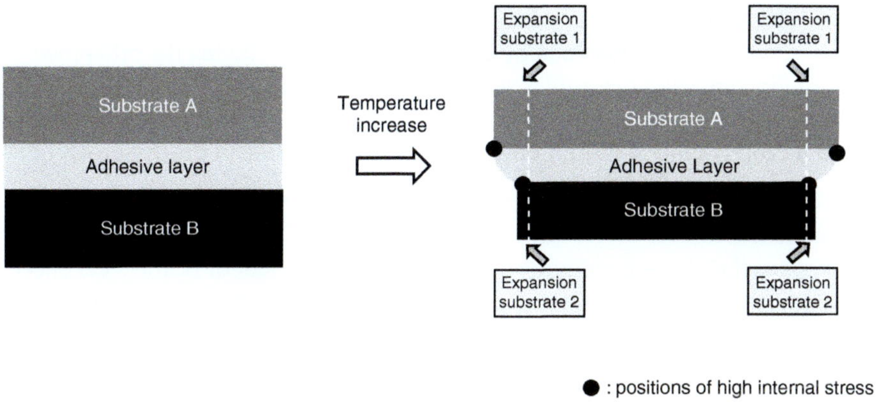

Figure 8.30 Internal loads due to thermal expansion of dissimilar substrates (schematic and not to scale).

stresses form in the bond, as these are easily relieved by the free displacement of the molecules. For this reason, such calculations for practical applications can only be carried out for adhesive layers with high glass transition temperatures.

8.4.3.2.2 Loads Caused by Dissimilar Substrates

If different materials are bonded in the component, as for example is the case with steel–aluminum composites (Figure 8.30), different coefficients for thermal expansion of the substrates lead to undesirable effects during curing that can damage the component. These relative movements of substrates A (aluminum) and B (steel) lead to high internal stresses at the overlap edges of the joint, which must be excluded of minimized by suitable measures.

Possible damage to the components includes:

- *Bulging of a flat bond* – in the direction of the joining part with the higher coefficient of thermal expansion,
- *fracture of one of the substrates* – if its strength is too low for the thermal stresses generated,
- *cohesion failure in the adhesive layer* – if both parts to be joined are of high strength and the adhesive layer is not able to compensate for the expansion paths that occur,
- *adhesion fractures in the bond* – in the case of inadequate surface treatment, and
- *microcracks in the adhesive and boundary layers.*

The linear expansion ΔL of a substrate with the length L_0 caused by a temperature increase ΔT can be calculated according to the following relationship:

$$\Delta L = \mathrm{L}_0 \cdot \alpha \cdot \Delta T$$

Such calculations are based on the assumption that the thermal length change of the substrates is a linear function of the original length, the coefficient of thermal expansion, and the temperature. For example, the above equation can be used to calculate the respective elongation of a steel and aluminum sheet in a steel–aluminum joint that is very commonly manufactured in the automotive industry. From these individual expansions of steel and aluminum, the relative movement in the bonded joint can be determined by simply taking the difference.

In the following example, this is done for a bonded component, which – as in many automotive applications – undergoes a significant temperature increase to dry the painted component and simultaneously cure the adhesive in the drying oven:

- With an overlap length of the steel–aluminum bond of 10 mm and a temperature increase from 20 to 170 °C, as is usual when the component is placed in a paint oven, the above equation calculates a change in length of the steel sheet ($\alpha = 12 \cdot 10^{-6}\ \text{K}^{-1}$) of 18.0 μm and of the aluminum sheet ($\alpha = 23 \cdot 10^{-6}\ \text{K}^{-1}$) of 34.5 μm. The difference results in the relative motion of these dissimilar substrates, which causes stress on the joint. Thus, to avoid damaging or compromising the appearance of the joined component, the cured adhesive layer must be able to relax the stresses generated by this relative motion of 16.5 μm.

Reduction of such internal loads is possible by the following measures:

- Production of adhesive layers with lower elasticity and shear moduli, which, however, is associated with increased creep of the adhesive and a reduction in adhesive strength,
- incorporation of toughening particles, which usually have no negative effect on the chemical and physical properties of the adhesive layer, and
- use of primers or adhesion promoters with a higher deformation capacity than the adhesive layers.

In some cases, stress relief can also be achieved by adjusting intelligent time–temperature profiles. For example, the cured adhesive component is heated to a temperature above the glass transition temperature of the adhesive. Thus, it is possible to relieve the stresses that have arisen in the bond as a result of the reduced elastic modulus of the adhesive in this state.

8.4.3.2.3 Loads Caused by Shrinkage of the Adhesive Layer

These loads are caused by the shrinkage behavior of adhesive layers and occur during curing or drying of adhesive layers. The effects are particularly noticeable when bonding large surfaces. Because of its influence on the strength, stress formation, and visual appearance of the component, this load is particularly important in the automotive industry, e.g. in the bonding of roof structures, doors, engine hoods, and trunk lids.

In the case of heat-curable adhesives, the following stresses occur in the corresponding phases (Figure 8.31):

- *heating-up phase* – during this phase, the volume of the adhesive layer is increased, which is represented in curve A-B,
- *curing phase* – here, the volume is reduced by the start of the crosslinking reaction and the associated formation of covalent bonds, which is shown in curve B-C, and
- cooling down phase – in this phase, further shrinkage of the adhesive layer is caused by cooling if the cured adhesive layer, demonstrated in curve C-D.

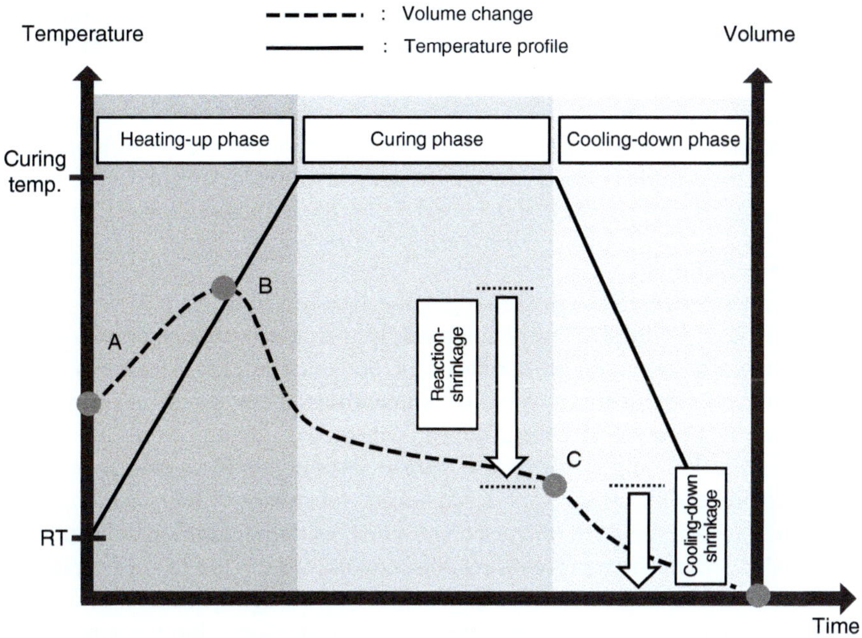

Figure 8.31 Shrinkage of adhesive layers during curing of heat-curable adhesives (schematic and not to scale).

Other reasons for the shrinkage of the adhesive layers are:

- *release of reaction products* – such as those formed during crosslinking by condensation reactions,
- *release of low-molecular-weight substances* – such as monomers and additives being part of the adhesive formulation, and
- *evaporation of solvents* – the process needed for drying of adhesive layers.

Residual stresses frequently occur in bonded joints due to shrinkage of the adhesive layer and, as discussed, they can have various causes. These generate more or less large initial stresses. However, the good news is that the residual stresses caused in this way can usually be absorbed by the intermolecular forces in the boundary layer.

8.4.4 Environmental Loads

8.4.4.1 Temperature Changes

Stresses due to temperature changes are particularly critical in the presence of water or moisture, whereby the adhesive layer is subjected to exceptional stresses. This combination very often leads to the failure of a bonded component. The reason for this is the fact that water diffused into the adhesive or the boundary layer undergoes significant expansion when the temperature increases, and the bond is blown away, so to speak. This does not happen suddenly, of course, but over a certain period of time, caused by the pressure created in the bond by the expanded water. To simulate this effect, climate change tests have been developed and repeatedly optimized over many years, particularly in the automotive industry. In such tests, laboratory samples or prototypes are subjected, for example, to repeated temperature changes from −40 to 80 °C over a longer period of time with simultaneous influence of humidity. After a specified period of time, the drop in adhesive strength in percentage is used to assess the behavior of the bond under the corresponding conditions.

8.4.4.2 Water Diffusion

Depending on the degree of crosslinking, adhesives can absorb up to 5 % water. Its diffusion into the adhesive layer causes swelling processes that negatively affect the strength and deformation properties of the bonded joint and, as already mentioned, this effect is intensified by elevated temperatures. Diffusion of water into the adhesive layer is a process in which the water molecules attempt to reach a static equilibrium state. In doing so, they move from areas of higher to areas of lower concentration, a process that is slow and temperature dependent. If there is damage to the adhesive layer in the form of cracks and pores, water penetration via capillary forces can also occur, which proceeds relatively quickly and is further accelerated if the water is in the liquid phase. This explains the different aging behavior under the influence of moisture of mechanically unstressed and stressed adhesive bonds.

8.4.4.2.1 Testing Water Resistance

To test the water or moisture sensitivity of a bonded component in the laboratory, a defined moisture–heat load is applied to the bond by exposing the specimens to water or to climatic chambers set to defined relative humidity and temperature. To achieve a state of equilibrium with a constant moisture load on the specimens over time, long exposure times of about 1000 hours must be scheduled. If necessary, the exact exposure times to reach the equilibrium state must be determined experimentally. The influence of water or moisture exposure on the quality of the composite can then be verified by bond strength tests (lap shear), generation of stress–strain curves, or dynamic mechanical thermal analysis (DMTA).

8.4.4.3 Corrosive Media Impact

Corrosion is the reaction of a metallic material with its environment. In the case of steel, elemental iron is oxidized by oxygen in the air to iron (II) and iron (III) ions. In the presence of water or moisture, insoluble iron (III) hydroxide and iron (III) oxide are formed, which remain on the surface (Figure 8.32).

In general, moisture and the simultaneous attack of corrosive media can lead to detectable damage to an adhesive bond in a relatively short time, which is further intensified by heat. These effects, which can be detected with the salt spray test according to DIN 50021, can be observed much more strongly with unalloyed steels than with aluminum alloys.

The following types of corrosion are particularly relevant in adhesive-bonding technology:

Bond line corrosion – This type of corrosion, caused by the influence of moisture on the boundary layer, is strongly influenced by the substrates used. It starts at the

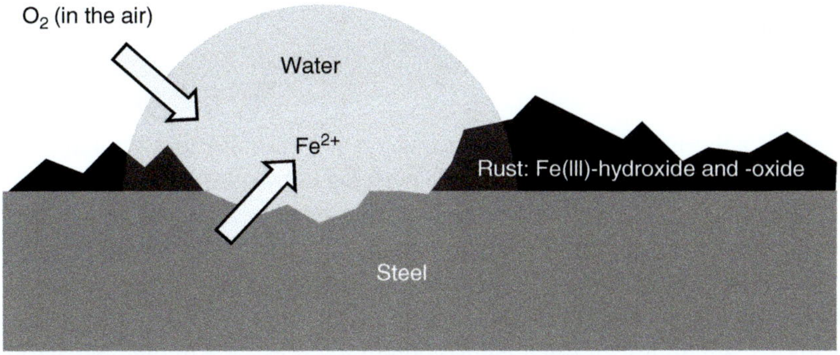

Figure 8.32 Formation of rust on a steel surface by reaction of atmospheric oxygen with iron under the influence of water (schematic and not to scale).

edges of the bonded joints and proceeds without or with simultaneous corrosion of the base metals. Bond line corrosion usually leads to a total loss of strength and it proceeds slowly due to the influence of moisture alone. However, the loss of strength occurs much more quickly if corrosion processes on the base material take place at the same time. The good news is that bond line corrosion can be partially or completely avoided by sealing the adhesive edges.

Creep corrosion – This type of corrosion is based on destruction of the substrate by corrosive media and subsequent infiltration of the adhesive layer. The resulting fracture of the bonded component is therefore not due to a direct failure of the established bonding forces, but to material destruction initiated outside the adhesive joint. This type of corrosion can be avoided by optimized surface treatment of the substrate.

Contact corrosion – When metals with different electrochemical behavior come into contact via an electrolyte (water) in a bonded component, contact corrosion occurs. Thus, in pairings of noble and base metals such as copper/iron, copper/zinc, or iron/aluminum, electrons can be transferred from one metal to the other, leading to electrochemical reactions at the surfaces and thus to corrosion. Compared to mechanical fastening methods such as screwing, riveting, and welding, adhesive-bonding technology is in an excellent position to prevent this type of corrosion. The adhesive layers between the metals have an insulating effect and are able to prevent electron transfer and thus corrosion.

Stress crack corrosion – This type of corrosion, which is not caused by electrochemical processes, frequently occurs in plastics such as PC, PS, or PMMA. It is caused by the action of stresses on the bonded joint with simultaneous exposure to media. These can occur by external loads, but also by internal stresses in the material, where the resulting cracking in the adhesive layer depends on the degree of stress, the temperature, the type of media, and the pH value. Residual stresses in plastic materials are often relaxed by solvents or monomers diffusing into the material, which leads to the formation of microcracks in the polymer and thus negatively affects the quality of the bonded component.

8.5 Adhesive Specification

8.5.1 Introduction

When selecting adhesives for a bonding process, it is important that, in addition to assessing their adhesive properties on the selected substrates, the production-specific conditions and the stresses acting on the bonded component in use are also strongly taken into account. The modern adhesive systems offered by adhesive suppliers generally exhibit the general adhesive properties required for industrial applications very well in terms of adhesion and cohesive forces. However, they differ primarily in their suitability for the intended bonding process and in their behavior under mechanical and environmental stresses such as temperature, moisture, UV radiation, and chemical media. They also vary as to whether and what type of surface treatment is necessary for the formation of a high-quality bond.

For the successful preparation of the adhesive specification, it is therefore essential to define the following parameters in sequence (Figure 8.33):

- technical properties of the substrate,
- expected loads acting on the bond,
- performance requirements for the adhesive,
- surface treatment method, and
- parameters of the manufacturing process.

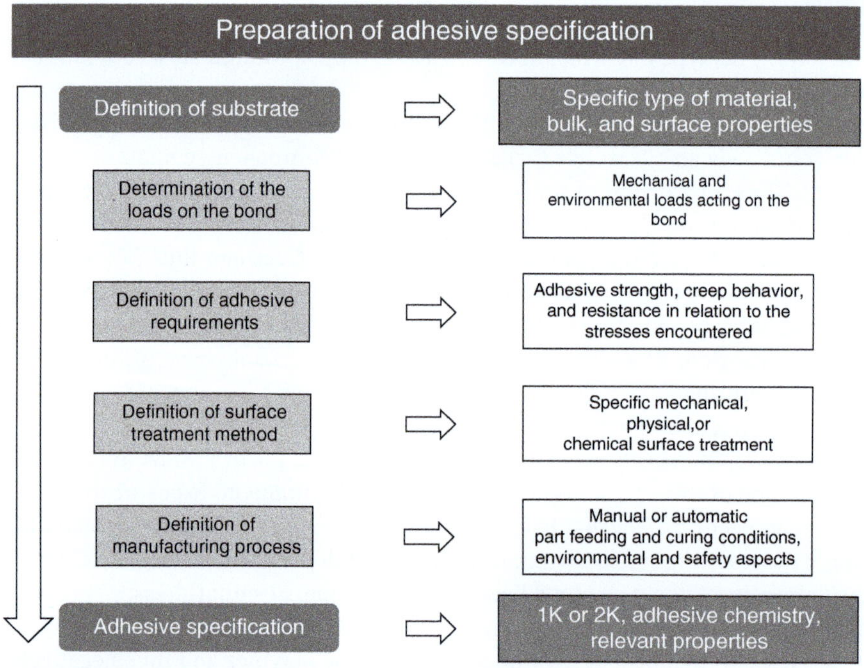

Figure 8.33 Definition of the parameters relevant for the adhesive specification.

8.5.2 Definition of Substrates

To obtain an optimal match with the adhesive, the exact type of substrates used with size and geometry as well as its surface chemistry (treated or untreated) and surface condition must be known. Also important are the storage and the manufacturing conditions of these materials if they influence the quality of the bond and thus the adhesive selection. At the beginning of the planning and the start of the project, the materials to be joined are usually available and have been specified beforehand, taking into account a wide variety of requirements for the component in the application, such as strength, long-term resistance, appearance, application, and processability. Even if adhesive bonding has already been defined as the joining technique prior to the selection of the parts to be joined, the assessment of the suitability of the selected substrates in terms of their adhesive-bonding capability is rarely given much attention. It is also often the case that the material has been *designed* for a mechanical joining technique, and adhesive-bonding technology later comes into play due to its many advantages, such as ease of processing, weight savings, cost savings, and appearance. Therefore, for the developer responsible for adhesive selection, the starting position in most cases is as follows: The substrates are fixed and the adhesive and the type of surface treatment must be adapted to the existing conditions.

For adhesive selection, the substrate materials should first be clearly defined and described according to the following criteria:

- *Specific material type* – For metals, the alloy should be known; for plastics, the chemical composition of the polymer and its structure (thermosetting or thermoplastic), which influence strength and thermal load-bearing capacity; as well as the additives used, such as plasticizers or/and processing aids, which can migrate to the plastic surface and influence adhesion.
- *Basic bulk properties* – Strength properties such as tensile strength, elongation at break, yield strength, elasticity, shear modulus as well as temperature resistance, thermal expansion, and thermal conductivity should be known.
- *Surface properties* – Because of their extreme importance to the quality of the bonded joint, the exact surface properties should be known. This includes information on surface roughness, the type or number of layers on the surface, and the presence of polar groups. Further information on already applied coatings, corrosion and wear behavior, fatigue resistance, diffusion, and dissolution behavior of the substrates would be useful.

8.5.3 Determination of Loads on the Joint

The load-bearing capacity of an adhesive bond is highly dependent on the adhesive used in the bond. Therefore, when selecting the adhesive system, it is important to know exactly the loads acting on the bonded component. In practice, the quality of an adhesive-bonded component is almost always influenced by jointly occurring mechanical and environmental influences that occur over the short or long term. For example, the joint is often subject to a long-lasting influence of temperature, weather, climate, and certain media, usually accompanied by mechanical impairments such as tensile, compressive, shear, peel, and/or torsional loads. The performance of an adhesive bond is therefore not only determined by the data obtained in static short-term tests, but according to the long-term impacts of the previously mentioned mechanical impairments and environmental influences. The aim of developing an adhesive bond is to minimize the strength loss caused by these influences and to create a component that retains the original strengths as far as possible. For this purpose, the selection of a suitable adhesive plays a paramount role and for the adhesive specification, it is important to know as precisely as possible the loads to be expected on the bonded component.

The following loads must be considered:

- *mechanical loads* – static: shear, peel, pressure, and tensile loads; dynamic: vibrations or impact loads,
- *humidity and stresses caused by media* – for example cleaning agents, solvents, or other chemicals such as coolants,
- *thermal loads* – for example temperature level and duration or rapid temperature changes, and
- *UV radiation.*

8.5.4 Definition of Adhesive Requirements

Once the loads that will be applied to the bonded component in service have been identified, the required properties of the adhesive layer must be designed to successfully cope with these loads. Knowledge of these properties provides a good basis for identifying the adhesive(s) that will be further evaluated in the feasibility phase. A unique feature of adhesive bonding technology is that a polymeric adhesive layer is present between the joined materials in the finished adhesive bond, which differs in many ways from metallic and other inorganic materials. Its strength is relatively low, and when a force is applied, its deformation is non-linear and also time-dependent. If the adhesive layer undergoes deformation of more than 1%, irreversible material damage can occur, which negatively affects the quality of the bonded joint. Under prolonged static (resting) load, the adhesive layers tend to undergo plastic deformation, the degree depending on the temperature, the intensity of the load, the properties of the joined parts, and the type of cured adhesive. This special behavior of the adhesive layer must be taken into account when designing the required adhesive layer and subsequently selecting the adhesive.

For this purpose, the following properties are of interest:

- *Adhesive strength* – This property is defined as the maximum stress required to break a bond in a shear tensile test and a very important physical property of the adhesive layer. Therefore, it is one of the first criteria for selecting an adhesive. By taking into account the mechanical loads that will act on the bonded component in use, a good estimate can be made for the required bond strength of the adhesive layer. For example, very high loads on the component require relatively stiff structural adhesives with bond strengths of, e.g. >10 MPa and relatively low loads on the bond allow semi-structural or elastic adhesives with bond strengths of, e.g. 1–10 MPa. Since the adhesive strength is strongly temperature-dependent, the corresponding values for the application temperatures must be strongly considered when selecting the adhesive.

- *Stress relaxation* – Especially when bonding different materials, temperature changes can lead to different dimensional changes of the substrates, resulting in functional limitations or even destruction of the bonded joint. The appropriate selection of adhesives as well as the optimal design of the adhesive bond can help to minimize the problem. Hereby, elastic adhesives with good stress relaxation properties are able to compensate for such effects. Non-toughened thermosets are generally poor at overcoming this problem due to their stiffness and low deformation capacity.
- *Creep resistance* – High creep resistance is a basic requirement for structural bonded joints that are subjected to constant mechanical stress and must last over a long period of time. Adhesive layers exhibit a more or less temperature- and mechanical-load-dependent visco-elastic behavior. They tend to creep (or flow slowly), which can lead to failure of the bonded component. This behavior must be taken into account when selecting the adhesive. Elastic adhesives are soft and flexible and therefore more prone to creep, whereas thermosets with high crosslink densities allow low creep under such conditions.
- *Corrosion resistance* – The influence of moisture, temperature, and corrosive media, such as salt solutions, can attack the substrates and adhesive layers in the resulting composites, leading to functional limitations or structural failure. The right choice of the properties of the adhesive layer, in addition to the selection of suitable substrates and their surface treatment, plays an essential role and thus enables adhesive bonds with good corrosion resistance.

8.5.5 Definition of Surface Treatment Method

The type of surface treatment often plays a decisive role in the selection of the adhesive. For example, there are some special adhesive systems that are capable of bonding low-energy plastics such as PE and PP without surface treatment. Others exhibit special strength properties with certain selected treatment methods, which is not the case with other adhesives applying the same treatment method. Also, there are treatment methods for certain applications that allow the use of less expensive adhesives and still achieve the required properties. Therefore, the type of surface treatment should be determined prior to adhesive selection. Always, the treatment of the material surface must be preceded by a cleaning and degreasing process, which is absolutely necessary for the preparation of a high-quality adhesive bond.

To improve the bonding strength, mechanical treatment is recommended in any case when bonding metals and for example, a significant improvement in adhesion to these surfaces can often be achieved by simply grinding the surface

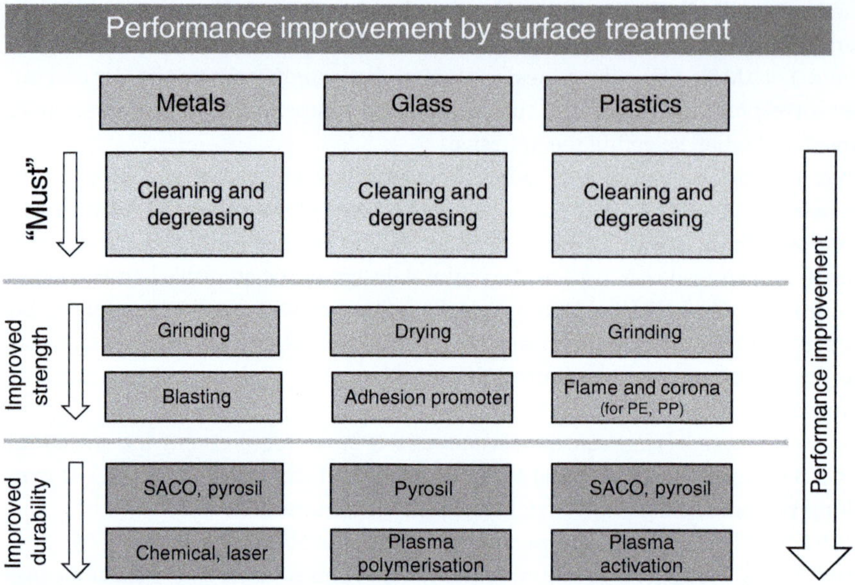

Figure 8.34 Improvement of bonding performance and aging resistance of metal, glass, and plastic joints by mechanical, physical, and chemical surface treatment.

with abrasives. This is the simplest method of mechanical surface treatment, and very often provides bonded joints with adequate strength and suitable moderate long-term aging behavior. The use of other surface treatment methods depends on the long-term requirements of the adhesive bond and the type of substrate. For example, blasting, SACO treatment, or the Pyrosil process is used for more demanding adhesive bonds, when the bonding process is performed on metal and plastic surfaces. For very high-performance bonding, laser and chemical treatment are used for metals, and plasma treatment for plastic surfaces. Low-energy plastics such as PE and PP in almost always require surface treatment, e.g. flame treatment, corona or plasma irradiation, to produce adhesive bonds with sufficient strength (Figure 8.34).

8.5.6 Definition of Manufacturing Process

When selecting the adhesive, the existing conditions in production relevant to the bonding process must be taken into account. If the current process capabilities are not sufficient, the selection of the adhesive must also be made under the aspect of the investment for process modifications or the acquisition of new equipment.

General process requirements – As a basic requirement, the highest standards of cleanliness must be met in the production of bonded components. Appropriate professional facilities must be prepared for the storage of the selected adhesives, bonded parts, and all auxiliary materials. The environmental conditions in the production facility must be adapted to the requirements of the adhesive curing process. Cycle times and the application process (manual or automated) must be taken into account, and suitable application equipment for adhesive application in the required thicknesses and widths must be installed.

Manual or automatic application – Manual application requires no or only low investment costs, but the quality of the process depends on the person carrying it out and their daily form. Automatic application methods require considerable investment costs, but lead to better quality of the component and further to shorter cycle times due to the smaller spreading width compared to the manual method. Depending on the method, the choice of adhesive will fall on the candidates, which, for example, have better compatibility with one or the other method or stand out due to fast cycle times particularly suitable for the automatic process.

Component feeding – When selecting the adhesive, it is important whether the component is fed manually by placing the workpiece in the working position or semi-automatically by placing the workpiece in a processing station outside the working area and then automatically moving it into the working position. These two methods result in different cycle or preparation times for adhesive processing, which affect the selection of the adhesive.

Curing conditions – The requirement for the type of cure and the curing time, as well as the equipment needed for the curing process, plays a major role in the choice of adhesive. For example, there are adhesives that cure quickly without additional process steps, such as the PSAs or the hotmelt adhesives, or those that need to run through an oven for some time to cure or dry, such as the 1K epoxy or solvent/water-based adhesives. These may require a larger investment in an oven. Moisture-curing adhesives, such as 1K PUs or silicones, cure slowly, but like the PSAs and hotmelts, do not require additional equipment. Thus, it depends entirely on the production requirements such as temperature, time and curing conditions at the user which adhesive is best suited for him.

Joining part fixation – The fixation of the component after joining, which is necessary for proper and reproducible curing of the adhesive, is achieved by various measures in the process or simply by the effect of gravity, for which no additional devices are required. Fixation can also be achieved by additional snap-fit joints or tongue-and-groove systems, or by the so-called hybrid process, in which adhesive-bonding is combined with mechanical joining processes such as spot welding, riveting, punch riveting, hemming, or screwing. It is obvious that, depending on the type of fixture, certain adhesive systems may or may not be suitable for the process step of joining part fixtures because of their property profile.

Environmental protection and occupational safety – If adhesives are classified as hazardous substances or dangerous goods due to their composition, they require certain protective measures with regard to environmental protection and occupational safety during their processing. These measures for the protection of people, the process facilities, and the environment are defined in more detail by laws, ordinances, and technical rules and must always be followed. These aspects play a very important role in the selection of the adhesive. In this respect, for example, dispersion, hotmelt (also reactive), and, above all, PSAs have major advantages over solvent-based or reactive adhesive systems. Corresponding information on occupational safety and environmental protection can be found in the adhesive supplier's safety data sheet. This includes, for example, the preparation of operating instructions specifying the protective equipment to be used, and, if necessary, medical care for employees. Further regulations to be observed concern the transport, storage, and disposal of the adhesive.

Quality assurance – A special feature of adhesive components is that they cannot be tested completely non-destructively. This is why this type of joining process is also referred to as a *special process*, meaning that quality must be *produced* and cannot be *tested*. This principle requires a functioning, practiced quality management system, appropriate quality assurance measures, and qualified personnel. The requirements regarding quality assurance in relation to the finished bonded component should also be taken into account when selecting the adhesive. It is obvious that more complex adhesive systems, which require additional steps such as mixing for 2K adhesives, ask for more effort than the easy-to-handle PSAs, hotmelt, or moisture-curing 1K adhesives.

8.5.7 Adhesive Specification

After intensive consideration and definition of the influencing variables relevant for the selection, the adhesive can be specified precisely. The following lists of general, performance, and process specification items serve as an example, which contains the essential components of a specification for the required adhesive and can be extended or refined for the respective application:

***General and performance specification*:**

- chemistry of the adhesive,
- 1K or 2K adhesive,
- physically or chemically curing,
- elastic or structural performance,
- minimum and maximum adhesive strength,
- elongation range,
- maximum service temperature, and
- short- and long-term resistance to the loads that occur.

***Process specification*:**

- storage conditions (temperature/ time),
- viscosity ranges,
- measures for adhesive preparation,
- open time and times for obtaining hand and final strength,
- adhesive application method,
- curing or drying conditions,
- waste disposal of adhesive residues, and
- measures regarding occupational safety and environmental protection.

The above list of general, performance, and process-relevant specification items serves as an important basis for discussions with suppliers and ultimately for selection of candidates for feasibility demonstration.

8.5.8 Some Practical Guidelines for Adhesive Selection

The list for the adhesive specification proposed above is a good basis for the exact specification of the adhesive requirements. For narrowing down adhesive candidates in terms of their suitability for certain bonding applications, here are some *practical guidelines* for cases, where certain properties are required for the bonding process as well as the use of the bonded component in practice:

- *Fast initial strengths* – Fast initial strengths give the user the opportunity to process the components quickly and thus make the process economical. The use of self-adhesive foam tapes, photoinitiated acrylates, fast 2K versions of acrylate adhesives, PU, or epoxy systems as well as hotmelt and cyanoacrylate adhesives is recommended for this purpose. Here, for bonding with higher (structural) strength requirements, acrylate, epoxy, PU, and photoinitiated systems are suitable. Foam tapes are an excellent choice for elastic bonding and are suitable for a wide range of materials. Cyanoacrylates are used for thin-layer bonding of small parts, and hotmelt adhesives for components requiring only low heat resistance, however fast subsequent processing.
- *High-temperature resistance* – For higher-temperature loads, highly crosslinked adhesives based on thermosets should be used, with epoxy resin adhesives achieving good results in many applications. These adhesives also show favorable creep behavior under static mechanical loading. The thermoplastic adhesives such as cyanoacrylates and non-reactive hot melts should be excluded at higher-temperature loads, with the reactive hotmelts showing better heat resistance due to subsequent crosslinking by moisture or heat. Silicones show excellent temperature resistance but low-shear strengths.
- *Bonding of large areas* – When bonding large areas with reactive adhesives, systems must be selected that have long open times. Contact adhesives, adhesive dispersions, some 2K systems, and self-adhesive films (PSA films) are suitable for these applications. These PSA products with almost unlimited open time are widely used for decorative film applications such as high-rise buildings, buses, and trains, and for retroreflective applications in the form of traffic signs.
- *Bonding of plastics* – Elastic adhesives with good deformation capacity should be used for bonding plastic parts. Adhesives based on PUs and self-adhesive foam tapes are very suitable for such applications. Products developed for these surfaces include special liquid acrylic adhesives that enable structural performance of bonded plastic parts, and double-sided adhesive tapes with excellent adhesion to low-energy plastics such as PE or PP. Both new product families require no surface treatment of the used PE or PP before bonding.
- *Bonding of materials with different thermal expansion* – Bonding materials with different coefficients of thermal expansion is often a major challenge for users.

In such bonded components, temperature changes lead to different dimensional changes, which result in limited function or even damage to the bond. To avoid this problem, elastic adhesive systems with good stress relaxation properties should be used. Candidates for this are PU, silicone, or elastic acrylate adhesives as well as the very elastic and stress-relaxing self-adhesive high-performance foam tapes.

8.6 Definition of Concepts

Now it is time to define concepts 1–3 for the verification work in the laboratory. To this end, we first learned about the basics of bonding – such as wetting, adhesion, and cohesion – studied substrate materials, their surfaces and possible surface treatment methods, acquired comprehensive knowledge about adhesives and their applications, and finally dealt with stresses that can affect bonded joints.

Building on all this knowledge, we are able to define:

- the substrate material,
- the surface treatment method,
- the adhesive(s) for the task at hand, and
- the process conditions for making the bond(s) in the lab.

Before we can start the laboratory work and produce samples for the upcoming feasibility work, the following items must be still worked out:

- *Estimation of required load-bearing capacity* – As a reality check, a theoretical estimate of the required load-bearing capacity of the proposed bond needs to be performed, taking into account the expected mechanical and environmental loads acting on the bond. This preliminary estimate is made without test data in the laboratory and on the basis of the characteristics of the treated substrate surfaces and the properties of the adhesive candidates, and – if available – taking into account reduction factors for the corresponding loads collected from data sheets and experience.
- *Theoretical proof: load < load-bearing capacity* – Using supplier specifications and data sheets, theoretical proof according to DIN 2304 that for the selected concept(s) the practically acting load is smaller than the load-bearing capacity of the selected candidates needs to be prepared. This verification is purely theoretical and preliminary. It must be refined after data from laboratory tests under load are available later in the feasibility step.
- *Specification laboratory specimens* – The dimensions and characteristic features of the laboratory specimens used for initial testing and for evaluating the performance degradation of the bonds after application of the loads encountered in practical use need to be determined. This is the basis for establishing reduction ratios for the appropriate loads.

As mentioned earlier, the theoretical estimate at this relatively early stage that the load on the composite component is less than the load-carrying capacity of the composite system is preliminary and must be performed for all 1–3 candidates, using information from, for example, data sheets, supplier information, and experience in similar applications. However, at the late stage of the feasibility assessment, it must be carefully demonstrated that the load-carrying capacity of the composite (including a safety and reduction factor) is greater than the actual assumed load. The evaluation is then based on real data generated at the beginning and after the application of the loads occurring in practice. This constant review of the load limits of the adhesive bond is a central requirement of DIN 2304, and suitable methods are used for the initial and long-term data. These are described later in the book.

In summary, the final specification of concepts before starting laboratory work should include the following:

- specification of substrates,
- specification of the surface treatment method,
- specification and exact name of the adhesive(s),
- process conditions for the production of the bond in the laboratory,
- specification of loads and time of their application for laboratory testing, and
- the dimensions of the laboratory samples for testing.

References

1 Oertel, G. (ed.) (1985). *Polyurethane Handbook*. New York: Hanser Publishers, Munich and Oxford University Press.

2 Saunders, J.H. and Frisch, K.C. (1987). *Polyurethanes - Chemistry and Technology*. Malabar, FL: Robert E. Krieger Publishing Company.

3 Estes, G.M., Cooper, S.L., and Toblsky, A.V. (1970). Blockpolymers and related heterophase elastomers. *Journal of Maromolecular Chemistry* C-4: 313.

4 Rave, A. (1995). *Principles of Polymer Chemistry*. New York: Plenum Press.

5 Gum, W.F. (ed.) (1992). *Reaction Polymers. Polyurethanes, Epoxies, Unsaturated Polyesters, Phenolics, Special Monomers and Additives; Chemistry, Applications, Markets*. Oxford: Hanser, Munich and Oxford University Press.

6 Barwich, J.D., Guse, D., and Brockmann, H. (1989). Härtung von Epoxidharzen mit Dicyandiamid und Monuron. *Adhesion* 5: 27–30.

7 Flegel, H. (2002). The future of adhesive bonding as a joining technique. *AutoTechnology* 5: 64–67.

8 Dahlquist, C. (1989). Creep. In: *Handbook of Pressure Sensitive Adhesive Technology* (ed. D. Satas), 97. New York: Van Nostrand Reinhold.

9 Chu, S.G. (1989). Viscoelastic properties of pressure sensitive adhesives. In: *Handbook of Pressure Sensitive Adhesive Technology* (ed. D. Satas), 158. New York: Van Nostrand Reinhold.

10 Pressure-Sensitive Tape Council (1994). Test Methods for Pressure-Sensitive Adhesive Tapes, Chicago.

11 Pocius, A.V. (2002). *Adhesion and Adhesive Technology*, 239. Munich: Carl Hanser Verlag.

12 Bauer, W. (1934). Adhesive. US Patent 1 982 946.

13 Ulrich, E.W. (1959). Pressure-Sensitive Adhesive Sheet Material. US Patent 2 884 126.

14 Butler, G.L. (1989). Natural rubber adhesives. In: *Handbook of Pressure Sensitive Adhesive Technology* (ed. D. Satas), 260. New York: Van Nostrand Reinhold.

15 Dahlquist, C.A. (1966). *Tack, adhesion, fundamentals and practise*. London: McLaren and Sons Ltd.

16 Ewins, E.E. Jr., Clair, D.J.S., Erickson, J.A., and Korcz, W.H. (1989). Thermoplastic rubbers: A-B-A block copolymers. In: *Handbook of Pressure Sensitive Adhesive Technology* (ed. D. Satas), 317. New York: Van Nostrand Reinhold.

17 Satas, D. (1989). Acrylic adhesives. In: *Handbook of Pressure Sensitive Adhesive Technology* (ed. D. Satas), 396. New York: Van Nostrand Reinhold.

18 Schlademann, J.A. (1989). Tackifier resins. In: *Handbook of Pressure Sensitive Adhesive Technology* (ed. D. Satas), 527. New York: Van Nostrand Reinhold.

19 Brockmann, W., Geiß, P.L., and Wagner, A. (2004). *Selbstheilungseffekte der Adhäsion in Klebverbindungen*. Proceedings of the 18th International Symposium Swiss-Bonding (ed. E. Schindel-Bidinelli), 17–25. Switzerland: Rapperswil.

20 Bennett, G., Geiß, P.L., Klingen, J., and Neeb, T. (1996). Structural bonding tape: an innovation in adhesive bonding. *Adhesive Age* 39 (10): 19–23.

21 Lutz, A. and Symietz, D. (2003). Innovative structural adhesive approach for car manufacturing. *AutoTechnology* 6: 42–46.

22 Habenicht, G. (2005). *Kleben - Grundlagen, Technologie, Anwendungen*, 616–617. Berlin: Springer-Verlag.

23 Bremont, M. (1994). Verbesserung der Beständigkeit von Klebverbindungen aus verzinktem Stahl und einem warmhärtenden Epoxidharzklebstoff. PHD Thesis. Technische Universität Kaiserslautern.

24 Voigt, D. (2008). Durability of adhesively bonded stainless-steel joints under accelerated ageing conditions. PHD Thesis. Technische Universität Kaiserslautern.

25 Neeb, T. (1999). Adhäsionsmechanismen an mechanisch vorbehandelten Metalloberflächen. PHD Thesis. Technische Universität Kaiserslautern.

26 Brockmann, W., Hennemann, O.D., Kollek, H., and Matz, C. (1986). Adhesion in bonded aircraft structures. *International Journal of Adhesion and Adhesives* 6: 115–143.

27 Rider, A.N. and Arnott, D.R. (2000). Boiling water and silane pretreatment of aluminum alloys for durable adhesive bonding. *International Journal of Adhesion and Adhesives* 20 (3): 209–220.

28 Brockmann, W. and Emrich, S. (2000). Model studies of the influence of enlarged surface areas due to increased adherent macroroughness on the adhesive strength and long-term durability of aluminum-epoxide joints. *Proceedings od EURADH 2000, Society Francaise du Vide, Le Vide Service, Technique et applications Paris, No. 296-2/4-2000*, pp. 311–319.

29 Brockmann, W., Emrich, S., Eicher, C., and Geiß, P.L. (2004). Dominant effects of nano scale surface structuring onto long-term durability of bonded metal joints. *Proceedings of EURADH 2004, DECHEMA*, Frankfurt, pp. 156–160.

30 Lucke, H. (1966). Kunststoffe und ihre Verklebung. *Kunststoff-Rundschau* 12: 9–13, 67–71 and 188–193.

31 Qu, J. (2000). Untersuchung zur Langzeitstabilität adhäsiver Verbindungen aus Floatglas und Kunststoffen. PHD Thesis. Technische Universität Kaiserslautern.

32 Lotz, S. (1995). Untersuchungen zur Festigkeit und Langzeitbeständigkeit adhäsiver Verbindungen zwischen Fügepartnern aus Floatglas. PHD Thesis. Technische Universität Kaiserslautern.

33 Graf, M. (2004). Die Langzeitbeständigkeit von Klebverbindungen aus Floatglas. PHD Thesis. Technische Universität Kaiserslautern.

34 Radovic, B. and Rothkopf, C. (2003). Eignung von 1K-PUR Klebstoffen für den Holzbau unter Berücksichtigung von 10-jähriger Erfahrung. *Bauen mit Holz* 6 (3): 24.

9

Feasibility (Step 3) – Demonstrating Feasibility and Validation of the Preferred Concept

9.1 Introduction

In the concept phase, we have developed, specified, and outlined 1–3 concepts for the planned bonding. The substrates to be bonded are clearly defined, but the concepts may vary in terms of adhesive selection, surface treatment methods, and processing methods. However, these should be clearly defined for each concept. Now the laboratory work begins, starting with the fabrication and testing of small laboratory samples of all the planned concepts initially and after exposing them to the most critical loads. To this end, Sections 9.2 and 9.3 will deal intensively with the production of bonded joints in the laboratory and the test methods used to evaluate them, including accelerated aging methods.

Based on the first laboratory results, the most suitable candidate is then selected, manufactured as a laboratory prototype and then intensively tested for the effects of the mechanical and environmental stresses encountered in practice. To this end, tests are carried out initially and after accelerated aging due to influences such as temperature, humidity, UV radiation, and corrosive media. If the tests are passed, practical components as used in real applications are designed, manufactured, and then tested for validation. The tools necessary to evaluate the properties of the bond and the subsequent design of the bonded component for validation are also comprehensively covered in the following Sections 9.4 and 9.5.

Thus, at the end of this step – the feasibility step –, a concept for the bonded component validated for the loads occurring in service is available, described in principle by:

- the substrates,
- the surface treatments,
- the adhesive,
- the process conditions, and
- the dimensions/design of the component.

The goal of the next step – the development step – is to develop a robust manufacturing process for the validated concept. The proof of feasibility that is now pending

Adhesive Bonding in Five Steps: Achieving Safe and High-Quality Bonds, First Edition. Jürgen Klingen.

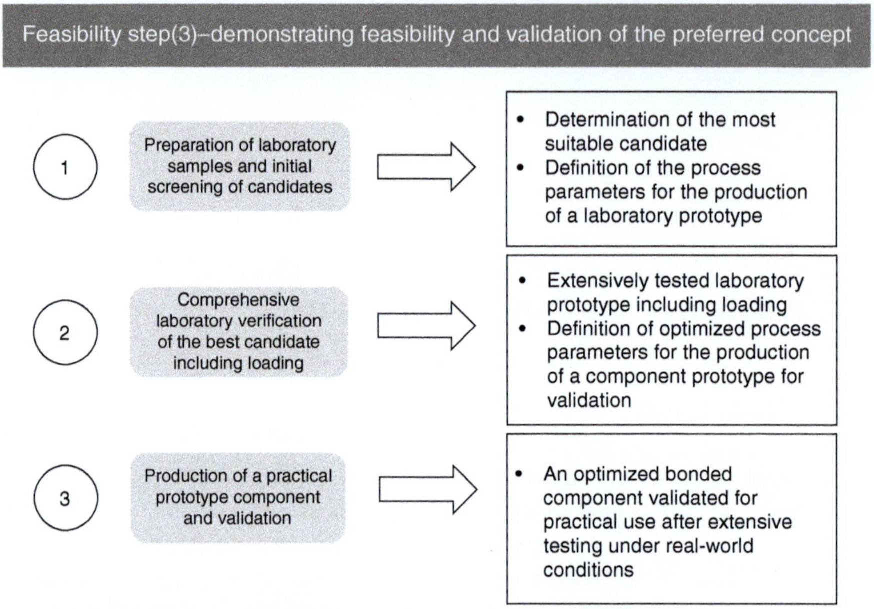

Figure 9.1 Major tasks and accomplishments of Step 3 (feasibility).

and includes the following process steps with the corresponding deliverables that must be in place for the project to move into the development step. These are demonstrated in Figure 9.1.

9.2 Production of Bonded Joints in the Laboratory

Now the laboratory work starts with the aim of verifying the concepts, selecting and checking the best candidate for bonding and, at the end, validating it by testing a prototype under practical conditions. But before that, we will go into the necessary measures before the actual execution. For this purpose, it makes sense to divide the upcoming laboratory work into

- *preparation* before making the bond and
- *execution* of bonding (Figure 9.2).

9.2.1 Preparations Before Starting Work in the Laboratory

To produce a bonded joint properly, a number of preliminary steps have to be carried out before starting the practical work for bond production.

These must ensure that the following requirements are met:

- safe working in all process steps,
- good organization of the work to be performed,
- set-up and cleanliness of the workplace,
- provision of the necessary tools, containers, and materials, and
- optimum preparation of the adhesive used.

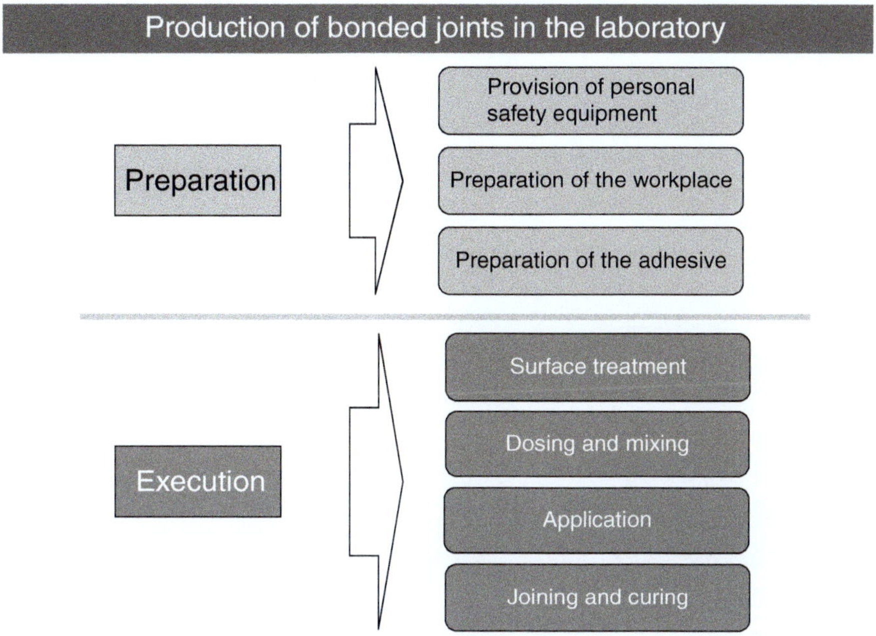

Figure 9.2 Preparation and execution of the production of an adhesive joint in the laboratory.

9.2.1.1 Provision of Personal Protective Equipment

An important task at the start of work is to ensure that occupational safety and health requirements are met. If injuries or health impairments cannot be prevented by technical or organizational measures, personal protective equipment must be used or worn by employees to protect themselves from risks to their safety and health. Thus, suitable protective clothing, chemical-resistant gloves, and laboratory safety glasses must be provided for the production of adhesive joints in the laboratory (Figure 9.3).

9.2.1.1.1 Safety Glasses

Since eyesight can be irreparably damaged by external influences, eye protection is immensely important during work. When preparing adhesive bonds, there is a risk of eye injury, e.g. during surface treatment due to flying particles, when applying the adhesives due to resin or solvent splashes, or because of UV radiation hitting the eye during the curing of photoinitiated adhesives. In Europe, all protective eyewear used must comply with the specific requirements of European standard EN 166 and be certified to this standard by an independent testing institute (Figure 9.4).

The safety glasses should have the following characteristics:

- low weight,
- low pressure force on the face,
- no image distortion,
- no color distortion (especially with signal colors),
- no restriction of the field of vision, and
- no fogging on the lens.

Figure 9.3 Personal protective equipment required for the production of an adhesive bond in the laboratory. Source: Dr. Jürgen Klingen, Aboso-Consulting.

Figure 9.4 Safety glasses according to EN 166 required for adhesive-bonding relevant work in the laboratory. Source: Dr. Jürgen Klingen, Aboso-Consulting.

Protective eyewear should be tested in the field to minimize wearer interference or disability at work.

9.2.1.1.2 Protective Gloves

If the worker comes into contact with chemicals or/and solvents during the work to prepare a bond, protective gloves classified according to EN 374 must be worn that protect against chemicals and solvents. Here, the European standard EN 374 (*protective gloves against chemicals and microorganisms*) specifies the requirements for gloves that protect the user against chemicals and/or microorganisms. In addition, these gloves are also tested according to EN 388 (*protective gloves against mechanical risks*), which specifies test methods and labeling against mechanical risks such as abrasion, cutting, tear propagation, and puncture.

With regard to chemical protection, EN 374 distinguishes between two classification levels that define the requirements for gloves for protection against chemicals:

- *Gloves for full chemical protection* – are marked with an Erlenmeyer flask and have breakthrough times of more than 30 minutes for three of the test chemicals used.
- *Gloves for basic chemical protection* – are marked with a beaker, are liquid-tight, and offer little protection from chemical hazards.

For work with chemicals and solvents in adhesive bonding, lightweight silicone-free polymer protective gloves made of nitrile rubber are very suitable (Figure 9.5).

These are for example characterized by:

- a wall thickness of about 120 μm and a length of about 240 mm,
- good puncture, abrasion, and tear resistance,
- protection against alkalis, oils, many solvents, lubricants and animal fats,
- excellent wet and dry grip,
- antistatic properties, but are
- not suitable for working with ketones or other strong organic solvents.

For work on bonded joints where no chemicals or solvents come into contact with the hands, special very light, breathable, knitted gloves are very suitable (Figure 9.6). For better grip, their areas of insides of the fingers and hands are coated with PU. This makes mechanical surface treatment by sanding the material over a longer period of time much more pleasant due to the breathability and also safer due to the excellent grip of these gloves.

9.2.1.2 Preparation of the Workplace

The clearly demarcated and well-lit bonding work area should be clean, dry, and free from sources of contamination. The ambient conditions should meet the requirements for adhesive-bonding work and the curing conditions of the adhesives used. This applies in particular to the temperature and to the humidity if moisture-curing

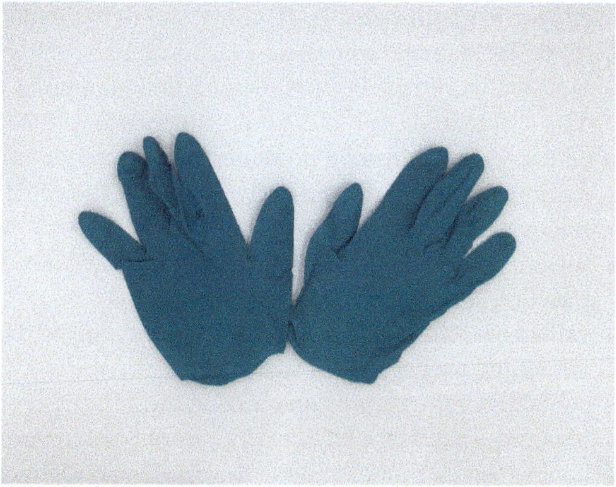

Figure 9.5 Protective nitrile rubber gloves according to EN 374 required for adhesive bonding relevant work in the laboratory. Source: Dr. Jürgen Klingen, Aboso-Consulting.

Figure 9.6 Breathable knitted gloves according to EN 388 with PU coatings for better grip. Source: Dr. Jürgen Klingen, Aboso-Consulting.

systems are used. Smoking, drinking, and eating are prohibited at the adhesive workstation, and it must be ensured that no silicone materials are stored or used in the vicinity. This is very important because silicones on the substrates repeatedly cause significant problems in the production of adhesive bonds due to the very low surface energy of these materials. The reason for the high frequency of this cause of failure is the extremely high presence of silicones in industrial production and their use in a wide variety of everyday items. Due to the risk of contamination, dusting activities such as grinding, blasting, drilling, or cutting must not be carried out in the vicinity of the adhesive workstation. Painting work in the vicinity should also be avoided, as adhesion-damaging additives are often released during such processes.

Before starting work on the manual production of a bonded joint, the following items should be available or ensured at the workplace:

- personal protective equipment such as suitable protective clothing, chemical-resistant gloves (according to EN 374), safety glasses (according to EN 166), and respiratory protection if necessary,
- functioning safety equipment such as chemical hoods, exhaust systems, eye showers, fire extinguishers, wash basins, and first-aid kits,
- fixed places for adhesives, auxiliary materials, and disposal containers,
- required work instructions, preparation standards, technical data sheets, safety data sheets, and templates for documentation,
- adhesives and adhesion promoters in sufficient quantities (at room temperature) within the minimum shelf life, indicating the mixing ratios and the date of first removal from the containers,
- solvent for cleaning the equipment used and for removing any dirt or spilled residues,
- sufficient cloths made of lint-free paper or undyed fabric for various cleaning operations and for degreasing surfaces,
- containers for weighing and mixing the adhesive components with suitable spatulas and stirrers as well as cartridge application guns (if provided), and
- application equipment and tools in a clean and functional condition.

9.2.1.3 Preparation of the Adhesive

The adhesives used are usually supplied in various containers often in cans, drums, or for smaller applications in tubes and must be stored according to the manufacturer's specifications. To form a functional adhesive layer, 1K systems can be used as supplied without any upstream mixing and require only moisture, heat, oxygen exclusion, UV light, or just pressure, as with PSAs. Epoxy, PU, silicone, and acrylic adhesives are available as 2K systems. In this cases the two components are stored separately and have a limited pot life after mixing, which can range from minutes to hours, depending on the adhesive. They are offered as user-friendly on-demand mixing and dispensing systems using dual-chamber cartridges, cans, drums or, for smaller applications, in tubes. The shelf life of the adhesives is specified by the manufacturer and is usually 3–12 months when stored properly in the original unopened container. Storage at low temperatures extends the shelf life. Filled adhesives tend to sediment and must therefore be stirred before processing to ensure a uniform consistency.

Before starting work, care should be taken to allow the adhesives used in the closed containers to adjust to the ambient temperature at the workplace for a few hours. Before use, the identity and expiration date should be verified and the lot number noted for traceability purposes. For new containers, the date and first opening should be clearly marked. To ensure the quality of the materials, a visual inspection focusing on the condition of the containers and the appearance or consistency of the contents should be performed, leaving the containers open only as long as necessary. If it is necessary to transfer into other containers, only those that are suitable should be used and marked in accordance with the legal requirements.

9.2.2 Execution of the Work for the Production of a Bonded Joint in the Laboratory

After the preparations described above, the practical work of producing the bonded joint can begin. Before performing surface treatment, the substrates used must be visually inspected, with attention paid to damage and contamination. In addition, after acclimatization, the accuracy of fit must be checked and adjusted if necessary. Subsequently, the surface treatment is carried out.

9.2.2.1 Surface Treatment

Surface treatment prior to manual production of a bonded joint focuses on ease of handling, suitability on a small scale, ecological traceability, and cost (Figure 9.7). It is very important that the substrate surfaces are always cleaned and degreased before surface treatment. The type of subsequent surface treatment depends on the requirements placed on the adhesive bond in terms of strength and long-term aging resistance. Here, adhesion to both metal and plastic surfaces can be improved by easy-to-perform mechanical surface treatment methods such as grinding or blasting.

9.2.2.1.1 Grinding

This surface treatment method, which increases the effective surface area of substrates, is the simplest type of mechanical surface treatment. It can be used very

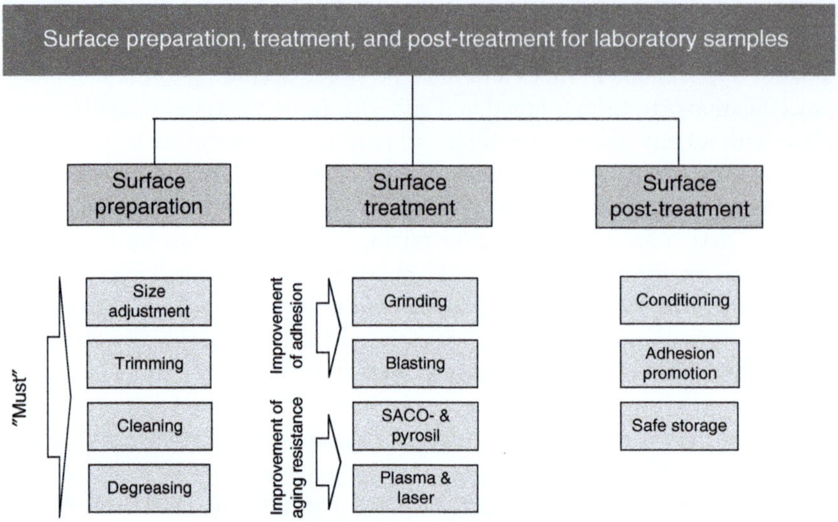

Figure 9.7 Manual surface preparation, treatment, and post-treatment suitable for production of bonded joints in the laboratory.

well for manual processes and very often provides adhesive bonds with good enough strength. However, the adhesion properties of the sanded surfaces are not as good as those of the blasted ones, especially when improved longer-term resistance is required. In addition to abrasive papers, three-dimensional abrasive fleeces, available as hand pads, strips, or discs, are also used for manual cleaning or treatment of surfaces (Figure 9.8). These three-dimensional nylon nonwovens with aluminum oxide or silicon carbide particles embedded in resin are very adaptable and easy to grip, have a long service life, and are very suitable for processing metals and plastic surfaces. Due to their adaptability, hard-to-reach places on the surface can be ground very well. The classification of these nonwovens is not by grain size, but as coarse, medium, fine, very fine, and ultra-fine.

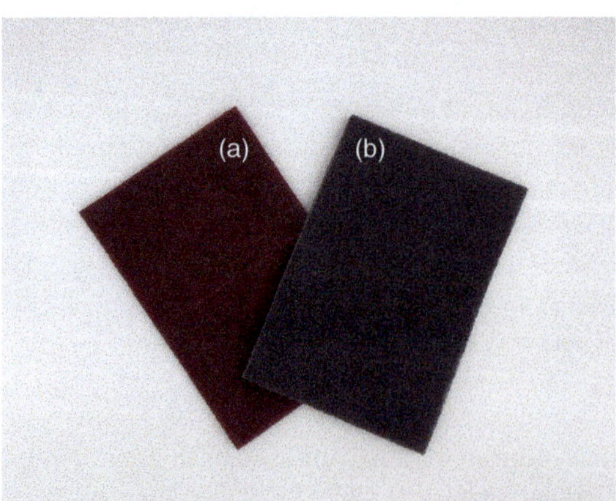

Figure 9.8 Abrasive fleece hand pads in the grades very fine (a) and ultra fine (b) for manual mechanical surface treatment of laboratory samples. Source: Dr. Jürgen Klingen, Aboso-Consulting.

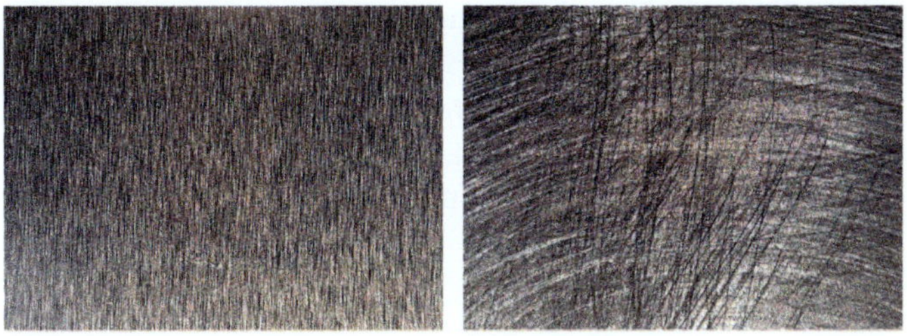

(a) (b)

Figure 9.9 Aluminum surface before (a) and after grinding (b) with an abrasive fleece hand pad. Source: Dr. Jürgen Klingen, Aboso-Consulting.

Before grinding, it is essential that the surfaces are carefully cleaned and degreased to prevent the grinding process from distributing impurities over the entire area or incorporating them into the surface. The mating parts should preferably be ground transversely to the direction of stress or crosswise (Figure 9.9). The process is carried out by hand or by machine with a medium grit of 120–180, which is most suitable for bonding purposes. After grinding, the surfaces must be cleaned of grinding dust, e.g. with solvent-soaked cloths.

9.2.2.1.2 Blasting

The application of this process, like grinding, results in clean surfaces and, by increasing the effective area, improves adhesion, especially to metal and plastic surfaces. This process also makes it possible to improve the aging resistance of bonded metal parts. This is particularly the case with the SACO process, which, in addition to material removal and surface enlargement, involves partial coating of the surface with functional silanes.

9.2.2.1.3 Pyrosil® Process

Applying this simple process for the manual treatment of substrates, the work is carried out by pyrolysis, whereby the surface of the substrate is touched with a gas flame containing a silicon-based substance. The result is an amorphous silicon dioxide layer on the surface, which significantly improves the adhesion and aging resistance of adhesive composites with, for example, metals, plastics, glasses, and ceramics. This easy-to-use process is effective, inexpensive, and easy to scale up.

9.2.2.1.4 Atmospheric Plasma

Atmospheric plasma activation of a surface is a physicochemical surface treatment process in which ionized air is generated in a plasma nozzle using high voltage. After leaving the nozzle, the plasma impinges on the material and creates a micro-blasted active surface containing polar groups. Plasma activation is a particularly efficient surface modification that leads to significantly improved wettability of the surfaces of plastics, metals, and glass. The process is particularly effective and helpful for nonpolar (low-energy) materials such as PE and PP, since no suitable adhesive bonds can be made with these materials without treatment.

After the surface treatment of the substrates is completed, the processing of adhesives takes place. After their proper preparation, 1K systems are applied to the substrate without any further upstream processing step. Then joining takes place and the adhesive bond is cured, requiring proper fixation. In the case of 2K systems, the adhesive components still need to be mixed before application.

9.2.2.2 Bonded Joints with 1K Adhesives

Since the substances that initiate the curing reactions are already part of the adhesive formulations and thus do not have to be mixed, 1K adhesives are generally much easier to process than 2K systems. A very advantageous feature for the user of 1K adhesives is therefore that, due to the presence of only one component, very simple dispensing equipment can be used.

1K adhesives are activated by:

- moisture, as in the case of cyanoacrylate and room-temperature curing PU or silicone adhesives,
- UV radiation or visual light, as in the case of photoinitiated free radical curing acrylic and cationic-curing epoxy adhesives,
- air exclusion in combination with metal ions on the substrate, as in the case of anaerobic adhesives,
- heat, as in the case of 1K epoxy adhesives,
- evaporation of organic solvents or water in the case of solvent and dispersion adhesives,
- solidification of polymer melts by lowering the temperature, as in the case of hotmelt adhesives, and
- exclusively applying pressure to the adhesive bond in the case of PSAs.

1K adhesives are available in tubes, bottles, cartridges, hobbocks, and drums, and they come in very low and very high viscosities. The shelf life of these products is usually 3–12 months when properly stored in the original unopened container. Extension is possible by storing at lower temperatures, but freezing of the adhesives must be avoided as the adhesives may be damaged. Before processing, the adhesives must be brought to room temperature. The simplest dispensing method for these adhesives is from tubes, which is very common for very smaller industrial, craft, and household applications. Often cartridges are used, supported by dispensing guns, which can be operated by hand or with compressed air. In this case, the control of the dosing quantity and speed of application is carried out by the worker. This method of operation has the advantage of great flexibility, but the dispensing results are highly dependent on the operator and their form on the day.

9.2.2.2.1 Cyanoacrylate Adhesives

Storage Since even traces of moisture are sufficient to start polymerization, these adhesives must be stored absolutely moisture-free, despite the built-in stabilizers. For this reason, they are packaged in relatively thick-walled plastic bottles made of PE or PP, which allow very low moisture diffusion. For small- or medium-consumption quantities, it is advisable to transfer the adhesive from a *storage* bottle to a smaller *working* bottle.

Dosage Form The adhesives are available in 10, 20, 50, and 500 g PE bottles in a wide range of viscosities. Usually the adhesives are colorless, but colored and fluorescent versions are also offered for better visual control of the applied adhesive layers. Viscosity ranges from very thin liquids with a viscosity of less than 20 mPa·s to high-viscosity gels with a viscosity of over 100 000 mPa·s. When processing the low-viscosity variants, it should be noted that the adhesive can be quickly absorbed by the material, thus reducing productivity. For such applications, as well as for gap bridging and applications on vertical surfaces, highly viscous thixotropic products (gels) are offered, which nevertheless cure very quickly.

Processing Cyanoacrylates cure extremely fast initiated by atmospheric humidity. They are applied in spots or in beads usually directly from the bottle onto a material surface. Then, the second part to be joined is applied and fixed, where due to the rapid curing reaction, repositioning is very often not possible. The subsequent joining pressure must be maintained until hand strength is reached. When the adhesive is applied, blooming occasionally occurs on the surface, caused by the evaporation of small amounts of the cyanoacrylate monomer and the subsequent deposition of the resulting polymer on the surface around the bond. The formation of these polymeric micro-crystallites can be avoided by reducing the application thickness and, if necessary, by extracting air in the area where the bonding process takes place. Control of the adhesive application by the user is possible by fluorescent systems, where the concentration of the added dye is in the ppm range and does not affect the storage stability and properties of the bond. Since cyanoacrylate adhesives cure extremely fast, direct contact between the adhesive and skin areas (e.g. fingertips) must be avoided at all costs. The moisture present on the skin causes immediate adhesion to the skin within a short time, and injuries may occur when the adhesive bond to the skin is subsequently detached.

Curing Cyanoacrylate adhesives cure by humidity, with optimum results at a humidity of approximately 40–60%. Below 30%, curing is greatly delayed. A spontaneous reaction occurs when damp and/or alkaline-reacting surfaces are bonded. Bonding with cyanoacrylate adhesives reaches hand strength after only a few seconds and final strength after approximately 24 hours.

Processing profile of cyanoacrylate adhesives:

- Cyanoacrylate adhesives are easy to process,
- they cure extremely quickly, thus enabling rapid further processing of the components,
- bond different materials without problems and are very suitable as fixing aids,
- are very economical, 1 g of adhesive yields up to 80 drops, depending on viscosity,
- curing takes place without additional process steps or energy input,
- for quality control, optical detection is possible,
- cure only in thin layers, since moisture from the ambient air or on the component surface cannot penetrate into thicker adhesive layers and thus initiate curing,
- bonding of parts larger than 50 mm · 50 mm is possible only to a limited extent,

- unmodified products have very low viscosities and therefore allow only very limited gap bridging,
- at a humidity of >80% or the application of the adhesives on strongly alkaline substrates, as is the case with glass, shock-like curing occurs, which leads to reduced strengths of the adhesive bond due to the stresses generated,
- acidic additives to the adhesive or acidic reacting surfaces delay or prevent the curing of the adhesive, and
- cyanoacrylates must not be metered using diaphragm metering valves, as small gaps in the valve can lead to local curing.

9.2.2.2.2 1K PU Adhesives

Storage Due to the spontaneous reaction of these adhesives with moisture, high demands are placed on the packaging of these products during transport and storage. Here, as during processing, it is essential to ensure that moisture penetration is prevented and that the tools and containers used are dry.

Dosage Form 1K PU adhesives are usually available as black, white, or gray paste-like masses and processed from tubes, cartridges (310 and 600 ml), tubular bags, or drums.

Processing 1K PU systems should be processed between +15 and +35 °C. Large-area applications are usually not possible with this adhesive, which is usually applied to a substrate in layer thicknesses of 2 to approximately 20 mm as a triangular- or round-shaped bead (Figure 9.10). For smaller applications in industry and trade, the adhesive is applied from tubes; for larger applications, the adhesive is applied from cartridges using dispensing guns operated by hand or with compressed air.

When manually applying the adhesive by a dosing gun, it should be taken care that:

- the first portion of the adhesive is always discarded,
- the application is carried out by slight contact of the nozzle with the component,
- when application as a triangular bead, the nozzle is positioned vertically to the substrate,

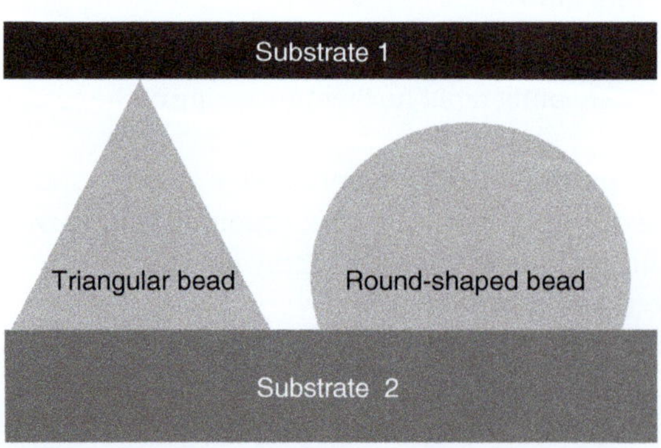

Figure 9.10 Bead application of high-viscosity adhesives in triangular- and round-shaped form.

- the adhesive application is carried out preferably without interruption or disconnection of the gun, and
- the bead end is exactly applied.

The bead application in triangular form has the following advantages compared to the round form:

- higher tolerance compensation,
- lower risk of air inclusions, and
- more uniform bead application.

Curing Since 1K PU adhesives require moisture from the environment to cure, it is important to ensure that the relative humidity in the processing rooms is between 30% and 75%, where humidity values <30% lead to a strong delay of the reaction. These adhesives can only be used to a limited extent for large-area bonding when moisture-impermeable substrates are used. As a rule, the adhesive layer cures from the outside to the inside at a rate of 2–3 mm per day. However, the curing time can be significantly reduced by using additional accelerators matched to the adhesive. These so-called *boosters* are available as attachments for the cartridges used or separately in the form of cartridges and mixers.

These accelerators cause:

- a complete curing of the adhesive within a short time,
- a curing speed that is independent of the humidity of the air,
- that the ambient temperature is an important factor for the curing speed, and
- that the adhesive is subject to the rules for 2K systems.

The formation of carbon dioxide, caused by the reaction of the isocyanates in the adhesive formulation with atmospheric moisture, can lead to undesirable bubble formation in the cured adhesive layer. This effect, which preferably occurs with porous substrates, can negatively affect the strength of the adhesive bond. Adhesives with reduced isocyanate content, the use of higher viscosity formulations, or extending the curing time by reducing ambient humidity are ways of minimizing this effect.

Processing profile of 1K PU adhesives:

- 1K PU adhesives are easy to process,
- inexpensive and can be used for various applications on many materials,
- the curing speed depends on the humidity and temperature in the working area,
- the adhesives react very slowly at 2–3 mm per day and often require an adhesion promoter,
- large-scale applications can only be performed to a limited extent, and
- high-moisture concentrations in the ambient air or water on the material surface cause carbon dioxide production during curing and thus bubble formation in the adhesive layer.

9.2.2.2.3 1K Silicon Adhesives

Storage Due to the spontaneous reaction of these adhesives with moisture, high demands are placed on the packaging of these products during transport and storage.

Here, as well as during processing, it is essential to prevent the penetration of moisture and to ensure that the tools and containers used are also dry.

Dosage Form 1K silicone adhesives are available in a wide range of colors as paste-like masses and are processed from tubes, cartridges (310 and 600 ml), tubular bags, or drums.

Processing 1K silicone adhesives should be processed between +15 and +35 °C. The adhesive is usually applied to a substrate in layer thicknesses up to 15 mm as a triangular- or round-shaped bead. Large-area applications are not possible with this adhesive. For smaller applications in industry and trade, the adhesive is applied from the tubes, but for larger applications cartridges with dispensing guns are used, which are operated by hand or with compressed air.

Curing Since 1K silicone adhesives require moisture from the environment for curing, care should be taken to ensure that the relative humidity in the processing rooms is between 30% and 75%. Humidity values <30% lead to a strong delay of the reaction. These adhesives can only be used to a limited extent for large-area bonding with moisture-impermeable materials. The curing of 1K silicones is initiated by the presence of hydrolysis-sensitive groups incorporated into the silicone molecule. As soon as the adhesive leaves the cartridge, and especially after joining, these groups react with the water in the air to create the adhesive network.

Here the following decomposition products are created:

- *acid cured* – acetic acid (smells like vinegar),
- *neutral cured* – alcohol, oxime, and
- *basic cured* – amine (smells fishy).

Since the curing rate of 1K silicone adhesives at a relative humidity of 50% and a temperature of 23 °C is about 1–2 mm per day, the time required for complete curing of the adhesive layer takes days. Acceleration of curing is possible by increasing the temperature of up to 150 °C or set higher humidity, the latter measure having by far the greater effect. Another means of accelerating the curing of the adhesive is the use of special accelerated systems, in which the acetic acid from the crosslinking reaction is neutralized by the use of, for example, alkali hydroxides, with the formation of water. This *internally generated* water is then available for curing the adhesive in addition to the *externally generated* water from the humidity in the air.

Processing profile of 1K silicone adhesives:

- 1K silicone adhesives are easy to process,
- the curing speed depends on the humidity and temperature in the working area,
- adhesives react very slowly at 1–2 mm per day (50% relative humidity, 23 °C), and
- the use of these adhesives on equipment on which other adhesives are processed is often not possible because of the risk of contamination by low-molecular-weight components of the silicones.

9.2.2.2.4 Anaerobic Adhesives

Storage Since the curing of the anaerobic adhesives is initiated by exclusion of oxygen, the commercially available products use containers made of air-permeable plastic. In addition, they provide for a large air space above the adhesive level in the container. After delivery, the adhesive should be stored in a refrigerator at 5–8 °C.

Processing and Curing These special 1K adhesives are applied to one side of the material and then the components are joined. In the joining zone, it is essential to ensure the exclusion of air and contact with metals, preferably iron or copper. Due to the presence of oxygen outside the joining zone, the adhesive does not harden and can therefore be removed there very easily. If no iron or copper ions are present on the material surface, the substrate can be activated by applying a metal spray. This additional treatment of the surface must be carried out for bonding with pure aluminum, precious metals, and plastics. Alternatively, the metal ions required for curing can also be transferred by brushing the substrates to be bonded with a copper or brass wire brush.

Processing profile of anaerobic adhesives:

- Anaerobic adhesives are easy to process,
- need component geometries, which enable air exclusion,
- need materials that provide metal ions for curing,
- are primarily used for metals and harden only after reaching the two curing conditions, presence of metal ions and air exclusion,
- in case of bonding with iron- or copper-free metals and plastics, need additional activation, and
- demonstrate low gap-bridging properties and are thus not suitable for components with higher manufacturing tolerances.

9.2.2.2.5 1K Epoxy Adhesives

Storage The storage time of these adhesives in closed containers is weeks to several months. These systems often require storage in appropriate refrigerators up to a maximum of +8 °C.

Dosage Form The adhesives are available in a wide range of viscosities from liquid to paste-like and are processed from tubes, cans, cartridges, hobbocks, or as prefabricated films.

Processing Before processing, the stored adhesive containers as well as the parts to be joined must be brought to the processing temperature (usually room temperature). Due to the reactivity of these adhesives, higher temperatures during processing should be avoided at all costs to prevent partial curing and thus a reduction in the strength of the adhesive bond. Manual adhesive application is carried out from tubes for small applications. For larger jobs, the adhesive is applied from cartridges using metering guns or from cans and hobbocks using spatulas. Accurate adjustment of adhesive thickness can be ensured by using glass beads with appropriate diameters in the adhesive or by using spacers during application.

Curing These adhesives are typically cured in ovens at temperatures of approximately 80–190 °C. The curing time is specified by the manufacturer and must be strictly adhered to. Humidity is not a factor with these adhesives, but the formation of condensation on the substrate surface must be avoided at all costs. The very long open times of the 1K epoxy adhesives allow very flexible planning of production of the bonded joint. In contrast to many other systems, the final strength of the component is achieved immediately after curing of the bonded component.

Processing profile of 1K epoxy adhesives:

- 1K epoxy adhesives cure at temperatures of about 80–190 °C,
- the curing time depends on the dimension of the component and the material used, and
- must be stored in refrigerator <8 °C to avoid partial curing during storage.

9.2.2.2.6 Hotmelt Adhesives

Dosage Form Hotmelt adhesives based on, for example, polyolefins, EVA copolymers, PUs, polyesters and PAs are available as powders, granules, films, rods (sticks), and blocks.

Processing and Curing Thermoplastic hotmelt adhesives are physically curing adhesives that are in solid form at room temperature and are melted before application to material surfaces. After joining, they cool again and return to their original solid state. Since adhesive composites with hotmelt adhesives are mechanically stable and functional after cooling and solidification, they allow very fast cycle times and immediate further processing in subsequent production processes. However, the melting or solidification behavior of these adhesives is reversible. As the temperature rises – e.g. during application of the bonded component – it softens again and is therefore only heat-resistant to a limited extent. Post-crosslinking variants of hotmelt adhesives, which are cured by moisture or heat after joining, are processed as described earlier in the book, but the end product is a crosslinked polymer with higher heat resistance that cannot be subsequently melted. Hotmelt adhesives are applied with hotmelt guns for manual processing of small quantities, with applicator heads from melting containers for larger quantities, and in the form of films for large-area applications.

After melting at temperatures of approximately 80–240 °C, the hotmelt adhesives are applied to a substrate and wet the surface very well due to their very low viscosity. Immediately afterwards, the joining process must be carried out very quickly, and after the two parts have bonded, the adhesive solidifies immediately. The open time of these systems can be very short and is determined by the cooling behavior of the melt and by the thermal conductivity of the parts to be joined. Therefore, when bonding metallic materials with hotmelt adhesives, preheating of the substrates is suggested and thus enables the production of high-quality joints.

For large-area applications, hotmelt adhesives are also used in the form of films. In this process, the hotmelt adhesive film is inserted between the bonded parts. Pressure and heat are then applied to approximately 20–30 °C above the softening range of the hotmelt adhesive system. In this way, the surfaces of the parts to be joined are very well wetted and, after cooling, the bond has sufficient strength for appropriate further processing.

Processing profile of hotmelt adhesives:

- Hotmelt adhesives allow easy and fast processing,
- soften at higher temperatures and wet the material surface in a thin liquid state,
- are reversible adhesives, which can be melted again after the bonding process has been completed,
- allow high production speeds due to the short consolidation time,
- allow only temperature-resistant materials due to the high processing temperatures involved,
- require a relatively high energy input for the melting process,
- do not represent hazardous materials,
- are stable and dry in their initial state, and
- allow very long storage times.

9.2.2.2.7 Photo-Initiated Adhesives

Storage Depending on the type of adhesive, products generally have a shelf life of 6–12 months when stored in an unopened original container. Before processing, the adhesives must be conditioned at room temperature.

Dosage Form Adhesives are offered in wide viscosity ranges from low viscosity to high viscosity or paste-like and are processed from bottles, cartridges, hobbocks, or canisters.

Processing When opening the adhesive container, no light of the wavelength required for curing should fall on the product. Manual dispensing of radiation-curing adhesives is done either by hand from small bottles and cartridges or semi-automatically using pneumatic dispensing equipment. It is important that the adhesive processed with the appropriate equipment is shielded from the light of the wavelength required for curing at all stages of the process.

Curing Polymerization of radiation-curing adhesives is initiated by irradiation with visible or UV light. Crosslinking occurs either by radical, as in acrylate systems, or by cationic polymerization, as in epoxy systems. This is done with the aid of photo initiators, which are already contained in the adhesive formulation. These substances decompose due to the incident radiation and initiate the curing reaction through the resulting reactive fragments. Curing is carried out with light sources that generate an intensity of about 100–1000 times that of normal daylight. Both area and spot sources are used for this purpose, usually equipped with LEDs or mercury vapor lamps. The major advantage of LED light sources, which are now market leaders, is their long service life of more than ten times that of mercury vapor lamps, better energy efficiency and controllability. During curing of photo-initiated adhesive systems the radiation intensity is highly dependent on the life of the light source and the distance to the adhesive. To ensure constant radiation intensities during adhesive curing, the radiation intensity at the adhesive application site should be measured continuously or periodically at intervals that are not too long.

With radical curing systems, complete curing is possible in a short time. However, these systems must be irradiated continuously, which requires the presence of at least one transparent joining part. If the irradiation of the adhesive is interrupted, the curing reaction is also stopped and the adhesive is not finally cured. The cationic-curing adhesives, on the other hand, allow pre-activation in which the entire bonding surface is irradiated. The joining process is then carried out within the *open time*. In contrast to the radical-curing systems, the crosslinking reaction takes several minutes to reach final strength without further irradiation. For this reason, these systems also allow the bonding of non-transparent substrates.

Processing profile of photocuring adhesives:

- Photo-initiated adhesives are easy to process and exhibit very fast curing,
- require relatively small amounts of energy for curing compared to the thermal systems,
- cure by visible or UV light in the wavelength matched to the adhesive,
- usually require at least one transparent joining part,
- radical curing systems require constant irradiation of the appropriate light for complete curing, and
- cationic-curing systems can be joined after pre-activation, in which case curing takes place without further irradiation.

9.2.2.2.8 PSAs

Dosage Form PSAs consist of a backing material coated on one side (single-coated) or on both sides (double-coated) with a dry adhesive. These products are also available as an adhesive layer without a backing material and are then referred to as adhesive transfer tapes. By varying the backing materials and adhesives, a variety of products with a wide range of properties are obtained, which are used in numerous applications. The products are offered in many widths and lengths in roll or sheet form and as die-cut parts. In the case of double-coated adhesive tapes and adhesive transfer tapes, the adhesive is protected by a release film that must be removed before joining.

Processing When processing these adhesive tapes, it is also essential that the surface is properly cleaned and treated if necessary, to ensure successful bonding. To avoid air entrapment, the tapes are applied to the material from one side under slight pressure using elastic squeegees or rubber-coated rollers. Wetting of the surfaces of the parts to be joined with the adhesive is achieved after joining by applying pressure again. Since the adhesive force is further increased, the component should not be loaded immediately after joining because final strength is often only achieved after approximately 72 hours at room temperature.

Processing profile of pressure-sensitive adhesives:

- PSA tapes are dry products that allow easy and fast processing,
- they have high initial strength, but usually require 72 hours at room temperature to achieve final strength, and
- because of the easily possible and often-invisible contamination of surfaces by fingerprints, gloves should always be worn when processing PSA tapes.

9.2.2.2.9 Solvent-based Adhesives

Dosage Form For small applications, the adhesives are offered in tubes, larger quantities in cans and hobbocks.

Processing and Curing Here, curing takes place by evaporation of the solvents, whereby the viscosity of the adhesive steadily increases due to the increasing solids content. After complete evaporation of the solvent, curing occurs through the formation of physical interactions between the polymer chains. Although solvent adhesives tend to be on the wane for ecological and occupational health reasons, they still have a good position in adhesive-bonding technology today due to their very good wetting properties on many materials and their suitability for diffusion bonding (cold welding) of thermoplastic polymers. Bonding with these adhesives requires at least one substrate with adequate permeability to remove the solvent. For the production of adhesive bonds with impermeable materials such as metals and glasses, these adhesives can only be used to a limited extent, since the removal of the solvent required to build up the final strength of these materials takes a very long time. In contrast, when bonding *porous* or *absorbent* materials, such as paper, cardboard, and wood, larger amounts of the solvents may still be present in the adhesive layer after joining, as they can quickly evaporate through or be absorbed by the substrates.

Natural and synthetic rubbers, PVAC, acrylates, and PUs are frequently used for solvent-based adhesives, the proportion of which in the solution is in the range of about 15–25% by weight. Esters, such as ethyl acetate, ketones, such as 2-butanone or tetrahydrofuran, are used as solvents. When processing adhesives containing solvents, a minimum drying time must be observed during which most of the solvent contained in the liquid adhesive evaporates. This should be followed in any case before joining, even if the substrates are solvent-permeable materials. The reason for this is the improved strength of the adhesive layer after evaporation of a significant proportion of the solvent, and thus easier handling of the adhesive bond in the manufacturing process. However, the drying time of these adhesives is limited by the open time (maximum drying time) during which wet or contact bonding is still possible without reducing the final strength. This is the case between the minimum and maximum drying times. If the maximum drying time is exceeded, bonding is no longer possible due to the lack of tack of the adhesive layer surface.

Wet Bonding In wet bonding, the substrates must be joined between the minimum and maximum drying times, while very solvent-permeable materials also allow the joining parts to be bonded immediately after adhesive application. As discussed, adherence to the maximum drying time is very important for the strength of the adhesive bond, since after this time limit the adhesives no longer exhibit tack and can therefore no longer be properly processed. Due to the fact that solvent is still present in the adhesive layer after joining, the strength of the adhesive bond is rather low at this point and final strength is achieved only after complete evaporation of the solvents.

Contact Bonding Here, the substrates are joined by bringing relatively dry adhesive layers together and then applying pressure. Also, in this case, joining must take place between minimum and maximum drying times, but very high initial strengths are achieved immediately after joining. The adhesives used here are mainly based on CRs and PUs, which change from the amorphous to the crystalline state after evaporation of the solvent and can thus build up greatly increased strengths. For this purpose, both parts to be joined are first evenly coated with the wet adhesive. Then, the solvent evaporates until the adhesive film feels relatively dry and no longer pulls threads when finger tested. In this state, the adhesive has only low tack. In the next step, the bonding surfaces are precisely positioned within the maximum drying time and joined under the highest possible pressure. Precise positioning is necessary because the very high initial strength of the adhesive layer does not allow subsequent correction of the bonded component position. In the case of flexible materials, pressing is carried out with the aid of rollers that exert a high line pressure on the adhesive bond. In the case of rigid substrates, suitable presses are used to exert surface pressure. This method can be used, for example, to bond metal plates over large areas. By exerting pressure on the resulting bond, the two still amorphous adhesive layers flow into each other to solidify by crystallization. Compared to wet bonding, contact bonding has the great advantage that the bonded components can be fully loaded immediately after joining.

Diffusion Bonding Another application for solvent-based adhesives is diffusion bonding (cold welding) of thermoplastics. The prerequisite for this is that only solvents capable of dissolving the surfaces of the two parts to be joined are used. In this process, both materials are coated with the wet adhesive and joined together under pressure after a short time. In the process, the polymer chains of the adhesive penetrate the dissolved substrate surface and intertwine with the polymer chains of the material. After evaporation of the solvent, a solid adhesive bond based exclusively on cohesive forces is formed after some time.

Processing profile of solvent-based adhesives:

- Solvent-based adhesives are easy to process, requiring special measures during storage and processing due to the presence of solvents,
- allow wet and contact bonding of various materials and diffusion bonding of plastics, and
- require compliance with the minimum and maximum drying time for good adhesive bonds.

9.2.2.2.10 Dispersion Adhesives

Dosage Form These adhesives are applied manually from tubes for small jobs and from cartridges using dispensing guns or from cans and hobbocks using spatulas for larger jobs.

Storage Dispersion adhesives must not be stored and processed at temperatures below freezing point, as in this case the structure of the dispersion is destroyed and the adhesive is no longer usable. Since dispersion adhesives are usually made with normal tap water, prolonged storage under certain conditions such as high humidity and elevated temperatures may cause mold growth, rendering the adhesive unusable.

Processing and Curing These adhesives are used for a wide variety of applications and are quantitatively among the most widely used adhesives in the world. Because they are not flammable or explosive and do not release harmful substances, they have been used for many years as substitutes for solvent-based adhesives. In dispersion adhesives, water represents the liquid phase in which the thermoplastic or elastomeric adhesive polymers are dispersed as the solid phase. Due to the size of the particles of approximately 500–2000 nm and the use of certain additives to stabilize the dispersion, these particles do not settle. When bonding with dispersion adhesives, at least one substrate must be permeable to water (analogous to solvent adhesives). Joining impermeable materials such as metals and glasses with dispersion adhesives is only possible to a limited extent, since it takes a long time to remove the water required to build up the final strength of the bond. PVAC, acrylates, rubbers, PUs, and CRs are frequently used for dispersion adhesives, the proportion of which in the dispersion is between 40 and 70 wt%. The curing of these adhesives takes place by removing the water from the adhesive layer, whereby the remaining polymer particles grow together to form a layer (film formation). A minimum temperature necessary for film formation must be maintained, above which the adhesive dispersion forms a crack-free film under certain conditions. Dispersion adhesives are much easier to process than solvent-based adhesives due to the absence of solvents and the associated much lower cost of safety precautions at the workplace.

Processing profile of dispersion adhesives:

- Dispersion adhesives are relatively uncritical from a safety point of view,
- use water as an inexpensive, non-flammable, and non-toxic solvent,
- require compliance with the minimum temperature required for film formation to produce good adhesive bonds,
- must not be stored and processed at temperatures below freezing,
- require longer periods of time or more energy to cure than solvent-based adhesives,
- are of limited use for water-impermeable materials due to the long curing times, and
- cannot be used with materials that corrode by contact with water.

(a) (b)

Figure 9.11 Common forms for manual processing of 2K epoxy adhesives in the laboratory from tubes (a) or dual-chamber cartridges (b). Source: Dr. Jürgen Klingen, Aboso-Consulting.

9.2.2.3 Bonded Joints with 2K Adhesives

Two-component adhesives are available in a wide range of viscosities from low-viscosity to paste-like. They consist of the two components, A (resin or also called binder) and B (hardener), which after careful homogenization by mixing result in a ready-to-use adhesive with limited open time. To ensure the appropriate strengths of the adhesive layers, the polyaddition 2K epoxy and 2K polyurethane adhesives require precise adherence to the specified mixing ratio, which should be maintained with an accuracy of approximately ±5%. Manual application of 2K adhesives is carried out by hand from tubes cans or drums (batch processing). For the production of laboratory samples, the adhesives are often processed from tubes (Figure 9.11a) or user-friendly dual-chamber cartridges (Figure 9.11b).

9.2.2.3.1 Mixing by Hand

When mixing components from tubes or cans by hand, the following points must be observed:

- mixtures over 50 g cannot be used, as high temperatures may occur with larger quantities due to the exothermic reaction of the components,
- the ratio is taken into account (based on volume or weight),
- the container size is chosen so that the filling height is only about 1/4 to 1/2 of the container height,
- only suitable disposable containers are used for mixing,
- suitable disposable stirrers made of wood or reusable stirrers made of glass or metal are used,
- adhesive components located at the bottom and at the edge of the container are sufficiently included in the stirring process to obtain a uniform mixture,
- to avoid the formation of bubbles, which can occur particularly with highly viscous adhesive components,
- the components are thoroughly mixed to obtain a homogeneous mixture,

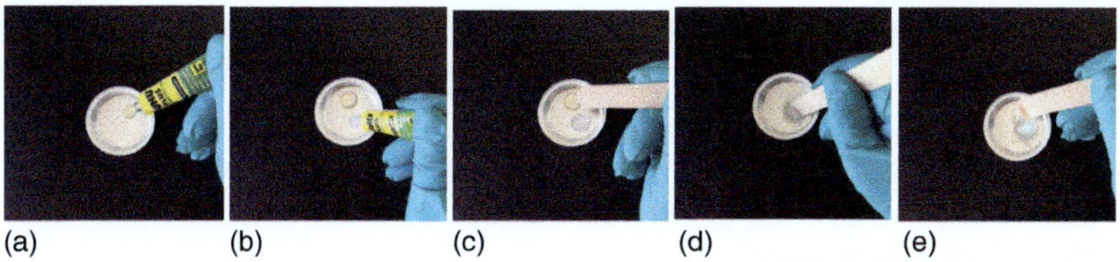

Figure 9.12 Production of very small quantities of 2K adhesive for laboratory samples from metering (a–c) and mixing (d) of the components to the finished adhesive mixture ready for application (e). Source: Dr. Jürgen Klingen, Aboso-Consulting.

- the pot life depends on the adhesive quantity and temperature,
- labeled reference samples are taken for quality control, and
- adhesive residues remain in the mixing container, which must be disposed of properly.

For small applications, a simple method of dispensing and mixing from the tube is possible, in which the two parts are filled into small aluminum trays and then mixed by hand using a wooden spatula (Figure 9.12). This method is suitable for applications where the bonding of relatively small parts is not regular, but rather sporadic. It allows the required quantities of adhesive to be supplied quickly and selectively and causes little waste. When using 2K systems, the manufacturer's specifications for the mixing ratios of the adhesive components may refer to the volumetric or gravimetric ratio. It is often necessary to convert from one form of representation to the other.

Use of Mixing and Dispensing Systems An easy way to process 2K adhesives quickly and safely is to apply the adhesive with user-friendly on-demand mixing and dispensing systems (Figure 9.13). These contain dual-chamber cartridges, which isolate the two resin components until they are needed. Mixing occurs as the materials are dispensed through the motionless mixing nozzle (static mixer) and is accomplished by hand tools or compressed air guns. Such systems allow dispensing, mixing, and application of the adhesive (spot or bead) to the substrate surface in just one step. These systems are available in various sizes of, for example, 50, 200, and 400 ml and are already filled by the manufacturer in the correct mixing ratio.

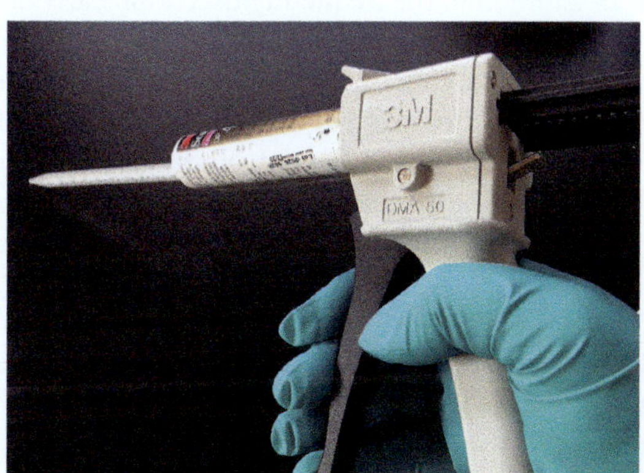

Figure 9.13 User-friendly on-demand mixing and dispensing systems for one-step operation using dual-chamber cartridges isolating the two resin components until they are mixed and applied. Source: Dr. Jürgen Klingen, Aboso-Consulting.

There are two options for processing the adhesive with such a system:

- *Mixing without the mixing tube* – The two parts of the adhesive are dosed from the dual-chamber cartridge into a suitable container, mixed by hand with a wooden spatula, for example, and then applied to the substrate. This process is carried out in three steps: dispensing by the dual-chamber cartridge system, mixing and adhesive application by hand.
- *Mixing with the mixing tube* – Dispensing, mixing, and application of the adhesive to the substrate surface is performed in just one step, and mixing occurs as the materials are dispensed through the motionless mixing nozzle.

The second option has the advantage that virtually no errors can occur in terms of mixing ratio and mixing quality. To ensure that the specified mixing ratio is present from the start, the metering gun should be actuated until material emerges from both chambers before attaching the mixing tube. After attaching the mixer, the first mixing tube contents should be discarded, as the unfilled mixing tube does not provide sufficient pressure for optimum homogenization. When using low-viscosity adhesives, longer mixing tubes should be used to ensure good mixing quality. In cases where the adhesive is applied by large dual cartridges and/or higher-viscosity adhesives, it is recommended to use pneumatic instead of manually driven dispensing guns. Also, when using such large cartridges and the associated long mixing tubes, exact positioning of the adhesive on the joining part can be made more difficult and problems can be caused by the relatively large distance between the worker's hand and the adhesive outlet at the nozzle tip.

When using disposable mixing tubes, the following should be observed:

- After a short break, the same mixing tube can be used again if the pot life of the adhesive has not been exceeded after processing the first complete mixing tube content after the break.
- In the case of longer-lasting breaks, the mixing tube should not be removed, as the cured adhesive seals the partially emptied cartridge well. Before a new mixing tube is then attached later, the old tube, including cured adhesive residues, should be removed. When attaching the new tube, some adhesive should be discarded to ensure a homogeneous mixture.
- In addition to the frequently used dual cartridge with the same diameters for the 1 : 1 mixing ratio, cartridges with different diameters are also available, allowing, for example, ratios of 2 : 1, 4 : 1, or 10 : 1. These cartridges are offered for a wide range of adhesives, but are usually significantly more expensive than the same adhesives offered in larger containers, such as cans, hobbocks, or drums. Therefore, when larger quantities are required by the user, double cartridges can be filled by external contactors, allowing a much more attractive price.

9.2.2.3.2 2K Epoxy Adhesives

Dosage Form The components of the 2K epoxy adhesives are offered from liquid to paste-like, depending on the type and application. The two parts are usually colored differently to allow easy control of the mixing quality and are available in tubes, cans, double cartridges, or hobbocks.

Processing Epoxy adhesives are among the most widely processed construction adhesives, and the reactants involved – epoxy resin and amine – must be made to react in a precisely defined weight ratio. In case of deviations, unreacted monomers of component A or B remain in the adhesive layer and have a negative effect on the strength of the bonded joint. The mixing ratios specified by the manufacturer must therefore be strictly adhered to, preferably within a range of about ±5%. The two-component adhesives are mixed and applied by hand from tubes, cans, or drums (batch processing) or with user-friendly on-demand mixing and dispensing systems described above.

When processing 2K epoxy adhesives, some conditions relevant for the successful production of the adhesive bond must be observed (Figure 9.14):

- *Pot life* – The so-called pot life is the time from the first contact of the two components until the mixed adhesive is applied to the substrate and thus describes the time available for successful processing of the adhesive. Depending on the adhesive, the pot life can range from a few minutes to several hours and is directly determined by the reaction rate between the resin and hardener. The joining process must be completed by the end of the pot life at the latest, as otherwise the viscosity of the curing adhesive would be so high that wetting of the material surface is only possible to a limited extent or not at all. The pot life depends on both the temperature and the amount of adhesive mixed. At higher temperatures, the chemical reaction is accelerated and thus the pot life is shortened.
- *Gel point* – The gel point describes the change in the physical state of the adhesive from liquid/viscous to solid.

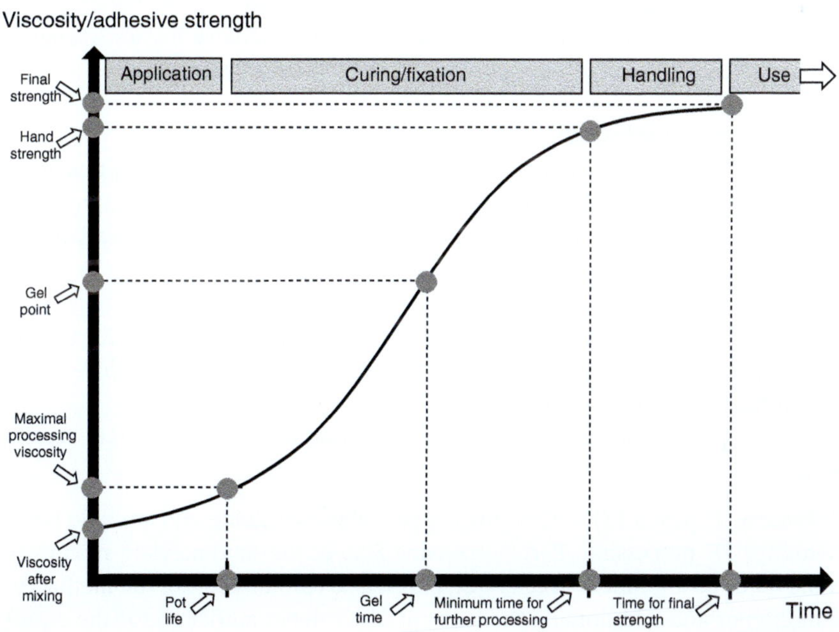

Figure 9.14 Course of viscosity/adhesive strength during application, curing/fixation, handling and use of a 2K adhesive.

- *Hand strength* – Hand strength is reached when the adhesive has not yet fully cured, but has already reached a strength that allows further processing, but without putting too much stress on the bond.
- *Final strength* – The final strength is reached when all chemical reactions possible under the curing conditions have been completed and thus the maximum possible strength of the bonded joint has been achieved.

Curing Curing of the 2K epoxy adhesives takes place at room temperature immediately after the two components are combined. Therefore, substrates should be joined immediately after application of the adhesive. The curing process takes several hours to days, but can be accelerated by increasing the temperature. Thus, adhesive layers that are additionally cured at, for example, 80 °C for one hour show higher strengths and, at the same time, lower elasticities than adhesives cured exclusively at room temperature. An important feature of 2K epoxy adhesives is the possibility of pre-curing to achieve hand strength by briefly raising the temperature for a few minutes using infrared radiation or induction equipment.

Processing profile of 2K epoxy adhesives:

- 2K epoxy adhesives consist of two components and can be processed very easily and quickly using user-friendly on-demand mixing and dispensing systems with dual-chamber cartridges, from which they can be precisely metered, mixed, and applied to the parts to be joined,
- react very sensitively to deviations from the specified quantity ratios, which must be maintained with a tolerance of approximately ±5%, and
- cure at room temperature in a few hours to days, but achieve higher strengths (with a decrease in elasticity) by applying higher temperatures.

9.2.2.3.3 2K Acrylate Adhesives

Dosage Form The 2K acrylic adhesives are available in various mixing ratios, with the base component often being a high-viscosity paste and the hardener component also being a paste or provided as a powder or varnish. Adhesives are available in tubes, cans, dual cartridges, and hobbocks.

Processing and Curing Compared to the other two-component systems, 2K acrylic adhesives have the advantage of relatively short curing times to achieve hand strengths already in about 5–120 minutes.

Three main processing methods are used to produce bonds with 2K acrylic adhesives:

- *A–B Process* (Figure 9.15) – This process provides two stable and storable components for 2K processing. Part A contains 50% of the methacrylate monomer and the required amount of accelerator, part B also contains 50% of the methacrylate monomer and additionally the amount of hardener sufficient for the entire adhesive batch. The application can be carried out either by dosing mixtures of

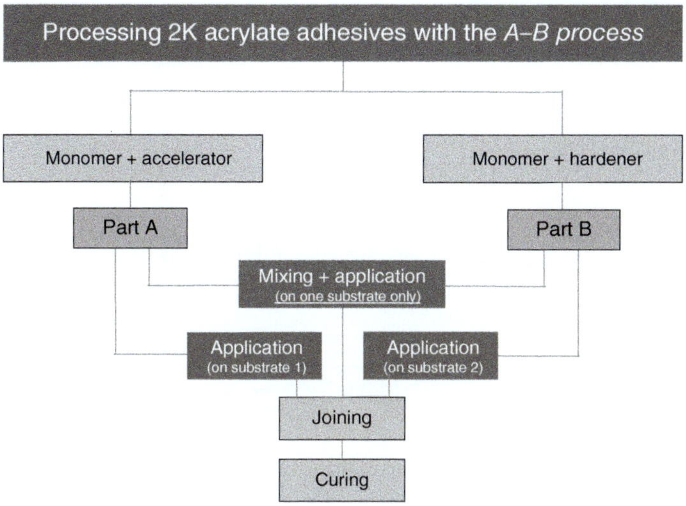

Figure 9.15 Processing of 2K acrylate adhesives using the *A–B process*.

parts A and B in equal amounts onto both materials or by dosing part A onto one joining part surface and part B onto the other. Going with the second option, mixing of the two parts is accomplished by the contact pressure used to join the parts. The subsequent curing to a sufficient initial strength takes place in only a few minutes. The *A–B Process* offers the user the advantage of dispensing the two parts of the adhesive in a 1 : 1 volume ratio. Since the hardener has already been incorporated into component B by the adhesive manufacturer, the user no longer needs to add small quantities of hardener, which is difficult to control. A particularly user-friendly and very popular application option for this process (as with the 2K epoxy adhesives) is to provide components A and B in dual-chamber cartridges from where the adhesive is processed by user-friendly on-demand mixing and dispensing systems.

- *No-mix process* – Here, the hardener is dissolved in a volatile organic solvent and applied to only one of the two material surfaces. After evaporation of the solvent, the material can be stored and further processed at any time later. This is done at the desired time by applying the already-compounded monomer accelerator mixture to the second substrate and then joining the two parts. A short time later, the compound is cured. The disadvantage of the *no-mix process* is that this method is limited to adhesive layer thicknesses of 0.3–0.4 mm, since the hardener must diffuse through the entire adhesive layer for complete curing.
- *Mixing method* – In this variant, the hardener is added to the monomer accelerator mixture in an amount of 1–3% immediately before processing, either as a paste or as a powder. Subsequently, joining must be performed immediately within the specified pot life. This method cannot be used with fast-reacting monomers and requires good process efficiency when mixing the relatively small amount of hardener with the adhesive matrix.

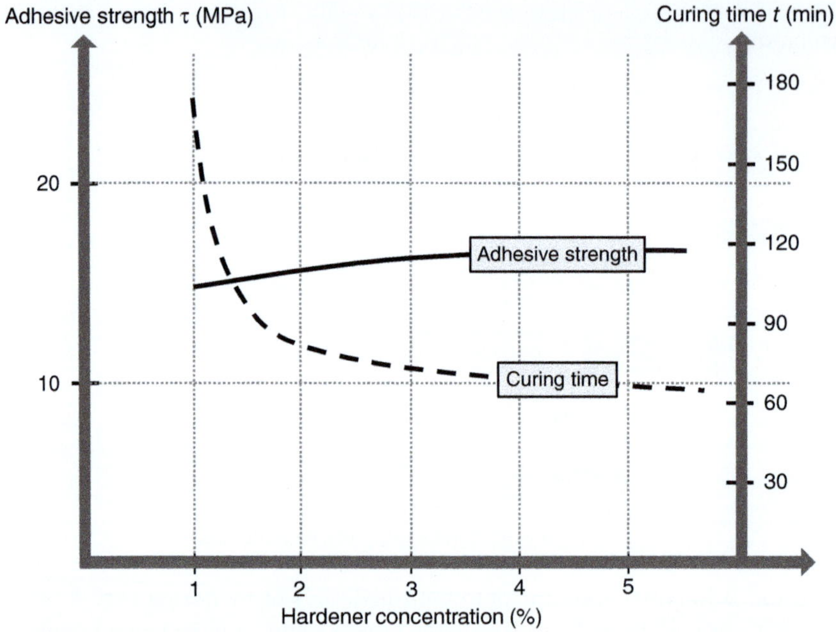

Figure 9.16 Dependence of adhesive strength and curing time on the hardener concentration in 2K acrylate adhesives.

Figure 9.16 shows the dependence of the bond strength and the curing time on the concentration of the hardener of a 2K acrylic adhesive. As can be seen, the adhesive strength remains relatively constant over a very wide range of hardener concentrations, while the curing time initially drops sharply and then also remains constant over a wide range. Thus, in the case of hardener concentration fluctuations, the 2K acrylic adhesives behave much more tolerantly with regard to the effects on strength and curing time than the 2K epoxy and 2K PU adhesives, where small changes in the amount of hardener cause large strength losses, making strict adherence to the specified resin/hardener ratios necessary for these systems.

Processing profile of 2K acrylate adhesives:

- 2K acrylate adhesives can be processed very easily and quickly in the A–B process, especially when using user-friendly on-demand mixing and dispensing systems with dual-chamber cartridges, from which they can be precisely metered, mixed, and applied to the parts to be joined,
- in contrast to the 2K epoxy and PU adhesives, smaller deviations from the specified mixing ratios are permissible,
- cure very quickly at room temperature in a few minutes to a few hours and therefore often have very short open times, in the minutes range,
- often require unusual mixing ratios (e.g. 1 : 10) and hardener additions of only a few percentage,

- allow only very limited large-area applications due to the very fast curing speed,
- must be stored in suitable refrigerators at about 8 °C, and
- often have a very strong odor.

9.3 Test Methods for Adhesive Bonds

9.3.1 Introduction

In adhesive-bonding technology, it is not possible to describe the performance of a bonded component solely from the values of static or dynamic tests, without the influence of the conditions occurring in practical use. For this reason, test results obtained after longer-term application of the loads occurring in the application should always be used to assess the quality of the bond, which then only allows statements to be made about the expected practical aging behavior of the bonded component over a longer period of time. For the comprehensive characterization of bonded joints, it is necessary to examine both the bonded joint and the materials involved – the pure adhesive layer and the substrates – with the appropriate methods. The tests of the adhesive bond are performed by so called technological tests, which are carried out both directly and after appropriate aging. For testing the pure adhesive layer and the substrates physical tests are applied (Figure 9.17).

9.3.1.1 Technological and Physical Tests

For example, determining the glass transition temperature of an adhesive layer or the tensile strength of a metal substrate are physical tests. They are used to characterize

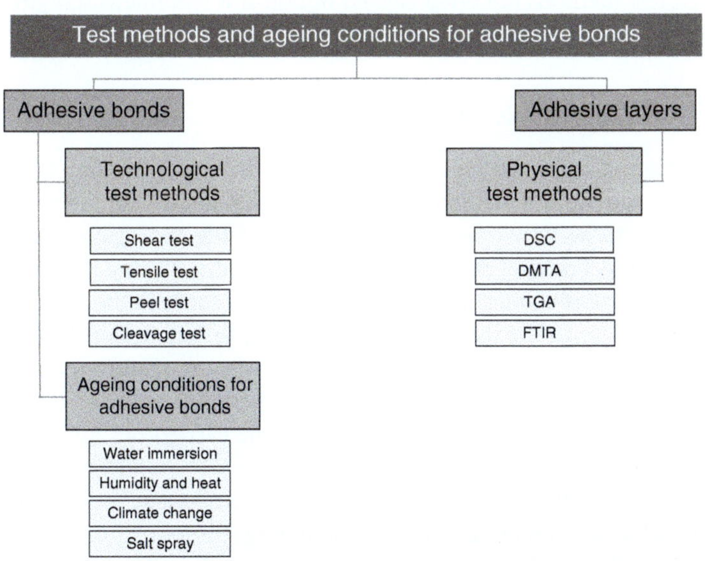

Figure 9.17 Technological test methods for adhesive bonds, aging conditions, and physical test methods applied in adhesive-bonding technology.

both the adhesive layers and the substrates, which is of great importance for assessing the quality of a bonded component. On the other hand, the determination of for example the adhesive strength (overlap shear strength) of an adhesive bond represents a technological test in which the different loads occurring in practice are to be taken into account as far as possible.

Due to the interactions between adhesive layer and substrate, the properties of adhesive composites cannot be fully described by determining only the properties of the pure adhesive layers. Therefore, methods that include the influences of both the adhesive and the substrates must be used to analyze the overall properties of the adhesive bond. The technical description of the properties of adhesive compounds is based on defined characteristic values, such as strength, deformability or resistance to mechanical stress, measured directly and after harmful environmental influences.

The aim of the technological tests is to develop a good property profile while taking into account the strengths of both the substrates and the adhesive layer, which provides as much information as possible about the behavior of the bonded joint in the application. In this context, it is important that precisely defined test methods are strictly adhered to, as otherwise no comparability with, for example, manufacturer specifications or/and results from the past is possible. For quality control or monitoring of the bonding process, additional application-related tests are often carried out on the prepared bonded joint, which reflect the loads of the bonded joints in practice as well as possible. The technological test methods are applied in the feasibility phase, first on laboratory test specimens and later on bonded components. The best basis for evaluation, on the other hand, are component tests under operating conditions, which are, however, time-consuming and cost-intensive. For this reason, standardized test methods are often carried out on test specimens that are similar to components, usually simplified and reduced in size.

9.3.2 Technological Test Methods for Adhesive Bonds

For the characterization of adhesive bonds, standardized test methods are used that provide data from measurements performed on well-defined test geometries. This allows conclusions to be drawn about the behavior of bonded components and enables adhesives and bonding methods to be compared with one another. Although in practice one usually finds multidimensional stress states in the bonded joint, pure tensile or shear forces are usually set when testing the strength behavior of bonded joints in the laboratory. Since the bonded joints in the application are usually subjected to environmental stresses – such as temperature, humidity, and radiation – tests to assess the load-bearing capacity must be carried out after the bonded joint has been exposed to these conditions.

These tests are performed in the laboratory with standardized test specimens, usually single-lap joints, which are exposed to the corresponding environmental conditions for an appropriate time and then tested. The strength reduction after aging forms the basis for the formation of reduction factors, which are used for the design of the bonded component. These factors are calculated from the ratio of strength after and before application of the corresponding load. If bonded joints are stressed beyond the limit of their load-bearing capacity, the adhesive layer can detach from the surface of a joined part (adhesion failure) or break within the adhesive layer (cohesion failure). For this reason, when using destructive testing methods, it is useful to examine and document not only the measured strength value, but also the type of failure and the associated fracture patterns.

9.3.2.1 Shear Test

The very frequently used *dynamic overlap shear test* – also called *adhesive strength test* – for determining the shear properties of a bonded joint is probably by far the most frequently applied test for bonded joints (Figure 9.18), which is used to assess the strength of bonded joints in research, development, and quality control. This test is very well suited to evaluate the aging behavior of bonded joints after the action of chemical and physical loads.

In general, a specimen is loaded in a shear test such that it suffers a sliding fracture along a plane parallel to the applied forces at the end. Here, one substrate is fixed and the shear force causes the other substrate to move in one direction, so that the material is stressed in a sliding motion. In this test, applied force is increased until the bond breaks. So, the test allows the determination of the shear modulus G in the elastic region of deformation and the *adhesive strength*, which is the maximal stress for breaking the bond.

Shear tests differ from tensile and compression tests in that the forces are applied parallel to the two contact surfaces, whereas in tensile and compression they are perpendicular to the contact surfaces. The result of the test is the shear strength or

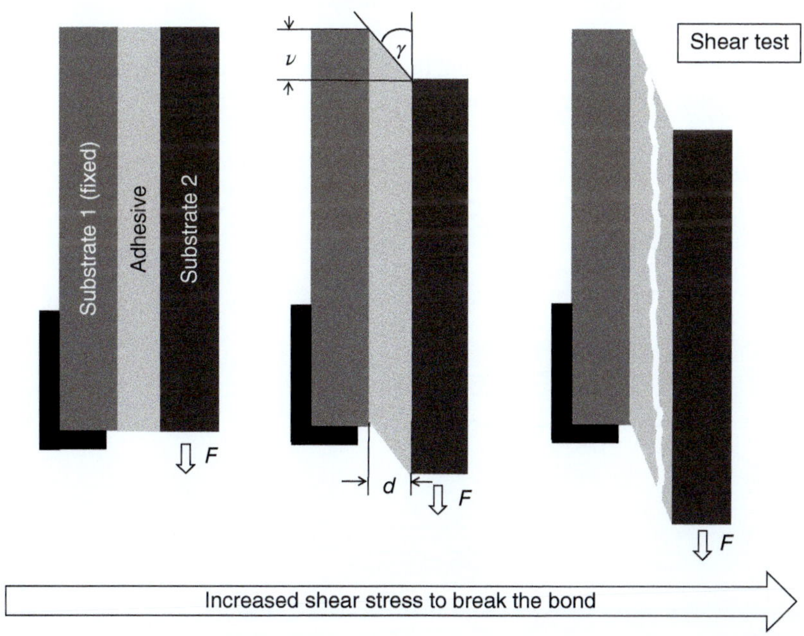

Figure 9.18 Overlap shear test to determine the shear stress (adhesive strength) and the shear modulus of a single-lap joint.

adhesive strength τ_{Bond}, which is calculated from the quotient of the maximum force F_{Bond} to break the bond and the bonded area A:

$$\tau_{Bond} = F_{Bond}/A$$

The unit of the overlap shear strength τ_{Bond} is N/mm^2 or MPa.

From the shear strain tan γ and the associated shear stress, the shear modulus G can also be calculated from the measurement as follows:

$$G = \tau/\tan\gamma$$

with the shear strain:

$$\tan\gamma = v/d$$

The unit of the shear modulus G is N/mm^2 or MPa.

For example, the description of the shear strain $\tan\gamma = v/d$ – with $v = 0.075$ mm and $d = 0.150$ mm – can be done in the following ways:

- as a quotient v/d: $0.075/0.150 = 0.5$,
- as a percentage v/d: $0.075/0.150 \cdot 100 = 50\%$, or
- as an absolute value of the deformation $v = 0.075$ mm.

Since the adhesive layer deformation increases proportionally with increasing adhesive layer thickness, the adhesive layer thickness should have no influence on the displacement angle. However, this is not the case, since larger thicknesses result in higher bending moments and greater transverse contraction, as well as higher shrinkage stresses and inhomogeneities in the adhesive layer.

Figure 9.19 shows the curve of the shear stress τ as a function of the applied shear strain tan γ, which is obtained by measuring the specimen with a tensile tester in the laboratory. Normally, only the τ_{Bond} – the value at break of the bond – is given, but

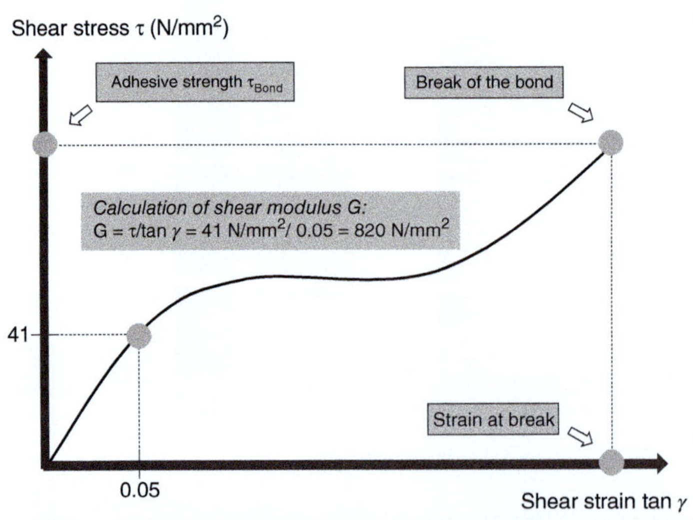

Figure 9.19 Course of shear stress as a function of the shear strain tan γ.

the course of the stress–strain curve often gives the designer of the bond additional information about the behavior of the joint under the applied load. It also allows the calculation of the shear modulus $G = \tau/\tan\gamma$, which is demonstrated in Figure 9.19, where G at a shear strain of 0.05 and a shear stress of 41 N/mm^2 result in the shear modulus $G = 820\,N/mm^2$.

The determination of the shear strength or adhesive strength τ_{Bond} of single overlap adhesive joints is described in DIN EN 1465 and given as the force/adhesive area in MPa (or N/mm^2) to break the bond.

The following test conditions are required for DIN 1465:

- specimen material: aluminum alloy $AlCuMg_2$ or a material with equal or higher modulus of elasticity than 68 000 MPa or higher yield strength than 290 MPa.
- specimen geometry: joining part: 100 mm · 25 mm · 1.6 mm, joining surface: 12.5 mm · 25 mm,
- surface treatment according to the specifications of the adhesive manufacturer or the relevant European standard ISO 4588,
- at least six measured values must be determined,
- the specimens must be stored and tested for a sufficiently long time in one of the standard climates specified in ISO 291 before testing, and
- the testing speed of the testing machine must be selected so that the bond is destroyed within 65 ± 20 seconds with increasing load.

9.3.2.1.1 *Substrate Dependence of Shear Test*

It is important to note that the adhesive strength τ_{Bond} determined according to DIN EN 1465 is not an adhesive-specific measure, but depends on the substrates used in the test. A disadvantage of the specimen geometry used in this test method is bending moments that superimpose the shear forces and lead to twisting of the bonding zone and thus to the formation of undesirable peeling forces. These can be avoided by using much thicker specimen plates in the shear test, which is described, for example, in DIN 54 451 [1]. Compared to the test method according to DIN EN 1465, this test is performed in a largely defined shear stress condition due to the improved specimen geometry. Here, the applied force does not cause strains of the specimen plates and bending moments, which normally lead to stress peaks and additional normal stresses in the composite. Thus, this standard can be used for substrate-independent determination of bond strength and also, for example, for analysis of the shear modulus and shear strain at break.

9.3.2.2 Tensile Test

The goal of this test method is to determine the tensile strength of a butt joint by applying perpendicular force to the bonded substrates (Figure 9.20), which is calculated from the quotient of the maximum force F_{Bond} to break the bond and the bonding area A [2]:

$$\sigma_{Bond} = F_{Bond}/A$$

The unit of the tensile strength test is N/mm^2 or MPa.

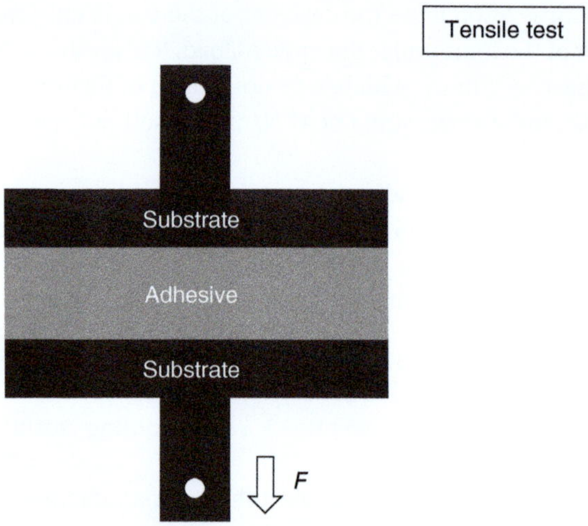

Figure 9.20 Specimen geometry and alignment of the force in the tensile test.

This test is described in DIN EN 26922 and since the stress proportions are relatively well defined by the geometric arrangement, this method is often used to assess the adhesion and cohesion properties of adhesive composites.

9.3.2.3 Peel Test

The goal of this test method is to determine the peel resistance of bonded joints by applying peeling forces to the bonded joint at different angles (usually 90° and 180°). DIN EN 28510 describes a 90° test of a flexible substrate bonded to a rigid specimen (Figure 9.21a), DIN EN 14173 a 180°T peel test for two flexible bonded substrates (Fig. 9.21b).

The result of the tests is the absolute peel resistance $p_{abs} = F_{peak}/\text{width}$ (N/mm) from the quotient of the first peel force peak divided by the bonded part width and the average peel resistance $p_{av} = F_{av}/\text{width}$ (N/mm) from the quotient of the subsequent average peel force divided by the bonded part width (Figure 9.22).

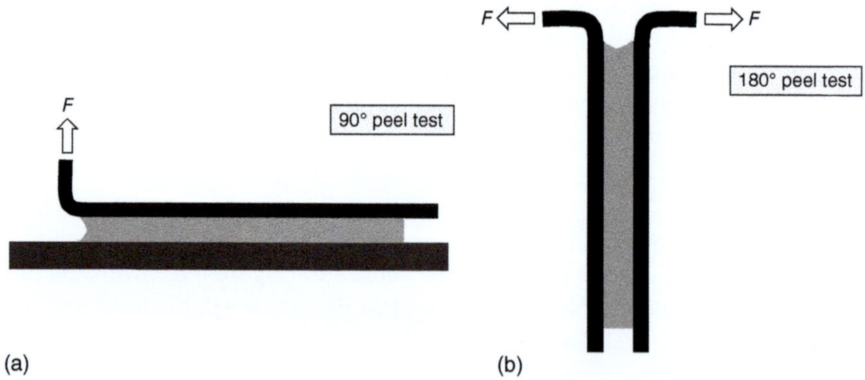

Figure 9.21 Specimen geometry and alignment of the forces during 90° (a) and 180° (b) peel testing.

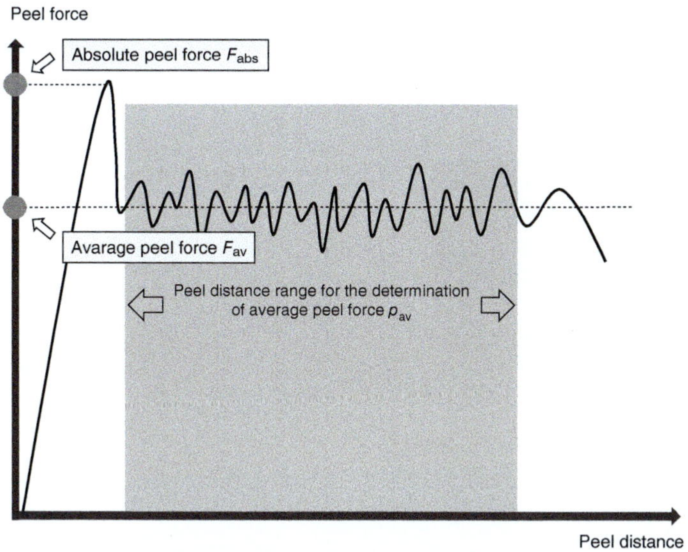

Figure 9.22 Course of the peel force as a function of peel distance and the determination of the absolute F_{abs} and average peel force F_{av}.

9.3.2.3.1 Relevance of Peel Tests

In contrast to the earlier described test methods, which produce a relatively uniform stress distribution over a relatively wide area of the specimen, peel tests produce predominantly localized normal stresses concentrated in one line. Although peel stresses should be avoided in the design of adhesive composites, the method is of great importance in assessing the adhesion and cohesion behavior of adhesive layers. This is particularly the case since, due to the complexity of the boundary layer, direct measurement of adhesion on bonded components is not possible and thus, specimen setups that force adhesive failure are needed. This is the case with peel tests, which makes this method very useful for determining the adhesion properties of surfaces and also the cohesion of adhesives. Because they are able to indicate differences with high sensitivity, they are often used for comparative evaluation of adhesives, surfaces, and surface treatment methods. Peel studies also enable brittle and tough adhesives to be distinguished by determining the forces required for peeling at relatively high peeling speeds and the corresponding records of the peeling curves.

Peel tests are also carried out to obtain initial indications of the water resistance of adhesive composites. For this test, 50% of the sample is first peeled and, after the bond has been relieved, water is injected into the boundary layer. When the peel test is continued, it is found that the presence of the water significantly reduces the peel strength in the case of a water-sensitive adhesive bond. In this case, the fracture pattern changes to partial or complete adhesive failure, which is associated with a significant, abrupt decrease in peel strength. However, this *wet peel test* qualitatively only provides initial indications of weak points in the interface and should not be used for the general determination of the long-term behavior of adhesive joints affected by moisture.

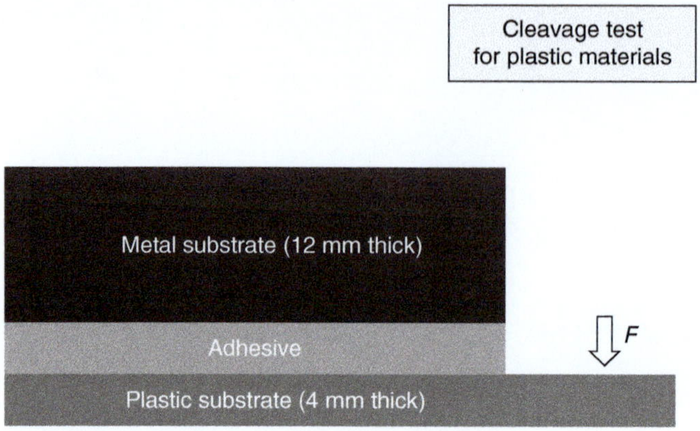

Figure 9.23 Specimen geometry and alignment of the forces during the cleavage peel test for plastics.

9.3.2.4 Cleavage Test

The objective of the related test methods is the determination of the bending peel resistance of the bond by applying forces vertically to the substrate to be tested to induce cleavage (Figure 9.23). The result of the *bending resistance test*, which is described in DIN 54461, is the so-called bending peeling resistance P_b as a quotient of the first strength peak occurring in the curve and sample width. Occasionally, the average peel resistance p_{av} from the quotient of the subsequent average peel force divided by the joining part width is recorded and documented.

This test is a variation of the peel test described above, using thicker mating parts, which can no longer be peeled due to their stiffness and thus induce splitting. The investigation of the cleavage peel strength by means of the geometry described allows the adhesion properties of plastic/metal joints in particular to be assessed much better than the overlap shear test can. The reason for this is that fracture in the plastic is avoided, which is very often observed in the shear tests of plastic materials. For such a bending peel test, a 4 mm thick and 25 mm wide plastic strip is bonded to a 12 mm thick steel substrate with the adhesive to be tested; any adhesive that escapes at the edges must be removed after curing. For testing, the protruding part of the plastic strip is pressed down at a speed of 5 mm/s.

9.3.2.5 3-Point Bending Test

In this specimen geometry, the thick substrate is bonded over a thinner substrate that is overlaid on both sides (Figure 9.24). For this test, which is described in DIN EN 1966, the load is applied in the center and on both sides of the supported specimen. Testing according to the three-point bending method is particularly suitable for plate-shaped adhesive composites and sandwich laminates.

9.3.2.6 Wedge Test

The wedge test was developed by Boeing to test the properties and aging resistance of joints bonded with surface-treated aluminum sheets. In this test [3, 4], a wedge is driven between two bonded test panels and the end of the crack that develops

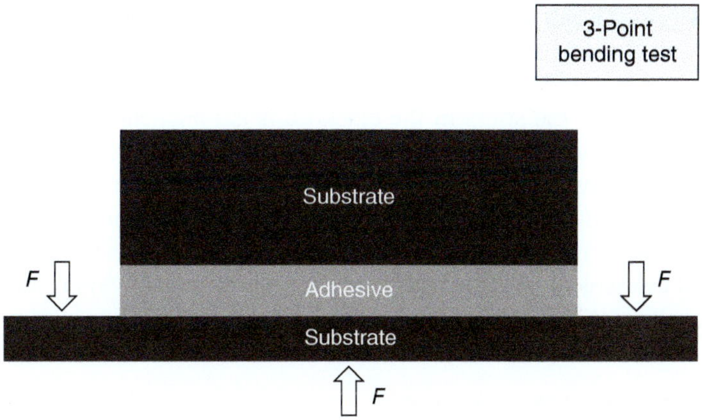

Figure 9.24 Specimen geometry and alignment of the forces during the 3-point bending test.

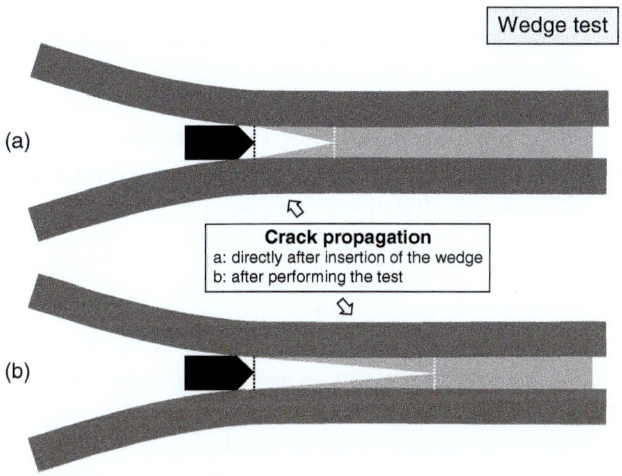

Figure 9.25 Specimen geometry for the wedge test with crack propagation directly after insertion of the wedge (a) and after performing the test (b).

after the wedge is driven into the adhesive layer is marked (Figure 9.25a). The crack growth under the influence of different climatic conditions is then measured, e.g. for 75 minutes at 50 °C and 95% relative humidity, and the crack enlarged by the influence of the loads is marked again (Figure 9.25b). To obtain further information on the aging behavior of the bond, the type of failure, cohesive or adhesive, is also checked and noted.

9.3.2.6.1 Significance of the Wedge Test

The wedge test is not an adhesive-specific test, but it does allow aging-determining influences on the overall behavior of an adhesive bond to be determined. An adhesive failure of the bond shows the significant influence of aging on the boundary layer, a fracture of the adhesive layer the corresponding reaction of the adhesive matrix. The popularity of this test arrangement compared to the shear strength test is explained by the quickly available results as well as the direct effect of the aging media on the adhesive layer under mechanical load. Also, no loading fixtures are required for these tests, since the stress in the crack tip is automatically set as a function of the stiffness of the specimen plates. However, this strong substrate dependence is also the main disadvantage of this method, which is the reason why the results of this experiment can only be discussed on a qualitative basis. Since certain degradation processes, as they occur in a long-term load test, are not observed in the wedge test, a general assessment of the aging behavior of the adhesive bond is not possible from the results of this experiment.

9.3.3 Aging Conditions for Adhesive Bonds

9.3.3.1 Immersion Test

The immersion test according to DIN 53287 *Resistance to liquids/Immersion* test is frequently used to determine the sensitivity of adhesive composites to water and moisture. Alternatively, Brockmann proposes a simpler test, whereby six identical samples are first placed in a 1-l wide-neck bottle. To each sample, 150 ml of demineralized water, which has a pH value of 7 at the start of the test, is added and storage in the water is carried out at room temperature for a period of up to arround 1 year. After aging, the samples are removed from the bottles and kept in a warming cabinet at 60 °C for 72 hours for reconditioning prior to testing.

9.3.3.2 Climate Test

This test – described in DIN 54456 – is usually performed for 4 or 8 weeks at 40 °C/95% relative humidity or 60 °C/95% relative humidity in a climatic chamber. After the specified aging time, the specimens are removed from the climatic chamber and stored in a heating cabinet at 60 °C for 72 hours for reconditioning. In this method, only the effect of water on the compound is tested.

9.3.3.3 Climate Change (VW PV 1200)

This climate change test follows a specification carried out by Volkswagen AG. Compared to tests with constant climate conditions, it is intended to simulate the influence of changing temperatures and the permanent condensation of moisture on the bonds caused by the cycle changes. The test is frequently used in the automotive industry on oiled test panels. The focus here is on adhesive applications in body-in-white and add-on parts where oiled steel and aluminum sheets are used. In contrast to the salt-spray test, this climate change test does not involve any corrosive attack on the bond.

The VW PV 1200 climate change test consists of the following cycles:

- 4 hours at 80 °C and 95% relative humidity,
- 2-hour cooling down to −40 °C,
- 4-hour holding at −40 °C,
- heating during 2 hours to 80 °C, 95% relative humidity.

The total duration of the test is 30, 60, or 90 days, depending on the system under test. After the specified aging time, the specimens are removed from the climatic chamber and stored in a warming cabinet at 60 °C for 72 hours for reconditioning.

9.3.3.4 Salt-Spray Test

The salt-spray test according to DIN 50021 aims to evaluate adhesive composites under highly corrosive conditions and is carried out at a specimen temperature of 35 °C. For this purpose, a test solution is prepared with a mass concentration of 50 g sodium chloride/liter of water, which has a pH value of 6.5–7.2. The solution is dispersed in the salt spray chamber with compressed air at a pressure of 0.7–1.4 bar, which is cleaned by three oil and liquid separators connected in series.

The test is performed as follows:

- the samples are clamped vertically in the stand and fixed in the chamber,
- aged in the test chamber for at least 500 hours,
- removed from the test chamber after ageing,
- rinsed with tap water at room temperature, and
- reconditioned in a heating oven at 60 °C for a period of 72 hours.

To protect specimens with corrosion-sensitive substrates from severe primary corrosion, the joints are coated in the area of the cut edges, e.g. with an e-coat.

9.3.3.5 Climatic Corrosion Test (KKT Test)

When bonded components are used in practice, there is often a superposition of different loads that damage the bonded joint. The various mechanisms can reinforce each other or cause a failure that does not occur or only occurs with a considerable delay when only one load is applied. This effect has led to the development of various combination tests, particularly in the automotive industry, with which a wide variety of component loads can be simulated at the same time. The aim of these tests is to represent aging loads that are as real as possible in a very short time.

The very aggressive climatic corrosion test is a combination of

- the VW PV 1200 test already presented,
- heat storage,
- condensation water test, and
- a salt-spray test.

One cycle of this test includes the following steps:

- 7-day climate change test VW PV 1200,
- 1-day room-temperature storage at 23 °C/50% relative humidity,
- 1-day heat storage at 80 °C/20% relative humidity,
- 21-day condensed water climate at 50 °C/100% relative humidity, with 2-hour salt spray at 35 °C every 84 hours.

To protect against very severe primary corrosion in the area of the cut edges, the specimens with corrosion-sensitive joining parts are additionally coated, e.g. with an e-coat.

9.3.3.6 Cataplasma Test

In this very simple test, the samples are in a closed atmosphere, which means that the corrosion products formed cannot be removed.

The test is performed as follows:

- wrapping the bonded sample with an absorbent cotton soaked in water,
- wrapping the sample package with an aluminum foil,
- airtight sealing of the aluminum wrapped package in a PE foil,
- storage for 21 days at 70 °C,
- storage for 16 hours at −30 °C, and
- determination of adhesion strength after two hours reconditioning at room temperature.

9.3.3.7 Outdoor Weathering Test

In the aging simulation methods discussed above, the desired time-lapse effect is achieved by significantly elevated temperatures or constant corrosive attack, either by condensation of atmospheric moisture or chloride ion attack. However, these extreme conditions can very easily induce processes to damage the bonded joint that will never occur in real component life. Therefore, to support the short-term tests, it is essential to perform additional long-term tests with only moderate degradation. Outdoor weathering tests and their variants should last at least one year (seasonal cycle), but several years are preferred. To intensify the conditions, the specimens can additionally be sprayed with a 3% NaCl solution once a week. This intensified outdoor exposure allows a combination of salt-spray exposure as well as exposure to climatic storage at widely varying temperatures – usually in the range of −20 °C to +30 °C – and natural humidity. Such tests might lead to severe corrosion of the bonded joint and thus to a drastic drop in strength. For the validity of the test, it is important that, in addition to the residual strengths, an exact analysis and documentation of the fracture surfaces is carried out.

9.3.3.8 Fracture Surface Evaluation

In addition to the evaluation of the mechanical characteristic values from the test methods already discussed, the evaluation of the fracture surfaces, especially after an aging simulation, is of decisive importance. From this evaluation, conclusions can be drawn about aging mechanisms and their temporal sequence, as well as about the causes of a change in the strength values. EN ISO 10365 differentiates the fracture behavior of bonded joints into fracture types, irrespective of the type of joining material and adhesive used. If more than one type of failure occurs in a fracture pattern, the approximate percentage must be given after each type of failure.

According to the standard DIN EN ISO 10365, the following fracture types are distinguished (Figure 9.26):

- *AF: Adhesion failure* – adhesion failure at one joining part,
- *AF: Adhesion failure* – adhesion failure on both parts of the joint,
- *CF: Cohesion failure* – fracture within the adhesive layer,

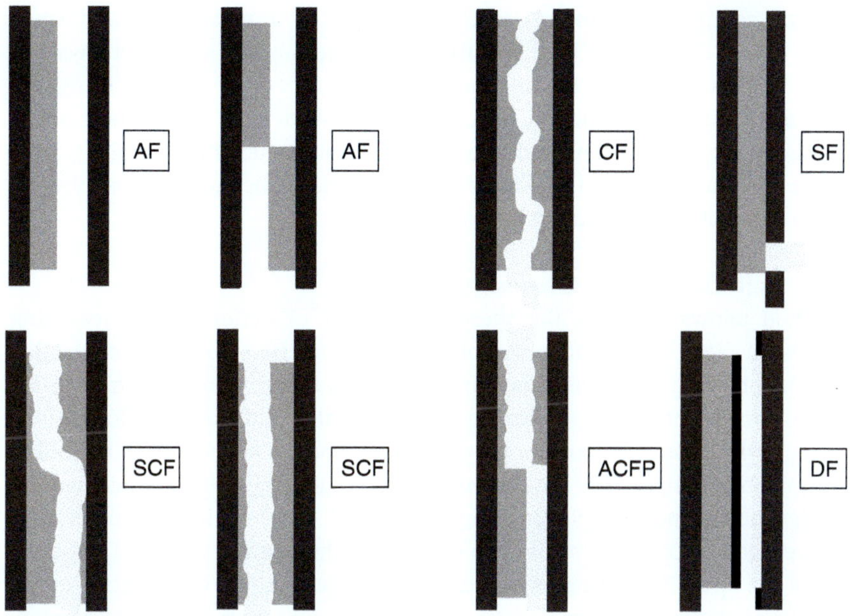

Figure 9.26 Types of composite fractures that can occur in bonded specimens before and after aging (according to DIN EN ISO 10365).

- *SF: substrate failure* – fracture within the substrate,
- *SCF: special cohesion failure* – special cohesion failure near the substrate,
- *ACFP: adhesion and cohesion failure* – mixed failure, and
- *DF: delamination failure* – delamination failure of an applied layer.

The fracture pattern depends very much on the adhesive type and the curing mechanism of the respective system. For example, unaged high-strength, hot-curing 1K epoxy adhesives usually show a near-boundary cohesive failure SCF with a very thin, friable polymer layer on one substrate side. After aggressive aging, there is then a change to adhesive failure AF. The failure behavior of room-temperature curing 2K epoxy adhesives is often very dependent on the film thickness. With unaged thin layers, cohesive failure also occurs near the interface SCF, but with thicker adhesive layers and good surface treatment, the fracture pattern changes to cohesive failure CF. 2K PU adhesives fail completely cohesively CF due to the usually very high mineral filler content. Changes in the fracture pattern are only to be expected in case of serious deficiencies in the surface treatment and very long aging times.

9.3.4 Physical Test Methods

The most important physical testing methods with relevance applied in adhesive-bonding technology are dynamic mechanical thermal analysis (DMTA), differential scanning calorimetry (DSC), and Fourier transform infrared spectroscopy (FTIR). DMTA and DSC are used for the commonly applied analysis of the thermomechanical properties of adhesive layers – especially the glass transition temperature T_g and the transition regions. Here, DSC detects the location of the transition regions based on the endothermic and exothermic effects, while DMTA determines the transitions by the change of the viscoelastic behavior of the adhesive layers as a function of temperature or frequency.

9.3.4.1 Differential Scanning Calorimetry (DSC)

The method allows studying the thermal behavior of materials by measuring differences in heat fluxes of a sample and a reference during a controlled temperature program. From these data, information can be obtained on phase transformations such as melting or solidification of materials, solid-state transformations in crystals, or the glass transition temperature of polymers. The method is thus based on a comparative temperature measurement of two substances at a defined heating rate, with the reference sample showing no thermal reactions in the temperature range investigated. DSC is also used to determine the degree of cure of adhesive layers and the optimum curing parameters. While one sample of an already-cured adhesive layer shows no thermal reactions in the temperature range investigated, heat is released in the other sample, which is not yet or only incompletely cured, due to the exothermic reaction. This results in a temperature difference between the two samples. A subsequent comparison of the two temperature–time curves allows statements to be made about the degree of curing of the adhesive layer and serves to determine the optimum reaction conditions in terms of temperature and time for curing.

9.3.4.2 Dynamic Mechanical Thermal Analysis (DMTA)

A common method for evaluating the viscoelastic behavior of adhesives is DMTA. The principle of DMTA is simple to describe: The method allows the determination of the reversible (elastic) and irreversible (viscous) response of a polymer to a deformation as a function of temperature or deformation rate. To do this, a sinusoidally oscillating force is applied to a material sample and the deformation of the material is measured. Both the amplitude and the phase shift of the deformation with respect to the applied force are determined. In the case of a viscoelastic material the resulting wave can be viewed as a superposition of two waves, one of which is in phase with the stress – representing an ideal elastic material – and the other of which precedes it by 90° – representing an ideal viscous material. It is therefore the sum of an ideal elastic component and an ideal viscous component with different amplitudes (Figure 9.27).

From the measured values, the viscoelastic properties of a sample can be determined as a function of time and temperature. The analysis can be performed in shear and tensile/bending modes, and in addition to the storage modulus (G' and E') and loss modulus (G'' and E''), the glass transition temperature T_g and the mechanical loss factor tan δ of the material can be determined. Furthermore, information on curing processes and on the rheological material properties can be obtained.

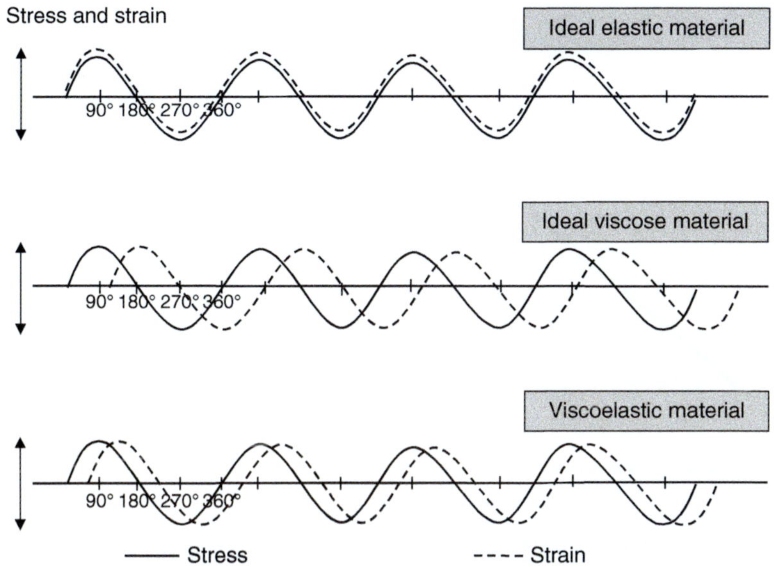

Figure 9.27 Deformation behavior of elastic, viscous, and viscoelastic materials in response to an applied sinusoidal oscillating stress.

In DMTA analysis, the deformations of the adhesive layers are usually kept so small that the linear viscoelastic range is not left. As discussed, the method allows the load to be applied in shear resulting in G^*, the complex dynamic shear modulus and in tension or bending mode with E^*, the complex dynamic strain/bending modulus. Since the shear modulus is more important than the others for the evaluation of adhesive layers, the former variant of the process is usually used in adhesive-bonding technology.

The following cases occur in the DMTA study of materials:

- *Ideally elastic material* – the deformation resulting from the load is in the same phase as the load. For materials with pure Hooke's elastic behavior: $G'' = 0$, and thus $\delta = 0$, the deformation field is exactly in phase with the stress field,
- *ideally viscous material* – the deformation resulting from loading is shifted by 90°, and
- *viscoelastic material* – the deformation resulting from the load is shifted by an angle <90°.

The phase angle δ between the stress and the responding strain is called the loss angle and the ratio of lost and stored energy per cycle leads to the loss factor

$$\tan\,\delta = G''/G'$$

The maximum of tan δ in the DMTA curve is an indicator of the dynamically defined glass transition temperature of the material under study.

Since it is not possible to describe the viscoelasticity of a material with a single material property, a complex modulus G^* is defined, which simply describes the complicated deformation mechanisms in the viscoelastic material (Figure 9.28).

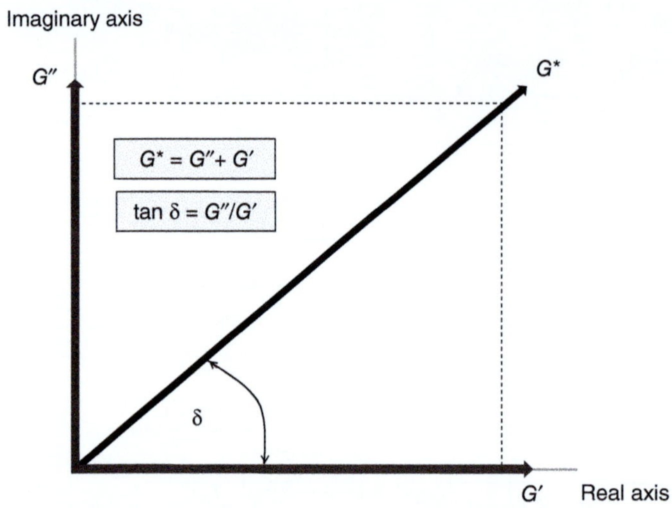

Figure 9.28 The representation and relationships between the complex modulus G^*, the storage modulus G', the loss modulus G'', and the loss factor tan δ.

The complex modulus G^* is obtained additively from the storage modulus G', which represents the elastic properties, and the loss modulus G'', which represents the viscous properties, to give $G^* = G' + G''$, where it can be divided into a real part and an imaginary part. The storage modulus G' is a measure of the mechanical energy that can be recovered, while the loss modulus G' indicates the amount of energy that is irreversibly converted to heat (dissipation). The mechanical loss factor $d = \tan\delta$ is the ratio of G'' and G' and indicates how much the deformation of the polymer lags the load. The curve around the tan maximum is called the damping peak and the tan δ maximum is at the temperature at which the damping of the polymer is greatest.

The tan δ curve versus temperature can be interpreted as follows (Figure 9.29):

- At low temperatures, where the polymer is in the glassy state, the amount of stored energy is large and tan δ, the ratio of internal friction as heat and stored energy, is small.
- As the temperature increases, the increase in micromovements of the molecules causes a decrease in the storage modulus G' and an increase in the loss modulus G'', resulting in an increase in tan δ.
- In the glass transition region, there is a rapid decrease in the storage modulus G' – which occurs in a relatively small temperature range – and the value of the mechanical loss factor tan δ, reaches a maximum.
- Subsequently, tan δ decreases with rising temperature, due to an increase in volume, which is caused by the higher degree of vibrations of the polymer molecules and, in addition, also due to interstices in the polymer matrix.

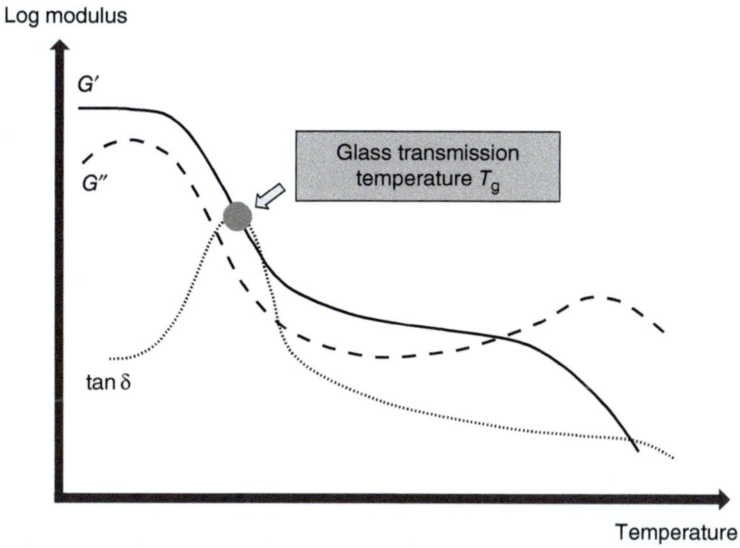

Figure 9.29 Typical DMTA curves of an adhesive showing the storage modulus G', the loss modulus G'', the loss factor tan δ, and the glass transition temperature T_g.

9.3.4.2.1 Time–Temperature Superposition

DMTA is not only capable of analyzing the viscoelastic behavior of polymers over a wide temperature range, but also allows investigation in a frequency range far beyond the scope of what is practically feasible of about 0.1–200 Hz. The principle of time–temperature superposition is applied, whereby a so-called reference curve is created by superimposing the DMTA results obtained at different temperatures. In this way, it is possible to evaluate the behavior of polymers either after very long mechanical stress or after exposure to very high stress frequencies. In combination with the classical methods for testing the strength of adhesive layers, the effort required to assess the temperature- and frequency-dependent behavior of bonded joints can thus be greatly reduced. However, the dynamic–mechanical analysis cannot provide strength values and is therefore no substitute for the classical mechanical testing methods.

9.3.4.3 Thermo Gravimetric Analysis (TGA)

This method enables the temperature- and time-dependent analysis of the weight losses of adhesives that occur during curing. The sample is placed in a crucible made of temperature-stable and inert material such as platinum or aluminum oxide in an oven at temperatures up to 2400 °C and the change in mass is determined using a microbalance. To prevent oxidation during analysis, the sample chamber is purged with, for example, pure nitrogen. During heating, the sample releases volatile components to the environment by decomposition reactions or evaporation. From the weight loss and the temperature at which the weight change occurs, conclusions can be drawn about the composition of the material. In addition, it is possible to determine the gases or fission products released using FTIR. In adhesives technology, TGA is used, for example, to control curing processes by detecting time-dependent weight differences.

9.3.4.4 Fourier Transform Infrared Spectroscopy (FTIR)

The FTIR spectrometer is a special variant of an infrared spectrometer, where, in contrast to dispersive measuring instruments, the spectrum is not recorded by a stepwise change in wavelength, but is calculated by a Fourier transform of a measured interferogram. In adhesives technology, FTIR spectroscopy offers the possibility to follow the curing mechanisms in adhesive layers and to perform evaluations for quality control of both the cured adhesive layers and the materials used in the adhesives, such as resins, crosslinkers, primers, adhesion promoters, and fillers. The characteristic so-called *fingerprint* spectra of the qualified *in spec* materials are recorded over a wider wavelength range and regularly compared with the spectra of the actual deliveries. In this way, quality deviations can be identified by comparing the *fingerprints*. FTIR spectra are vibrational spectra triggered by infrared radiation that reflect the translational and/or rotational vibrations of molecules or molecular groups in polymers. Depending on the atomic state, they show differently pronounced absorption bands located in specific and characteristic regions. For example, the absorption bands of C=C double bonds are in the range of 1640–1680 cm^{-1}. The curing of an acrylic adhesive leads to a continuous reduction of these double bonds in the adhesive. It is therefore possible to determine the degree of cure by measuring the amount of double bonds still present in the adhesive. Due to the need to determine many data points in a very short period of time, the special property of FTIR to provide a temporal resolution of 100 spectra per second is of great benefit for the analysis of the curing properties of adhesives.

9.4 Properties of Adhesive Bonds

9.4.1 Introduction

In the last section (9.3), the test methods were discussed – subdivided into technological and physical ones – which are necessary for the evaluation of bonded joints initially and after prolonged exposure to loads. The exact description of the shear, tensile, peel, and cleavage strength determinations as well as the accelerated aging conditions to which bonded joints must be subjected was given. Also, the reader learned what test methods in particular can contribute to the understanding of the properties of bonded joints. This section now deals with the properties of bonded joints and among other things, it discusses which ones are important, what they depend on, and what they mean for the bonded components in practical use.

9.4.1.1 Structural Adhesive Bonds

Thermoset adhesive layers, e.g. based on epoxy or crosslinked PU adhesives, have higher adhesive forces and a lower creep tendency than thermoplastics due to their crosslinked network. Such structural adhesive layers are therefore preferred for highly stressed applications. For their usually thin, low-deformation adhesive layers, the adhesive force depends on the adhesive layer thickness, with the highest adhesive force values being achieved at low thicknesses of around 0.1–0.3 mm. Due to the limited deformability of their adhesive layers, an increase in thickness does not lead to a reduction of stress peaks in the bond. It is very important to note that the maximum strength of bonds with thermosetting adhesives depends on the yield point of the substrates, and when the yield point of the substrate is reached, a further increase in strength by increasing the overlap length of the bonded joint is no longer possible.

9.4.1.2 Elastic Adhesive Bonds

This is different in elastic bonds, where 0.5–5 mm thick low-modulus adhesive layers are produced. Due to their uniform stress distribution in the bond and thus the lack of stress peaks at the overlap ends, the adhesive force can be increased linearly by increasing the overlap length. Thus, the elastic systems are able to compensate for the lower adhesive force compared to thermoset adhesives.

A property profile for an adhesive that meets higher requirements looks as follows:

- aging-resistant bond to the surfaces of the parts to be joined,
- high cohesive strength,
- suitable deformation capacity to reduce stress peaks,
- low creep tendency under static load,
- high thermal resistance, and
- high resistance to moisture and corrosive substances.

9.4.2 Thermomechanical Properties of Adhesives

The physical and mechanical properties of adhesives are dependent on temperature, although the degree of dependence differs for thermosets and thermoplastics. In contrast to the non-crosslinked thermoplastics, the thermosets do not show significant abrupt changes because of their close-meshed network.

9.4.2.1 General Thermomechanical Properties of Adhesives

In general, the temperature dependence of the adhesive strength τ_{bond} is very similar for a wide spectrum of adhesives used for industrial applications and can be divided into three ranges (Figure 9.30):

- *Range A* – At low temperatures in the glassy state, a slight increase in bond strength is initially observed, although the duration of low-temperature exposure has no effect on the strength of a bond. In this temperature range, PU and silicone adhesives demonstrate exceptional good deformation properties.
- *Range B* – With an increase in temperature, in the so-called rubber-elastic range, there is an increase in adhesive strength up to a maximum, which is due to the improved removal of stress concentrations at the overlapping ends. This characteristic behavior can extend over a wide temperature range, depending on the basic chemical structure of the adhesive.

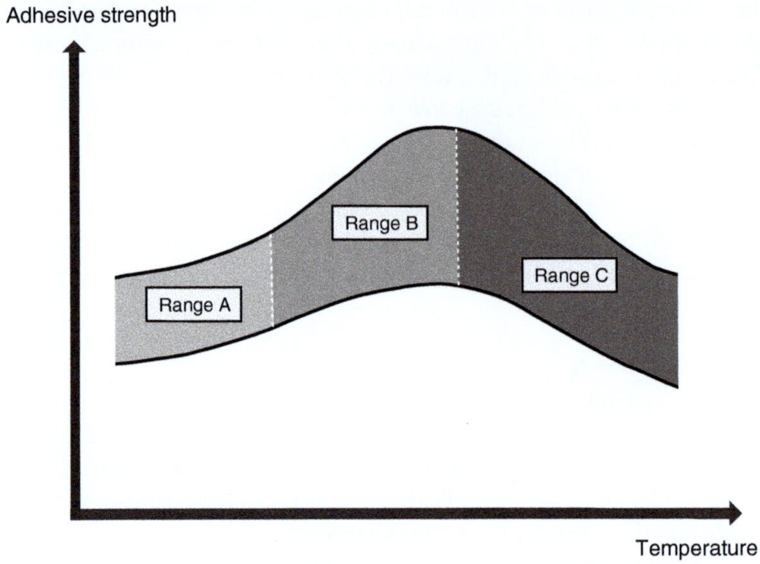

Figure 9.30 Ranges of dependence of adhesive strength on temperature.

- *Range C* – Subsequently, at higher temperatures, the adhesive force decreases due to the onset of flow processes in the adhesive layer and decomposition phenomena of the polymer molecules.

9.4.2.1.1 Glassy Region

The position of the glass transition temperature (T_g) is an important factor in the behavior of adhesive coatings. It is defined as the average temperature of the range at which the micro-Brownian motions – rotations and vibrations of the molecules of polymers – cease with appropriate cooling. Below this temperature, the adhesive is glassy and brittle and behaves completely elastically. The physical and mechanical properties are relatively constant in this temperature range and depend mainly on the chemical bonds in the molecules and the degree of crosslinking of the adhesive layer.

If thermosetting adhesives are used at room temperature, for example, they are below their T_g and therefore also exhibit the properties characteristic of this range (position A in Figure 9.31). If one increases the temperature from there, they undergo the transition to the elastic range, whereby their storage modulus decreases, which is characteristic for this transition. The T_g of thermoset epoxy adhesives, for example, are in the range of 80–100 °C. In contrast to thermosetting adhesives, elastic bonded joints, such as those produced with PU adhesives, are above their T_g when they are used in practice (position B in Figure 9.31). Thus, they are in the elastic region of the thermomechanical curve, which is expressed by the name of these adhesive systems and explains their lower storage modulus and

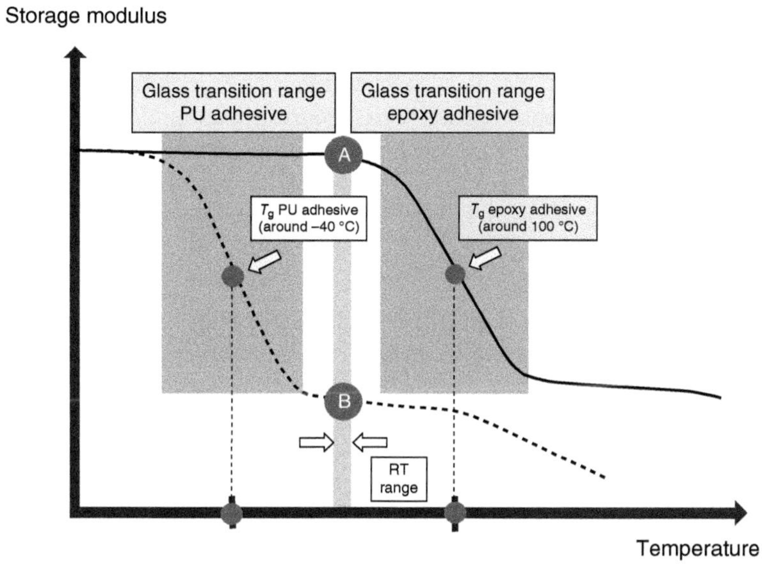

Figure 9.31 Course of storage modulus and position of glass transition ranges of a PU and epoxy adhesive depending on temperature.

thus their lower strength compared to thermoset adhesives. The glass transition of elastic adhesive systems is usually well below the operating temperature, in the case of PSAs, for example, often below −20 °C.

9.4.2.1.2 Elastic Region

Above the glass transition temperature, the behavior of the adhesive layer changes abruptly. Now rotational and vibrational movements – the micro-Brownian motions – are possible, which allow the free movement of whole molecules and molecular chains without any change in the position of the molecules. Due to the lack of crosslinking, these movements are particularly pronounced in thermoplastics, while they are more moderate in thermosets – due to their high crosslinking density. Thus, starting from the glass transition temperature, significant mechanical and physical property changes occur in the adhesive composite when the temperature is increased. This is caused by an increase of the mobility of the molecular chain segments and ultimately results in a thermostable rubber-elastic state of the adhesive.

9.4.2.1.3 Flow Region

If the temperature is increased further, the occurrence of macro-movements of the molecules gradually brings the adhesive system into the flow state. While in the micro-movements already discussed the molecules maintain their position, the macro-movements initiate the free mobility of whole molecules or molecule chains, which is particularly pronounced in thermoplastics due to the lack of crosslinking.

Thus, the flow temperature – the limiting temperature between the solid and liquid states of the adhesive – is approached, at which the intermolecular forces are overcome to such an extent that the molecular chains can move freely and slide toward each other. These irreversible displacements of neighboring macromolecules cause viscous flow and are thus responsible for the intensive irreversible creep processes in adhesive layers. Upon further heating, the decomposition temperature of the adhesive polymers is reached, at which the chemical bonds are destroyed.

9.4.2.1.4 *Relevance of T_g for Practice*

For the evaluation of adhesive systems with regard to their suitability for certain applications, knowledge of the position of their glass transition temperatures is of considerable importance. For example, a self-adhesive foam tape used for mounting profiles on painted materials must be above the glass transition temperature during processing to wet the surface and thus build up the necessary intermolecular forces at the surface of the substrate. On the other hand, for structural hem flange bonding of car doors, only adhesives with glass transition temperatures above the application temperatures can be considered because of the required strength.

9.4.2.2 Thermomechanical Properties of PSAs

These adhesives behave like viscoelastic materials and exhibit very different thermomechanical behavior over a temperature spectrum of, for example, −100 °C to +200 °C, represented by the following ranges (Figure 9.32):

- glassy state (1), which is determined by the chemical nature and structure of the base polymer,

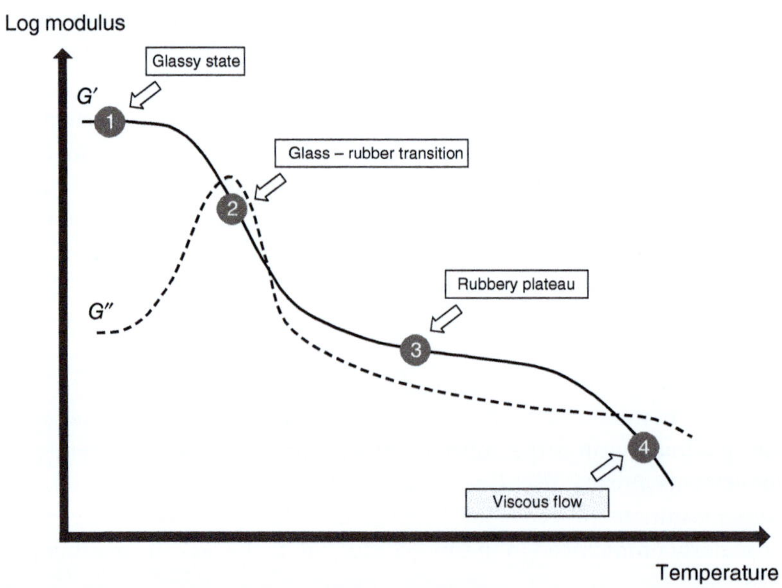

Figure 9.32 Thermomechanical ranges of PSAs.

- transition from glassy to rubbery behavior (2), which is determined by the glass transition temperature of the polymer composition,
- rubbery plateau (3), which is determined by the plateau modulus of the base polymer and the total polymer content of the adhesive, and
- viscous flow (4), which is determined by the molecular weight of the base polymer, the polymer content, branching, and crosslinking.

9.4.2.2.1 Dahlquist Criterion

A necessary condition for PSA performance is that the adhesive meets the well-accepted empirical *Dahlquist criterion*. It is the result of Carl Dahlquist's foundational work at 3M in the 1960s and defines the requirement for a PSA as follows: "*For measurable fast-appearing tack, the modulus of elasticity must be positioned below a certain fixed value that is completely independent of the type of adhesive, the part being bonded, and the pressure applied.*" According to Dahlquist, a G′ value (storage modulus) of $3 \cdot 10^5$ Pa or less in the bonding phase at RT is required to demonstrate tack (Figure 9.33).

The Dahlquist criterion requires that after a deformation of the adhesive of 1 second, the storage modulus must be equal to or below a value of $3 \cdot 10^5$ Pa (position 1 in Figure 9.33). For good peel behavior, an additional requirement must be met: from 1 to 0.01 second time scale, G' must increase significantly, accompanied by a high-energy loss expressed by high tan δ values. In other words, the G' value after 0.01 seconds contact time must be significantly higher than the G' value after 1 second contact time. The local deformation time of 0.01 seconds implies the peeling of a 40 μm thick adhesive layer with a peeling rate of 4 mm/second (deformation

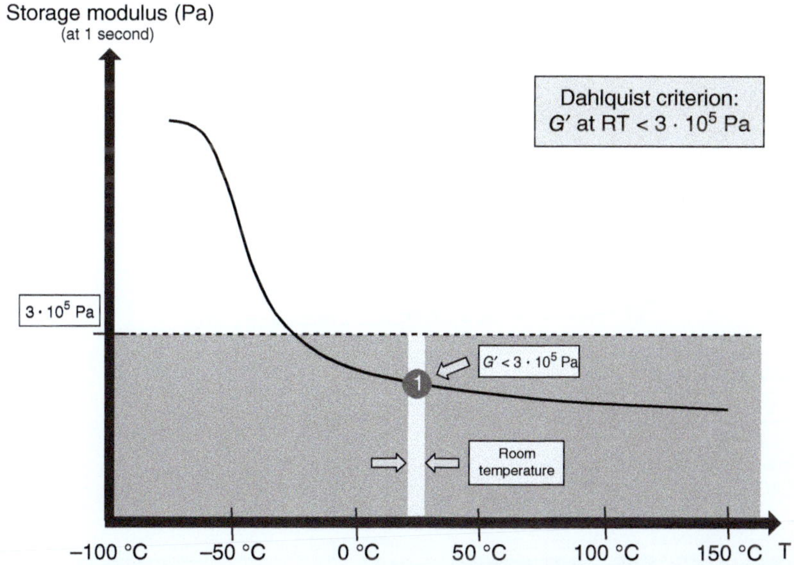

Figure 9.33 Course of the storage modulus of a PSA meeting the Dahlquist criterion for tack after being subjected to a stress at room temperature for 1 second.

time = thickness/peeling rate = 40 µm/4 mm/s = 40 µm/4000 µm/s = $1 \cdot 10^{-2}$ s). Creep occurs after long-term static loading of the bonded joint and therefore, for good creep resistance, G' must not drop too much on very long time scales.

9.4.2.2.2 Thermomechanical Effects of Tackifiers in PSAs

Tackifiers adjust the viscoelastic properties of an elastomer to make it a PSA, or they improve the properties of an adhesive that already meets PSA requirements. They have higher T_g's than the base polymer of typically 50–120 °C and basically act as diluents. As a result, they improve compliance (softness) and wetting. Figure 9.34 shows the effects of adding tackifier to an elastomer. So, the glass transition region (position 1) of the elastomer is shifted to higher temperatures, resulting in a higher T_g of the PSA. In the rubbery plateau, the storage modulus G′ is reduced (position 2), allowing the material to meet the Dahlquist criterion (position 3) and thus become a PSA.

9.4.3 Shear Properties

The goal in designing bonded joints is to allow the outside forces to act parallel to the substrates involved. This is in contrast to loads such as tension, compression, and peeling, which act perpendicular to the contact surface. The great advantage of forces acting parallel to the contact surface is therefore that much higher forces can be transmitted than would be possible with a load acting perpendicularly. It is therefore possible to scale the load capacity of an adhesive bond within limits via the size of the contact area, which is not possible with tensile loads. For this reason, the shear

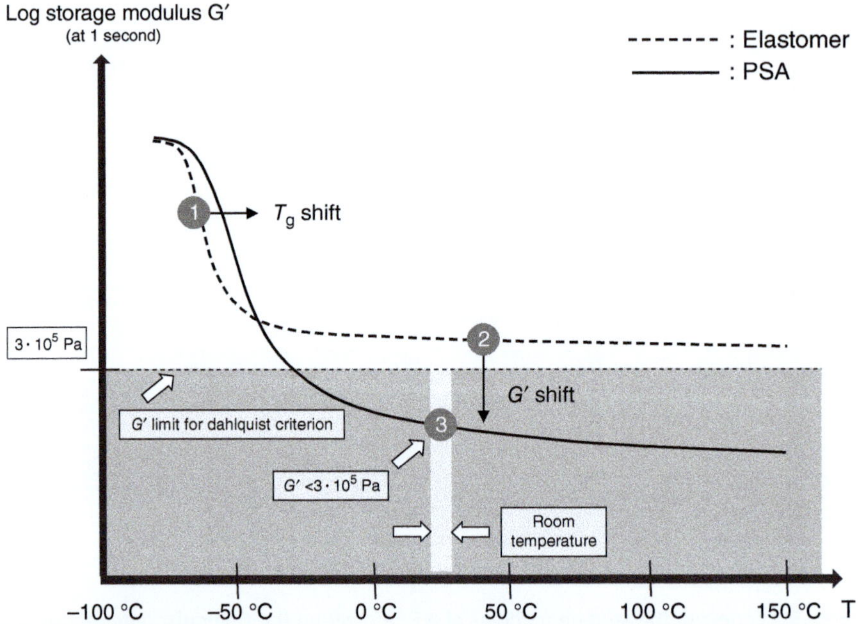

Figure 9.34 Effects of adding tackifiers to elastomers on the storage modulus in the glass transition and elastic region.

property of an adhesive bond is of great importance and is given high consideration in the design of bonded components. When selecting adhesives, the bond strength is the first indicator of the area-related holding capacity of an adhesive and therfore, this value is encountered wherever adhesives are offered, discussed, compared, or evaluated. However, since it provides the developer with important information about the behavior of an adhesive bond under internal and external loads in the application and its long-term behavior, the shear modulus as well as the stress–strain and creep-flow characteristics of the bond must also be taken into account for getting a more comprehensive picture.

9.4.3.1 Shear Modulus

The shear modulus is usually a material constant that provides information about the linear-elastic deformation of a component due to a shear force or shear stress (the unit is $N/mm^2 = 1\,MPa$). The shear modulus, as frequently applied in adhesive-bonding technology and measured using the substrates intended for the bonded component or relatively thin standard substrates according to DIN EN 1465, is not a material constant but depends on the substrates used for testing. For substrate-independent determination of the shear modulus, much thicker specimen sheets are used in the shear test, described for example in DIN 54 451. Compared to the test method according to DIN EN 1465, the test is performed in a largely defined shear stress condition due to the improved specimen geometry. Here, the applied force does not cause strains of the specimen plates and bending moments, which normally lead to stress peaks and additional normal stresses in the composite.

The shear modulus is the ratio of the shear stress τ and the elastic angular deformation $\tan\gamma$ (shear strain), which occurs at very small deformations in the linear viscoelastic range (Figure 9.35):

$$G = \tau/\tan\gamma \text{ with the shear strain } \tan\gamma = v/d, \text{ and thus } G = (\tau/v)\cdot d.$$

Shear stresses applied to the adhesive layer cause displacement of the individual surface segments relative to each other by the angle γ resulting from the displacement v of the layer thickness d. Low stresses produce elastic deformations of the adhesive layer, while their increase causes plastic deformations up to breakage of the bond. The higher the elastic component in the viscoelastic adhesive layer, the greater the shear stress, which can be explained by the dense crosslinked structure of such systems, as is the case, for example, with the highly elastic epoxy adhesives. Here, the shear stress in the adhesive layers cannot be relieved by free-flowing molecular chains, which is the case in adhesives with a large plastic content, such as hotmelt and PSAs. Crosslinked systems therefore generally have significantly higher shear stresses than thermoplastic or elastomeric adhesives, especially at higher temperatures. Shear moduli of adhesive layers are in the range of 100–10 000 MPa and are very commonly used to describe their thermomechanical behavior, especially to represent properties at different temperatures.

9.4.3.2 Shear Stress–Strain Behavior

Particularly in the case of high-strength, structural bonds, it is necessary to know the deformation behavior at defined stress conditions, from which the

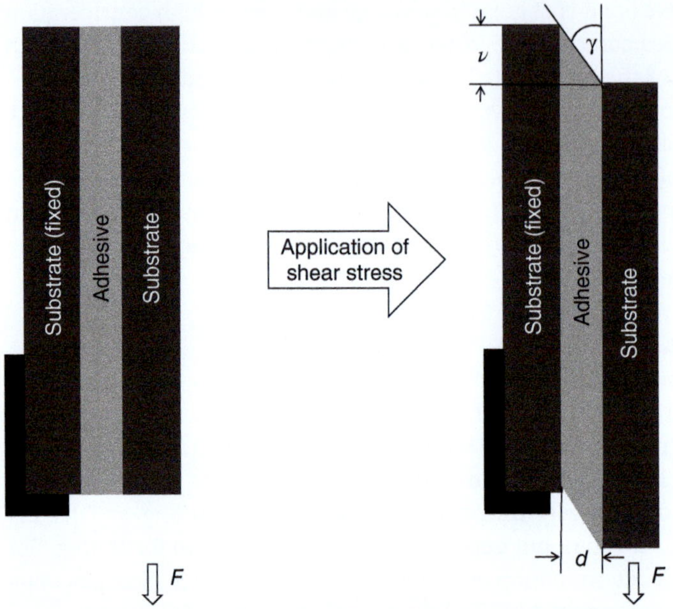

Figure 9.35 Determination of the shear strain tan γ of an adhesive from the response to a shear stress applied parallel to the bonding surface of a single-lap joint.

adhesive-layer-specific parameters of interest can be derived. Thus, the representation of the shear stress–strain behavior of a bonded joint allows the user to make a very good assessment of the possible stress states in the bonded component from the initial phase of elastic deformation to the fracture of the joint. Since the adhesive layer deformation increases proportionally with increasing adhesive layer thickness, the adhesive layer thickness should have no influence on the displacement angle. However, this is not the case, since larger thicknesses result in higher bending moments and greater transverse contraction, as well as higher shrinkage stresses and inhomogeneities in the adhesive layer.

The analysis of shear stress–strain behavior provides very good information both for an assessment of the strength of bonded components and, in addition, their long-term durability after exposure to appropriate loads. A major advantage of this method is that the load is applied in the shear direction during the test. This allows a good comparison with the practical application, in which bonded components are designed in the sense of an adhesive-compatible construction in such a way that mainly stresses occur in the shear direction.

When recording a shear stress–strain diagram, the behavior of adhesive bonds can be divided into three ranges (Figure 9.36):

- *Range A* – A linear progression is initially observed in the elastic range (1–2).
- *Range B* – A long flow range follows caused by the plastic deformation of the adhesive layer (2–3).
- *Range C* – The shear stress increases due to the orientation and interaction of the adhesive layer polymers until it finally breaks the bond (3–4).

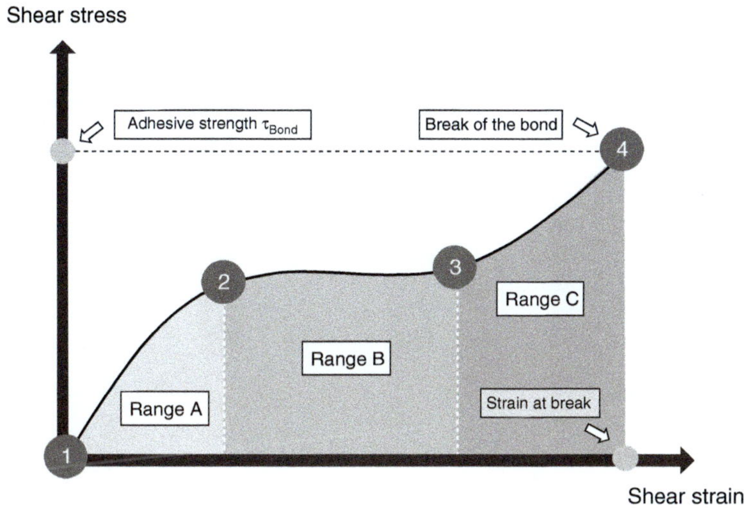

Figure 9.36 Shear stress–strain diagram of a bonded joint.

9.4.4 Creep and Flow

9.4.4.1 Creep

For the long-term behavior of an adhesive bond, the time-dependent mechanical behavior in the form of creep under load is of great importance. This property is defined as the time-delayed reversible deformation of viscoelastic materials under a constant, relatively small load, whereby a strain limit dependent on the applied stress is asymptotically established. Once this limit is reached, no more movement takes place at the molecular level, so that the creep process is terminated (Figure 9.37a).

9.4.4.2 Flow

If a high stress is now applied to the joint, irreversible deformation occurs, which is known as flow. The cause of this behavior is molecular motion that occurs after bonds are broken and only partially replaced by new molecular interactions. This leads to a progressive loosening of the polymer composite in the adhesive layer, and after reaching the deformation limit, the bond breaks (Figure 9.37b).

9.4.4.3 Influence of Creep on the Bonded Joint

For the evaluation of adhesive strengths under permanent load, values obtained after long-term tests should be taken into account in addition to the short-term evaluation of creep. This provides information on the stress state of the corresponding adhesive layer at the limit of linear viscoelastic behavior, which is necessary for the assessment of the strength of a bonded component or the structural design of an adhesive bond. In practice, it has been shown that permanently loaded components have a sufficient service life if this load is less than 50% of the short-term static bond strength. It can therefore be assumed that an adhesive bond that can withstand a short-term static load of 20 MPa, for example, can withstand a long-term static load of 10 MPa. In such

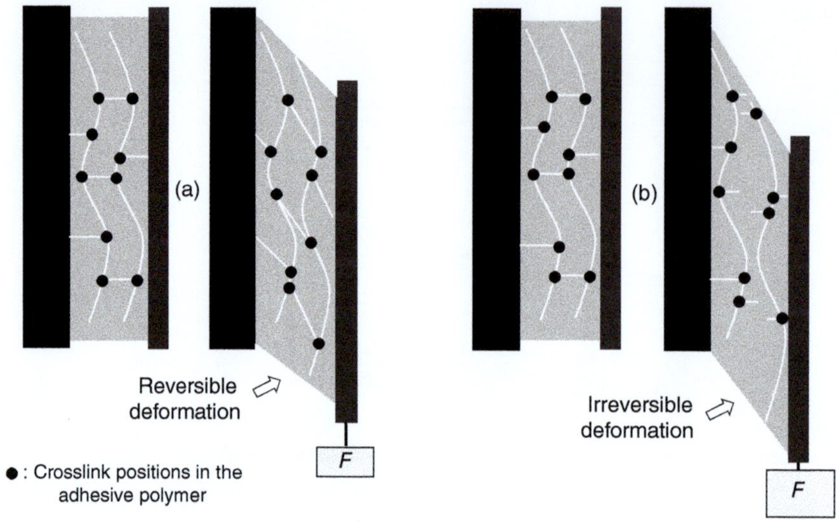

Figure 9.37 Simplified representation of reversible deformation (creep) by applying a low (a) and irreversible deformation (flow) by applying a high shear force to the adhesive layer (b).

a constellation, the strain to break the bond is unlikely to be reached and thus the application can be carried out successfully. However, this is a rough guide based on experience, which should be verified as best as possible in long-term studies.

The creep behavior in adhesive layers is influenced to a greater or lesser extent by various factors. In detail these are:

- temperature,
- level of load,
- properties of the joining part materials,
- chemical character of the adhesive polymers, and
- the degree of crosslinking of the adhesive layer.

9.4.4.4 Phases and Types of Creep Behavior in Adhesive Joints

According to Späth [5], the creep and flow behavior of adhesive joints under constant high load, which ultimately leads to joint failure, can be divided into three phases (Figure 9.38):

- *Phase 1* – Elastic creep in the first phase occurs at the beginning of loading in both crystalline and amorphous adhesive layers and leaves no plastic deformation. It can be explained by the release of molecular interactions and the displacement of chain segments (1–2).
- *Phase 2* – Then creep occurs at a constant rate and equilibrium is established, with chemical bonds being broken and new intermolecular forces being established at the new position (2–3).
- *Phase 3* – Finally, the deformation limit of the adhesive layer is exhausted and, after acceleration of the creep processes in the adhesive layer, flow is observed and subsequently the fracture of the bond is initiated (3–4).

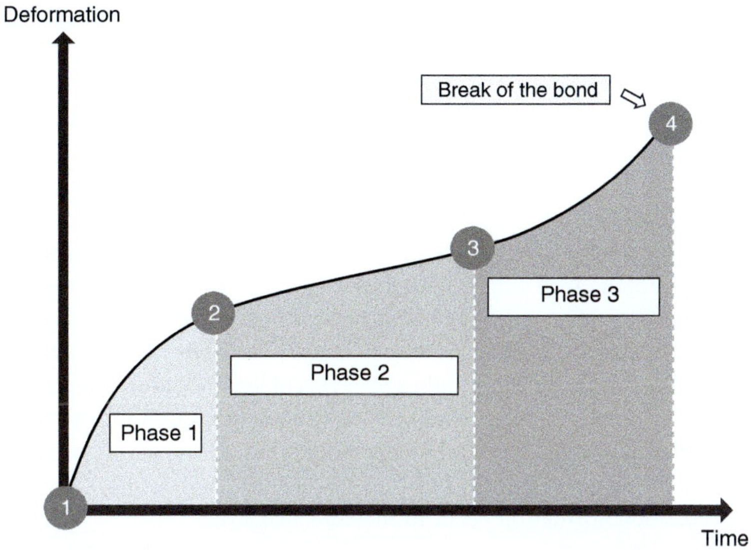

Figure 9.38 The three phases of creep and flow behavior of bonded joints under constant high mechanical loads (resulting in failure) according to Späth.

Based on extensive research, Brockmann [6] proposed to divide the creep and flow behavior of adhesive bonds under constant high loads into four types (Figure 9.39):

- *Type A* – shows steady sliding at high speed due to plastic behavior of the adhesive layer resulting in a rapid failure of the adhesive composite.
- *Type B* – has a gradual creep course due to successive molecular stretches.

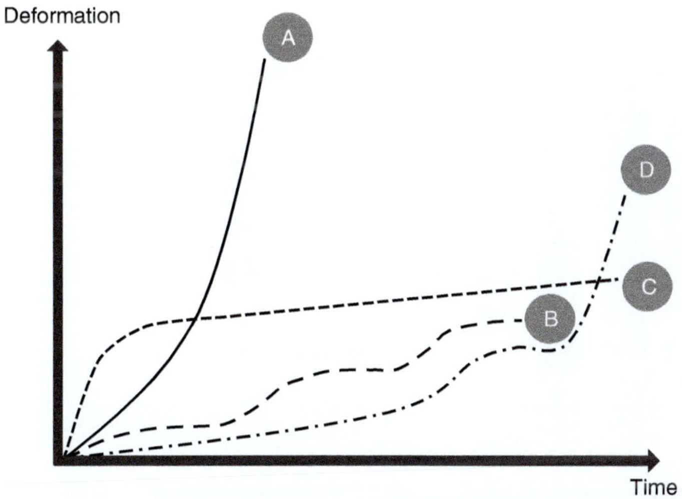

Figure 9.39 The four types of creep and flow behavior of bonded joints under high mechanical loads (resulting in failure) according to Brockmann.

- *Type C* – is characterized by uniform creep, which arises after an initially rapid increase of the creep.
- *Type D* – represents a combination of A and B, whereby after a stepped creep, a rapid sliding of the joining parts and failure of the bonding occurs.

9.4.4.5 Creep Curves

The creep of an adhesive layer is the time- and temperature-dependent deformation after application of a constant load, where both the loading and unloading phases are of interest. Time-dependent creep curves with a loading (Figure 9.40a) and unloading (Figure 9.40b) phase are very useful for assessing the long-term creep behavior of bonded joints and analyzing the elastic, viscoelastic, and plastic contribution to the strain occurring in adhesive layers. Here, adhesive layers of length l_0 are loaded with a constant force F over a period of time and the resulting deformation is observed. Then, the load is relieved and the residual deformation after a long period of time t_∞ is noted.

The following types of deformation occur in the adhesive layer:

- *elastic deformation* – which becomes effective immediately after loading and is completely reversible immediately after unloading,
- *viscoelastic deformation* – which takes effect time-dependently after loading and is completely reversible time-dependently after unloading, and
- *plastic deformation* – which becomes effective time-dependently after loading and is completely irreversible after unloading.

As discussed, time- and temperature-dependent elastic, viscoelastic, and plastic deformations of the adhesive layer take place when a bonded joint is subjected to constant load. In this process, the macromolecules contained in the adhesive layer react spontaneously and basically strive to relieve the forced stresses by

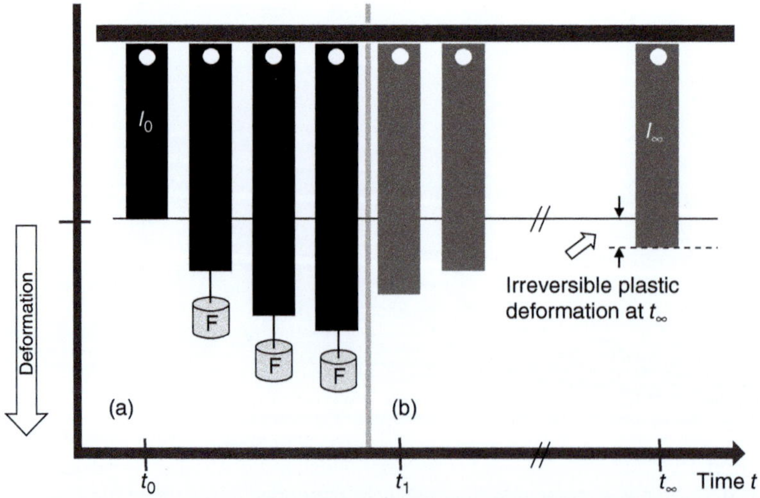

Figure 9.40 Simplified representation of the creep behavior of an adhesive layer under high mechanical loads, showing the loading (a) and the unloading phase (b).

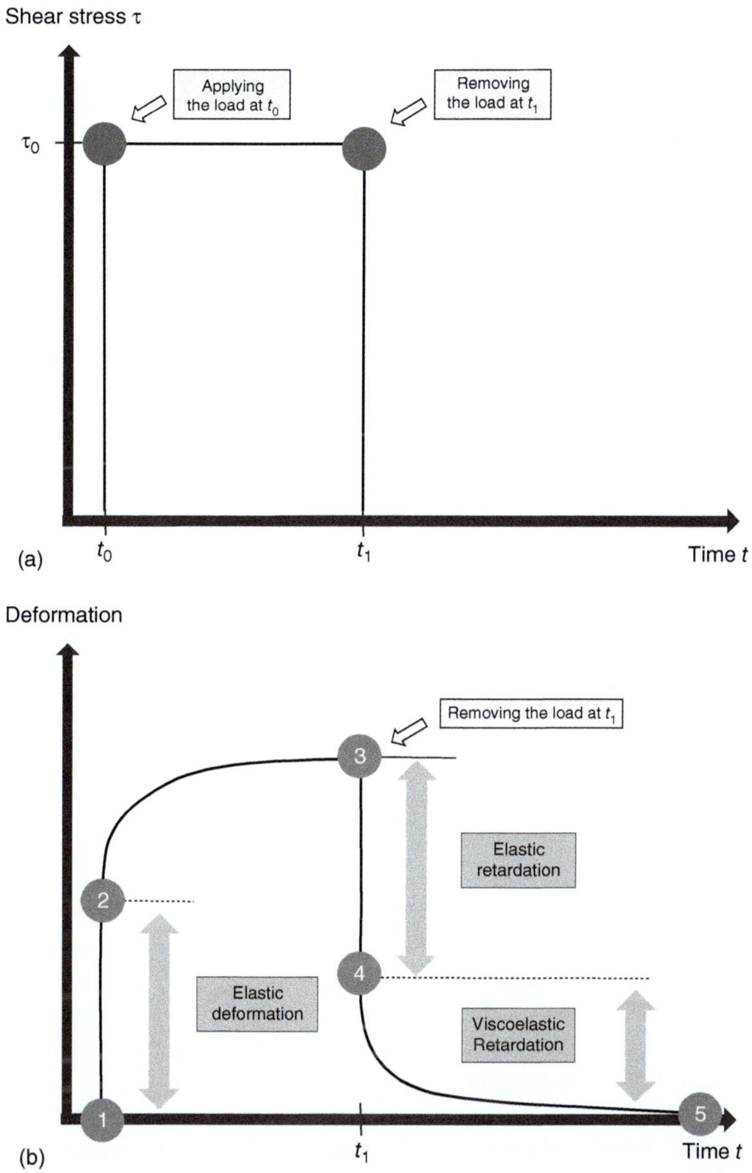

Figure 9.41 Creep diagram of an adhesive without permanent irreversible plastic deformation after long term stress application, represented by the time-dependent course of the applied shear stress (a) and the resulting deformation (b).

rearrangements. If the load is removed, the system tries to return to its original position. If only reversible elastic and viscoelastic deformations were generated, the system returns completely to its initial state, which is shown in Figure 9.41. Here, the load is applied at t_0 and removed at t_1 (Figure 9.41a), and after load removal the deformation of the adhesive layer decreases to zero (Figure 9.41b).

If irreversible viscous movements occur in addition to reversible elastic and viscoelastic deformations, a residual displacement is observed after the load is

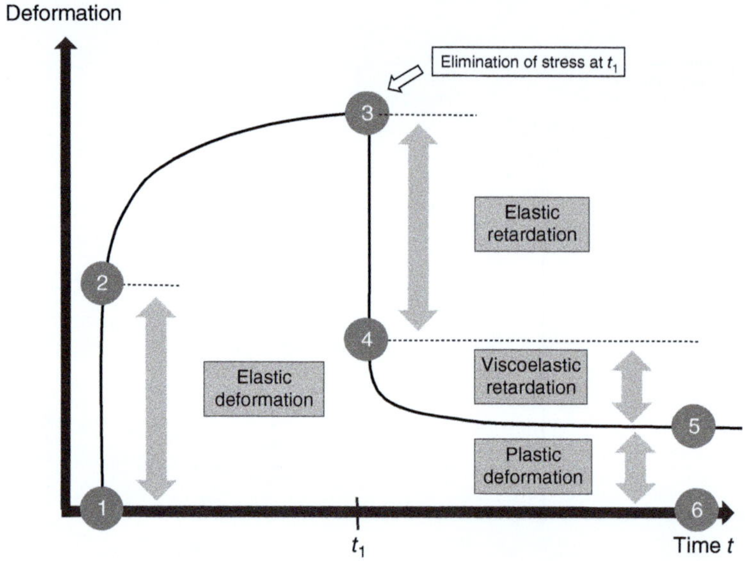

Figure 9.42 Creep diagram of an adhesive with permanent irreversible plastic deformation after long term stress application, represented by the time-dependent course of deformation.

removed. This is demonstrated in Figure 9.42, where the load is applied and released as shown in Figure 9.41a, but a residual deformation of the adhesive layer is observed, caused by irreversible plastic deformation in the adhesive layer.

9.4.4.6 Stress Relaxation Curve

The stress relaxation of an adhesive layer, on the other hand, is the time- and temperature-dependent stress relief after a constant deformation has been applied. It is an important characteristic of an adhesive layer that indicates its ability to relieve internal and external stresses that occur during processing and in practical use. These include local stresses at the overlap edges of the bonded joint caused by external loads, especially when the temperature changes, and stresses that occur during processing, such as during curing. If such a deformed layer is now held for a time, a spontaneous stress increase can first be observed. Then, depending on the holding time and the temperature, a time-dependent stress reduction occurs. This degradation is called stress relaxation and is very well pronounced in adhesives due to their viscoelastic property. This reduction of residual stresses over time takes place through relaxation processes in the adhesive layer and, in extreme cases, leads to a complete reduction of the stress peaks at the overlapping ends of the bonded component.

The stress relaxation behavior of an adhesive layer is shown in Figure 9.43. Here, the deformation is applied at t_0 and kept constant for a long time (Figure 9.43a). The response of the adhesive layer is shown in Figure 9.43b, where the stress does not decrease to zero after a long period of time, but remains at a certain value.

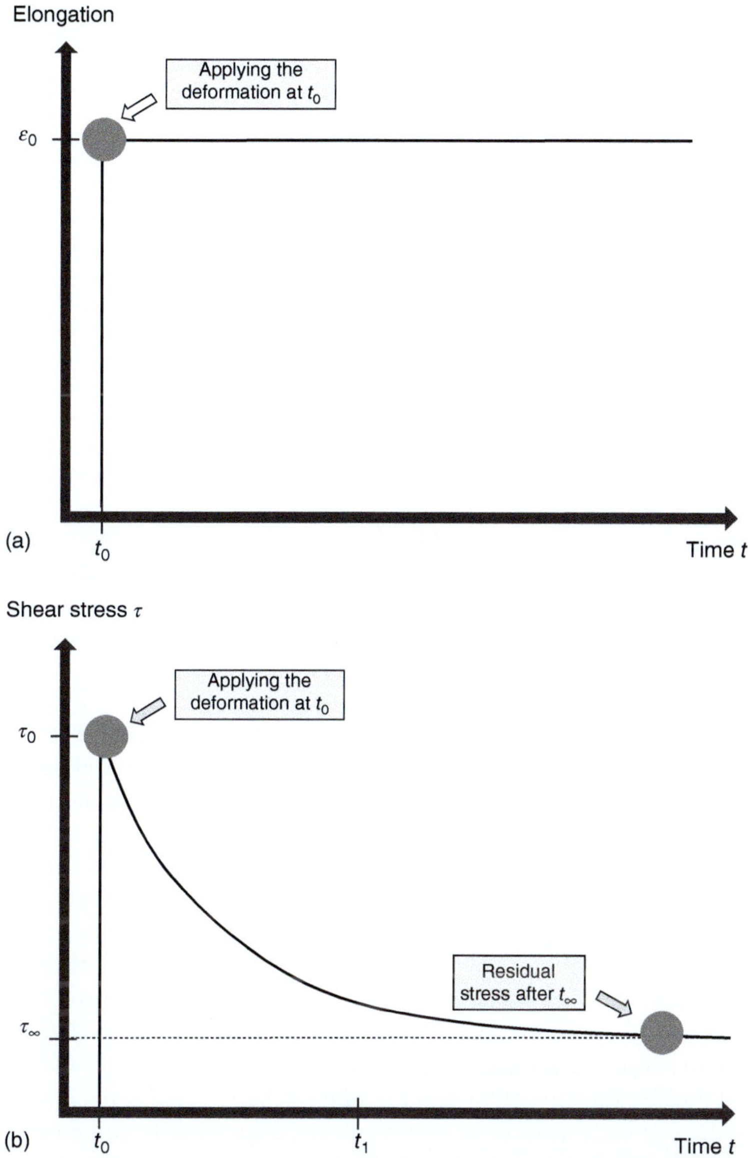

Figure 9.43 Stress relaxation diagram of an adhesive with residual stress after longterm deformation, represented by the time-dependent course of the applied strain (a) and the resulting stress (b).

9.5 Designing Adhesive Bonds

9.5.1 Introduction

The successful production of a high-quality bonded joint is achieved by a systemic joining process based on a large number of interlocking process steps that must

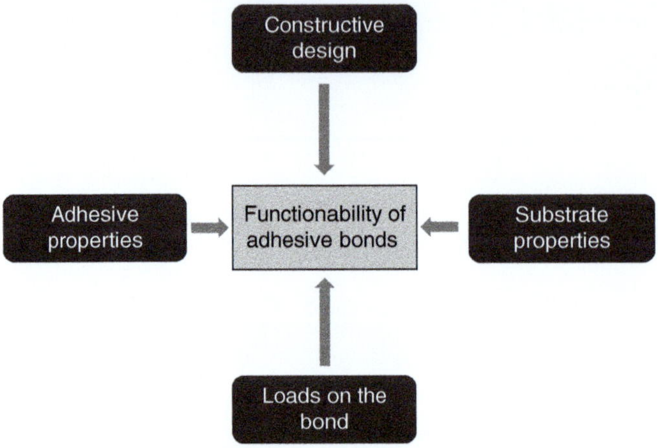

Figure 9.44 Factors influencing the functionality of an adhesive bond.

be optimally designed and coordinated. For the functionality of a bonded joint, the fundamental influencing variables are the properties of both the adhesive layer and the joining part material, as well as the load on the component and the constructive design of the bonded joint (Figure 9.44). Since errors in the design of the bonded joint are very often the cause of its failure, great attention must be paid to this influencing variable even in the early planning phase.

When designing adhesive joints, it is important to remember that adhesive bonding is one of several possible joining techniques, each with its own rules and peculiarities. However, if these are taken into account for the adhesive-bonding method, composites of the highest quality and often unique properties can be created compared to the alternative joining methods. It is often necessary to make compromises in the design of components, but these should never be at the expense of safety. For example, the design of a bonded joint should not only meet the requirements of the product designer, who wants an invisible joint, but also take into account the technical requirements for a suitable bonding area. Since the type of construction of a bonded component usually has a very large influence on the cost of a bonded joint, the aim is to design simple components with few construction elements in such a way that the best possible function of the bonded joint can be achieved at the lowest possible cost. In this context, measures for easy repair and disposal of the component after use must already be taken into account in the design.

The main task of a functioning bonded joint is the transmission of forces and the permanent absorption of all stresses occurring in the bonded joint.

Therefore, two important requirements must be met when designing adhesive bonds:

- *provision of sufficient bonding area* – to transmit the forces in the application, and
- *avoiding stress peaks in the bond* – caused by loads occurring in the application [7].

When designing and dimensioning bonded joints, it must therefore be ensured that the design can withstand the entire load spectrum occurring over the intended service life. To this end, the variables influencing the strength of the bonded joint are broken down into subunits, each of which follows this concept and is expected to function as a reliable system when interacting. It is very important for a reliable and efficient dimensioning of structural bonded joints to strongly consider the time-, deformation-, and temperature-dependent material behavior of the adhesive polymers. The required strength of an adhesive bond can be determined empirically on the basis of tests and empirical values or analytically by calculating the true stresses in the materials using the mechanical material constants [8]. For this purpose, finite element analysis methods and tools can be used to reliably predict the expected stress and deformation distributions in the bonded joint, provided that the relevant material parameters are available.

9.5.2 Provision of Sufficient Bonding Area

9.5.2.1 Introduction

When bonding metals, the substrates have about ten times the strength of the adhesive layers. If a butt joint were now provided, joining the transverse surfaces of the substrates and the adhesive, only a relatively small load could be transferred, which would only utilize the strength of the substrate to about 10%. With a given adhesive, a higher load-bearing capacity of the bonded joint would only be possible by increasing the joining area, but this cannot be achieved in a butt joint without the uneconomical increase in substrate thickness. Therefore, component geometries should be selected in which the stress of the adhesive layer is largely shear-oriented and where the load application takes place over a large joining area. Such designs allow the load-carrying capacity to be increased and take into account the different deformation and strength properties of the substrates and the adhesive layer. For this reason, simple lap joints or lapped geometries are very often used. Here, in contrast to the butt joint, the user has the option of freely selecting the size of the bonding area and adapting it within certain limits to the strength of the substrates, where their yield strengths $R_{p0.2}$ play an important role.

The yield strength $R_{p0.2}$ is a material property and denotes the stress value up to which the material shows no permanent plastic deformation under uniaxial and free tensile stress. If the yield strength is undershot, the material returns elastically to its original shape. If the yield point is exceeded, there remains a change in shape, e.g. an elongation in a tensile test. For materials with continuous yielding, the yield point cannot be clearly identified from the diagram. For this reason, in such cases a yield strength $R_{p0.2}$ is often given instead of the yield strength R_e, which can be read off unambiguously in each case. Here, $R_{p0.2}$ is the stress that leads to a small permanent deformation of 0.2% after unloading (Figure 9.45).

This advantage of choosing simple lap joints instead of butt joints becomes apparent when comparing the load-carrying capacity and substrate strength utilization of both geometries.

9.5.2.2 Butt Joints

If in a butt joint with for example aluminum-joining parts made of $AlCuMg_2$ – with the width $b = 25\,\text{mm}$ and the thickness $s = 5\,\text{mm}$ – an adhesive is used that has a tensile strength of $\sigma_{\text{Bond}} = 20\,\text{N/mm}^2$, the breaking force is: $F_{\text{Bond}} = \sigma_{\text{Bond}} \cdot b \cdot s = 20\,\text{N/mm}^2 \cdot 25\,\text{mm} \cdot 5\,\text{mm} = 2500\,\text{N}$, which is the maximum load capacity for

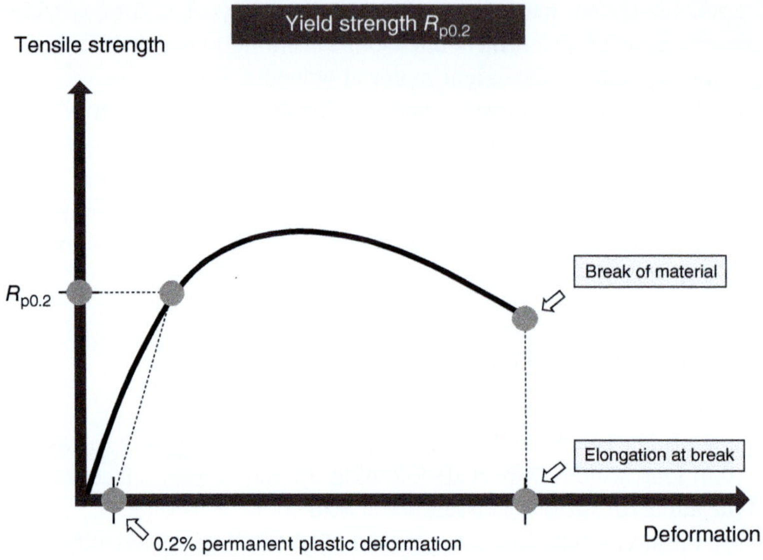

Figure 9.45 Position of the yield strength at 0.2% permanent deformation ($R_{p0.2}$) in the stress–strain curve.

the butt joint with these dimensions. The bonded surface area is 125 mm^2 and cannot be increased without using thicker substrate material, which would not economical. When the bond breaks, both the aluminum substrates and the adhesive layer are subjected to the same stress σ of 20 MPa, with only slight deformation of the aluminum sheets, which have a yield strength $R_{p0.2}$ of about 280 MPa. Thus, when the adhesive breaks, 100% of its tensile strength capacity is used up, with only about 7% of the tensile strength capacity of the aluminum substrate being reached at this point. Higher loads would be possible by increasing the substrate thickness, but at the cost of higher material usage, with the substrate strength capacity of about 7% still very poorly utilized. Thus, only forces which are limited by the tensile strength of the adhesive layer and the adhesive forces acting on the substrate surfaces can be transmitted in the butt joint geometry. This represents a major disadvantage of the butt joint–bonding geometry.

9.5.2.3 Single Lap Joints

When applying a single-lap geometry, the situation is very different. Using the same substrates and the same adhesive as in the first example and an overlap width of 25 mm and an overlap length of 12 mm, the breaking force in shear mode $F_{\text{Bond}} = \tau_{\text{Bond}} \cdot b \cdot l_{\ddot{u}} = 20\,\text{N/mm}^2 \cdot 25\,\text{mm} \cdot 12\,\text{mm} = 6000\,\text{N}$, which is the maximum load capacity for the single lap joint with these dimensions (Figure 9.46). The bonding area here is 300 mm^2 and can be further increased within limits, which would lead to an increase in load-bearing capacity. The tensile stress σ in the joint parts is $6000\,\text{N}/125\,\text{mm} = 48\,\text{N/mm}^2$. Compared with the butt joint geometry, the single-layer joining arrangement in this example increases the strength utilization of the aluminum substrate to about 17% (compared to 7% in the butt joint).

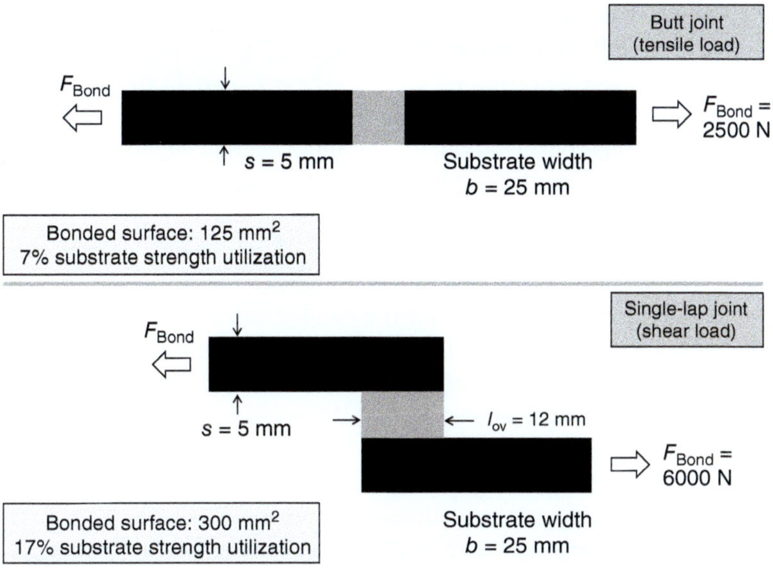

Figure 9.46 Increase the load-bearing capacity of a joint by using single-layer joints instead of butt joints.

9.5.2.3.1 Optimization of Single Lap Joints

In the case of relatively low-loaded single lap joints, the stresses occurring are well compensated by elastic and plastic adhesive layer deformations in the relatively short overlaps. Due to the short overlap length and thus small bonding area, the transmissible load in the event of fracture is so low that the stresses occurring in the joined part are still below the yield strength of the substrate. Thus, in this case, essentially only the shear stresses caused by a part displacement occur, which are supplemented by relatively small stress peaks from the bending moments. For optimization, further utilization of the strength can be achieved in single-lap joints by increasing the overlap length and thus the bonding area. The optimum is reached when the forces acting on the bonded joint cause deformations that are just within the elastic range of the substrate material. Plastic deformation does not yet take place because the material has not yet reached its yield point. Here, the stress peaks at the overlap edges of the adhesive layer reach a value characteristic of elastic deformation, and since there is a balance between the load on the substrate and the strength of the adhesive, the optimum material utilization for the bond is achieved.

The optimal overlap length can be determined for metal bonding by equating the forces present in this equilibrium state:

- Occurring force on the substrate (F_S): $F_S = R_{p0.2} \cdot b \cdot s$
- Occurring force on the adhesive layer (F_{Bond}): $F_{Bond} = \tau_{Bond} \cdot b \cdot l_{ov}$
- Equating the forces: $F_S = F_{Bond}$
  $R_{p0.2} \cdot b \cdot s = \tau_B \cdot b \cdot l_{ov}$
  $R_{p0.2}/\tau_{Bond} = l_{ov}/s$
- This gives in the optimum overlap length: $l_{ov} = R_{p0.2} \cdot s/\tau_{Bond}$

For bonding 1.5 mm thick aluminum sheet ($AlCuMg_2$, $R_{p0.2} = 280$ MPa) with a 2K epoxy adhesive of the adhesive strength of $\tau_{Bond} = 28$ MPa, the optimum overlap length of $l_{ov} = R_{p0.2} \cdot s/\tau_{Bond} = 280\,\text{MPa} \cdot 1.5\,\text{mm}/28\,\text{MPa} = \mathit{15\,mm}$. In the optimum, the ratio $R_{p0.2}/\tau_{Bond}$ is equal to the overlap ratio $r_{ov} = l_{ov}/s$. In practice, it has been shown that according to this consideration, the optimum overlap lengths are 10–20 times the thickness for metal bonding and 2–5 times the thickness of the substrate for plastics.

If the component has to withstand higher loads, one could assume that larger overlap lengths (with constant joint width) would allow higher forces to be transmitted to meet the higher strength requirements for the bonded structure. However, this is not possible due to the special circumstances of an overlap joint with increasing mechanical stress. Above the yield point of the substrates, increasing overlap lengths result in plastic deformation of the bonded parts at the overlap ends. Thus, the load can no longer be absorbed leading to fracture of the adhesive layer starting from the overlap ends. In general, increasing the mechanical load beyond a certain limit increasingly forms stress peaks and thus reduces the transmissible load.

This behavior is demonstrated in Figure 9.47 where the adhesive strength τ_{Bond} initially remains constant depending on the overlap length, but drops sharply after the elastic limit of the substrates is exceeded, ultimately leading to fracture.

9.5.2.3.2 Adhesive Bond Utilization Factor δ

In terms of economic material use, the adhesive bond utilization factor δ is of considerable interest to the user. This is the quotient of the tensile stress σ_{pres} present

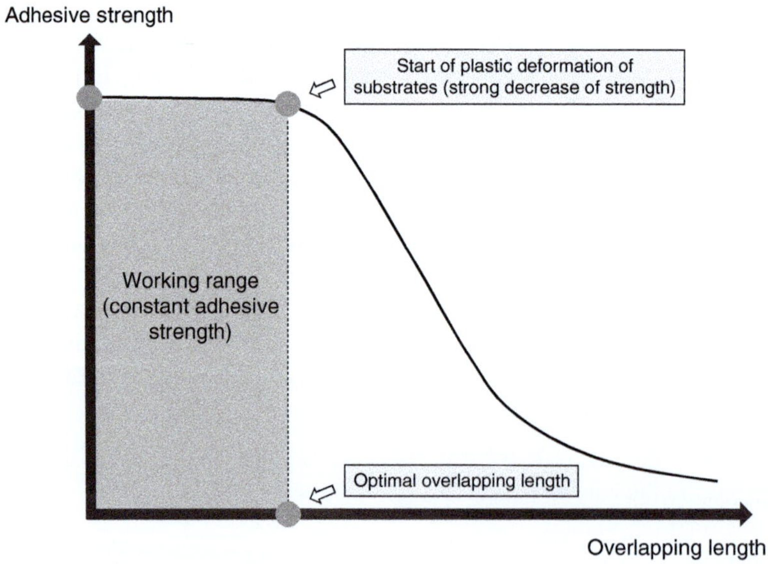

Figure 9.47 Dependence of adhesive strength on the overlapping length in the critical deformation range of the substrate.

in the bonded part at the break of the bond and the substrate strength, which in the case of metal bonds is represented by the yield strength $R_{p0.2}$:

$$\delta = \sigma_{pres}/R_{p0.2}$$

This value is chosen as the basis for calculation in order to keep the loads occurring in practice within the limits of elastic joining part deformation. It supports the user in avoiding plastic deformation of the substrates, which would lead to visible deformation of the joint and subsequently to its fracture. Also, with regard to the economical production of metal joints, it is important to know to what extent the strength of the joining part materials is actually utilized by the applied loads.

For this purpose, the bond strength utilization factor δ can be used as follows:

- *δ is far below 1* – the substrate is in the elastic deformation range of the material, but the utilization of the material is poor,
- *δ approaches 1* – the deformation of the substrate is still exclusively elastic and the utilization is already good,
- *δ = 1* – the optimum utilization of the material is achieved, and
- *δ is above 1* – the joined part is already in the plastic deformation range of the material and thus the quality of the joint is worse, provided with undesirable stress concentrations.

9.5.2.3.3 Influence of the Overlap Length

The influence of the overlap length on the failure load F is shown in Figure 9.48. Here, the values for the failure load increase as the overlap length increases, but the slope decreases. This leads to decreasing adhesive strength τ_{Bond}, which is represented by the dashed black lines in the illustration. For small overlap lengths generally up to approximately 5 mm, the failure load F – the maximum force that can be applied – is determined by a largely homogeneous stress distribution in the adhesive joint, whereby the values are not yet influenced by deformations of the substrates. Although the failure load still increases, the curve flattens out with increasing overlap length, which represents increasing stress concentrations at the overlap ends and a reduction of the adhesive strength, which is the average stress occurring in the bond.

Above a certain overlap length, a significant drop in failure load is then observed. The cause of this drop is stress peaks whose magnitude locally exceeds the adhesive strength with increasing overlap. As a result, the adhesive layer breaks from both sides toward the center, which then leads to failure of the component due to the reduced bonding area. When bonding metals, the yield strength $R_{p0.2}$ of the substrates is regarded as the upper limit of the load and, as described above, the optimum of the overlap length l_{ov} is calculated by:

$$l_{ov} = R_{p0.2} \cdot s/\tau_{Bond}$$

9.5.2.3.4 Influence of the Substrate Thickness

A larger thickness of the substrate has a positive effect on the strength of the bonded joint, since it positively influences its stiffness and bending moment. Thus, when the

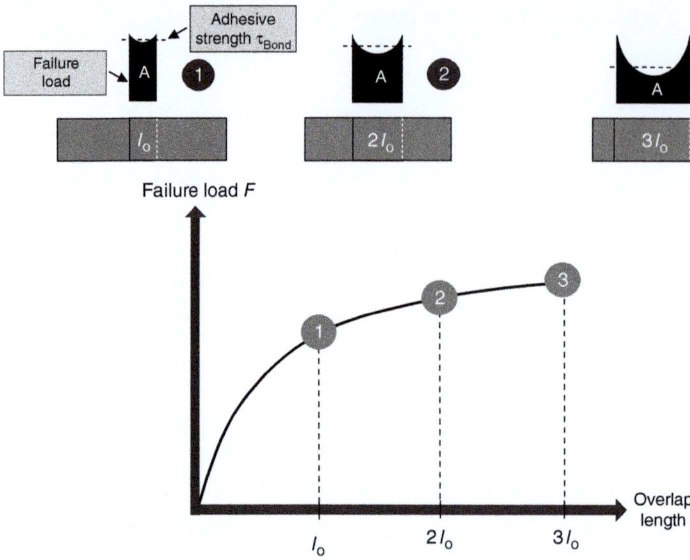

Figure 9.48 Influence of the overlap length on the failure load (F) represented by the areas A and the development of adhesive strength τ_{Bond}, represented by the black dashed lines.

bonded joint is loaded, less elongation occurs with thicker bonded parts than with thinner ones, which means that the undesirable stress peaks in the adhesive layer are reduced with thicker bonded parts, and thus larger portions of the adhesive layer are used for load transfer. The overlap ratio $r_{ov} = l_{ov}/s$ takes into account the dependence of the adhesive strength on both the overlap length and the thickness of the bonded part. In the optimum, the ratio $R_{p0.2}/\tau_{Bond}$ is equal to the overlap ratio $r_{ov} = l_{ü}/s$ and for metal joints, the economically justifiable overlap ratios are in the range of 10–20. Of particular importance is the dependence of the adhesive strength on the overlap ratio with simultaneous knowledge of the degree of adhesive bond utilization.

9.5.2.3.5 *Influence of the Substrate Width*

The load F that causes a bonded joint to break increases approximately proportionally with the overlap width (Figure 9.49), where the fluctuation range of the strength values is greater for smaller than for larger widths. The stress distribution in the bonded joint parallel to the overlap width is constant, which is a major advantage of the adhesive-bonding process over other joining methods such as riveting and screwing.

9.5.2.3.6 *Influence of Bonding Area*

In general, lower adhesive strengths are assumed on average for large bonding areas than for small ones. Since a defined area can vary in length and width, the influence of the bonding area on the adhesive strength is not necessarily proportional. Accordingly, a single-layer joint with an adhesive area of 100 mm^2 can, under

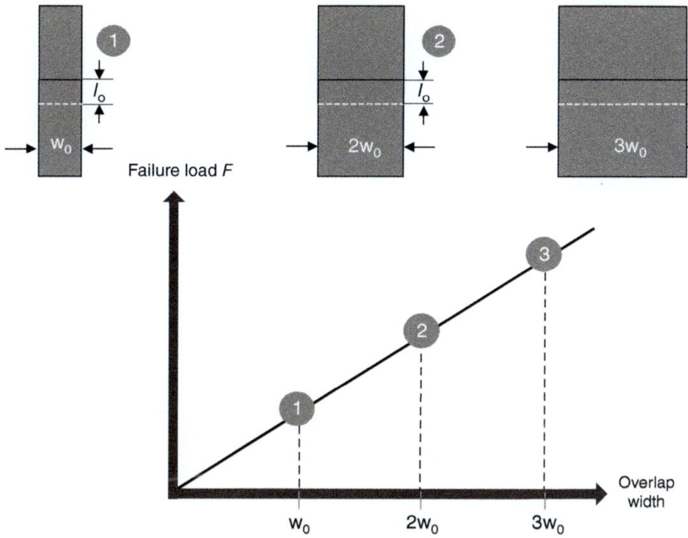

Figure 9.49 Influence of overlap width on the failure load F.

otherwise identical conditions, transfer higher loads with a length of 5 mm and a width of 20 mm than with a length of 20 mm and a width of 5 mm. The reason for this is the already-discussed increased formation of stress peaks at longer overlap lengths.

There are other points that can lead to a reduction in the adhesive force and must be taken into account when increasing the bonding area:

- the uniformity of the adhesive layer may no longer be guaranteed with a larger bonding area,
- the possibility of geometric variations of the substrates is more likely with larger areas, and
- different temperature distributions and uneven contact pressure over larger areas may affect the uniformity of cure.

Therefore, to ensure optimum strength of the bonded part with large bonding areas, special precautions are necessary to ensure exact adherence to the manufacturing parameters, especially with fast-curing adhesives.

9.5.2.3.7 Influence of the Adhesive Thickness

In practice, adhesive layer thicknesses in the range of 50–200 μm have proven to be favorable for surface roughness ranges of approximately 30–70 μm. If the adhesive force is plotted against the adhesive thickness, a relatively steep increase in the curve can initially be observed. The maximum adhesive force is formed from 50 μm to about 200 μm, which is confirmed by a wide range of empirical values and studies. From 200 μm, the bond strength decreases due to increased residual stresses in the bond up to a final thickness of approximately 500 μm. After that, it remains constant

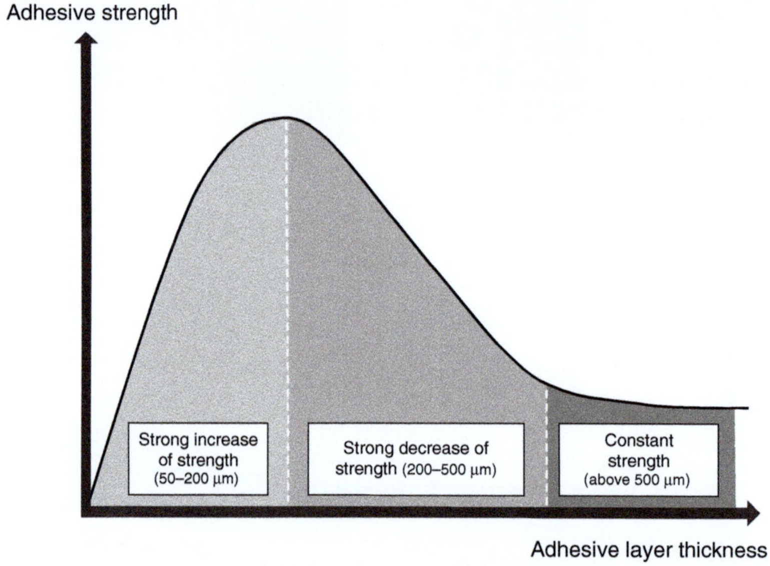

Figure 9.50 Influence of adhesive layer thickness on the adhesive strength τ_{Bond}.

and is now determined by the significantly lower strength of the adhesive layer compared to the substrates (Figure 9.50).

9.5.3 Avoiding Stress Peaks

When a load is applied, the strength of bonded single lap joints is limited by the stress peaks that form at the overlapping edges. These must be minimized or avoided in the interest of optimum design of the component.

Adhesives are essentially subject to the following types of mechanical load:

- shear,
- tensile,
- compressive,
- peel,
- cleavage,
- bending, and
- torsion.

To be able to respond to an increase in mechanical load, adhesive joints must be designed for shear loads. In other words, during application, the bonded parts are loaded parallel to the surface so that the entire bonded surface is subjected to stress. As a result, bonded joints can usually be dimensioned within limits by enlarging the bonding surface accordingly so that any forces that occur are distributed over the entire surface and punctual stress peaks are thus largely avoided or minimized. Tensile forces acting perpendicular to the bonding plane should only be permitted in exceptional cases, since the increase in mechanical load-bearing capacity by enlarging the bonding surface – as used in shear arrangements – cannot be realized by this geometry. Compressive loads largely have no influence on the quality of the bonded joint and torsional loads are treated like shear loads.

In the interests of optimum utilization of the bonding area and avoidance of stress peaks, the following types of load on the component must be avoided:

- peel,
- cleavage,
- bending, and
- plastic deformations of the substrates.

9.5.3.1 Peel Loads

Peel loads result in near-linear, very high stress effects on the bonded joint, often exceeding the ultimate strength of most adhesives. This is because during peel operation, only a fraction of the total adhesive area is used for load transfer. Therefore, if peel loads are to be expected, their negative influence on the quality of the bond must be avoided by appropriate design measures. Additional force- or form-fit joint types are available for this purpose and peeling forces can also be avoided by stiffening the joint parts at the overlapping ends. Figure 9.51 shows the stress distribution in a bond at fracture during peel and shear loading. In this illustration, the force F_{Bond} required for fracture is represented by the volume and determined by:

- $F_{\text{Bond}} = \delta \cdot l_0 \cdot b_0$ in the case of peel loading (a), and
- $F_{\text{Bond}} = \tau \cdot l_0 \cdot b_0$ in the case of shear loading (b).

Figure 9.51 nicely illustrates why fracture occurs at a higher force application under shear loading than under peel loading, and why peel loading should be avoided at all costs when designing a bond.

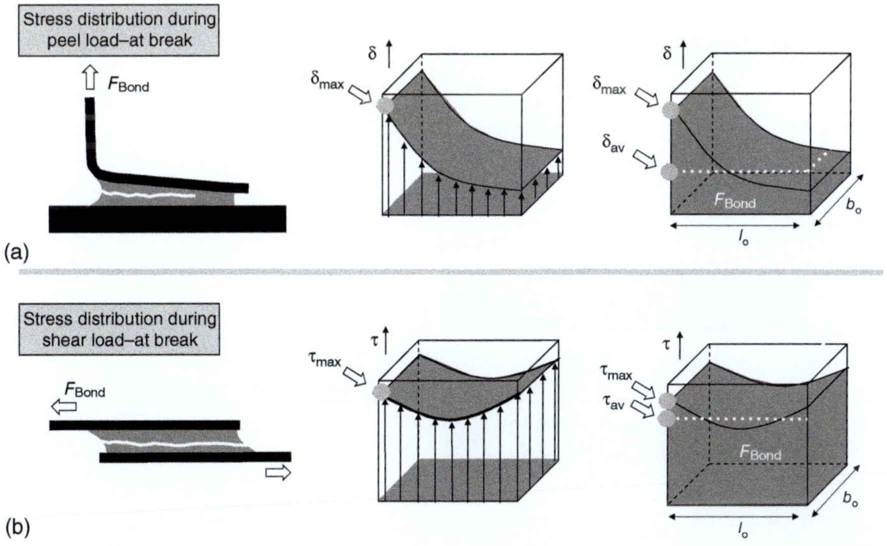

Figure 9.51 Stress distribution in the adhesive bond during peel (a) and shear loading (b).

The reason for this is the occurrence of significant local stress concentrations during peeling in a highly confined line-shaped zone, represented by δ_{max}. At this point, fracture sets in and the remaining composite area is only involved in force transmission to a small extent. Thus, it is not able to support the composite, which is expressed in low values for the fracture load F_{Bond} and indicated by the relatively small volume in Figure 9.51a. This is different when shear forces act on the composite. The stress is much more evenly distributed during loading, and thus a much higher force is required to destroy the bond, indicated by a much larger volume in Figure 9.51b compared to the peel test.

9.5.3.2 Bending Loads

In the case of bonded single-lap joints, undesirable bending moments occur due to eccentric force introduction into the component, which lead to normal stresses perpendicular to the adhesive layer and increase the stress load of the bond in this area. The influence of the bending moment decreases with increasing overlap length and can be minimized by reinforcing the substrate in the force-introducing area using bonding geometries in which force is introduced centrically (Figure 9.52).

Particularly in the case of thin substrates, care must be taken to ensure that the bonded joint is designed to be resistant to bending, which can be achieved in the case of overlap joints by using reinforcing adhesives at the beginning and end of the overlap. In the case of bonded joints subjected to tensile stress, the occurrence of such stress peaks can only be prevented by using an appropriate cardan joint.

9.5.3.3 Cleavage Loads

When joining parts with particularly high stiffness, cleavage loads must be avoided (Figure 9.53). If this type of load occurs in the application, it results in a very uneven

Figure 9.52 Avoidance of bending moments by a two-shear connection, allowing centric shear loads.

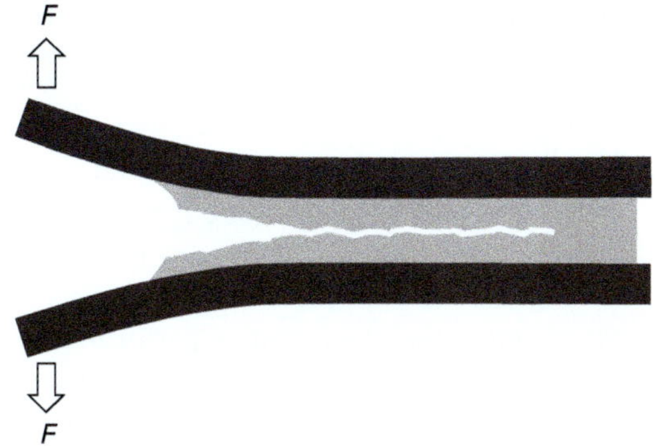

Figure 9.53 Cleavage load acting on an adhesive bond.

load distribution with high stress peaks on the adhesive layer. If this rather rare form of load occurs, the application of additional mechanical fastening elements such as rivets or screws helps to avoid negative influences on the bonding quality.

9.5.3.4 Plastic Deformation of the Substrates

Plastic deformation of the substrates was discussed earlier in connection with the limits of adhesive length extension to accommodate higher loads. Here we found that deformations of the bonded parts beyond the elastic range have a negative effect on the quality of the bond. Thus, these deformations must be taken into account in the design of the bonded part, and to avoid them, the optimum overlap length should be selected in the design of the part. This can be calculated by considering the bonded component and adhesive strengths, and the bond strength utilization factor $\delta = \sigma_{pres}/R_{p0.2}$ is of great help in this respect. Values well below 1 should be avoided because of the poor economics of such a design. Values close to 1 are very suitable because they represent both good economy and ensure that the substrates are in the elastic deformation range. Values for δ above 1 must be avoided because of the negative influence on the bond quality.

9.5.4 Dimensioning of Bonded Joints

Due to the manifold influencing factors, some of which are difficult to determine, an exact calculation of bonded joints with respect to their load-bearing capacity is only possible to a limited extent. As an alternative to finite element analysis (FEA), however, it is possible to dimension bonded joints for simply designed joints according to the usual design principles using corresponding table values for the various designs. It should be noted, however, that the loading in the application is composed of both the applied mechanical loads and the thermally induced residual stresses in the bond. These depend on the adhesives, substrates, and temperature range in which the component is used. Particularly in the case of structural (high-strength) bonding, where thin substrates and rigid adhesives are generally used, uneven stress distribution in the adhesive joint must be expected. In contrast, the occurrence of stress peaks in elastic bonding is avoided by using elastic adhesives, high-performance foam tapes and setting relatively thick adhesive layers.

To characterize the mechanical behavior of an adhesive or adhesive bond, the following test methods are usually used:

- *Tensile test* – this test is performed with a cured adhesive specimen (or cured adhesive bond) stretched axially to break and is used to determine the mechanical properties of an adhesive layer (or adhesive bond), such as tensile strength σ = force/cross-section, elasticity (change in length/original length) and Young's modulus (strength/deformation).
- *Shear test* – the shear test is performed with a cured adhesive joint sheared axially to failure and is used to determine shear strength – also called adhesive strength – (τ = force/bonding area), shear strain (tan γ), and shear modulus (G = shear strength/tan γ).

Tensile and shear strengths of adhesives or bonded joints are the basis for adhesive joint calculations. They can be easily determined with a tensile tester in the laboratory and should be listed in the data sheets of each adhesive, indicating the method used, the test temperature, and the test speed. The shear modulus is a characteristic value for each adhesive that indicates, among other things, its suitability for high-strength or elastic bonding.

When determining the dimension of the bond, the following influencing factors are of decisive importance:

- acting mechanical loads,
- residual stress of the bonded joint,
- strength and conformability of the adhesive,
- reduction factors for the various loads occurring in the application, and
- safety factors.

These parameters must be determined in the laboratory or taken from technical information or data sheets and then fed into a calculation.

A practical approach has become established for estimating or roughly calculating the dimensions of a bonded component. Here, the dimensions of the bonded joint are determined on the basis of the shear strength (adhesive strength) τ_{Bond}, which is measured in a shear test as described above:

$$\tau_{\mathrm{Bond}} = F_{\mathrm{Bond}}/l_{\mathrm{ov}} \cdot b$$

with the force F_{Bond} required to break the bonded joint, the overlap length l_{ov}, and the overlap width b.

However, the bond strength, which is easy to determine in the shear test, has the disadvantage that it does not take into account the local stress peaks in the most heavily stressed area of the bonded joint, which are located at the end of the overlap. These are usually responsible for the failure of an adhesive bond and not the shear stress τ_{Bond} determined in the shear test, representing the average of the individual stresses occurring in the adhesive bond. Thus, at these positions of stress concentration, both the shear stress and the normal stress have a maximum due to the different deformation behavior of the bonded part and the adhesive layer.

The bond strength τ_{Bond} is therefore only a *technological* value relevant for practical use, which – provided with reduction and safety factors – can be used for dimensioning bonded joints. However, these estimates/calculations based on adhesive manufacturers' technical data sheets and users' specifications must be verified by practical tests in the laboratory. Later, verification is carried out on a larger scale, taking into account the conditions occurring in the application, such as temperature, humidity, alternating climate, and mechanical loads. Since the influencing variables and thus the overall constellation usually change regularly on the way to developing an optimum bonded joint, the final definition of the dimensions usually follows an iterative process with many changes and fine-tuning of the parameters, corresponding trials in the laboratory and test runs on a larger or pilot scale. According to DIN 2304, an adhesive bond must be designed so that the load is less than the load-bearing capacity. This means that for an adhesive bond, the shear stress occurring in the application must not exceed the permissible shear stress (load-bearing capacity) τ_{Bond}, calculated from the adhesive test data and taking into account appropriate reduction factors and safety factors.

9.5.4.1 Reduction Factors

The reduction factors that adapt the results obtained in the laboratory to reality, taking into account the conditions of practical use of the component, consider, for example

- long-term static loads,
- dynamic loads,
- thermal loads,
- moisture loads,
- UV radiation loads,
- climate-changing loads,
- critical geometries (e.g. occurrence of bending moments), and
- variations in adhesive quality.

The reduction factors can be obtained from the adhesive manufacturers for the corresponding adhesives and indicate the number (<1) by which an adhesive strength determined under normal conditions must be multiplied to arrive at the value for the actual condition. In many cases, the bonded joint is not subjected to all the stresses listed here in practice, so that not all reduction factors have to be taken into account.

In Figure 9.54, for example the application of the reduction factors for heat and moisture loads to determine the load-bearing capacity is demonstrated. The initial adhesive strength τ_{Bond} (4), measured in the laboratory under normal conditions, is mathematically reduced by the factor for heat to obtain (3), and by the factors for both heat and moisture to obtain (2). Considering the loads occurring in the practical application (1), the selected adhesive in this example allows application under heat and moisture, since this load in practical application (1) is significantly smaller than the calculated load capacity of the bond (2). However, to prove feasibility, this calculation must be verified by comprehensive laboratory tests. Also, a safety factor

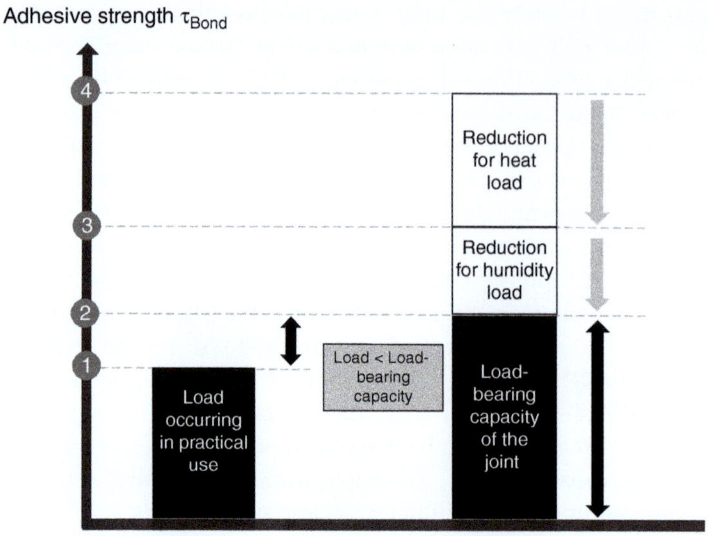

Figure 9.54 Load-bearing capacity of an adhesive bond after reduction of its initial adhesive force τ_{Bond} due to heat and moisture load.

might be used for the final determination of the load-bearing capacity of the joint, which was not done in the example.

Often the reduction factors are not determined individually, but one factor is used that represents all conditions. This puts the user on the safe side, but this approach often leads to *over-dimensioning* and thus to increasing costs. For example, in many applications UV or extreme temperature/humidity loads often do not play a role, e.g. in indoor applications. Therefore, it makes sense to carefully discuss and evaluate all loads occurring in the application and to critically scrutinize them to avoid *oversizing*.

As a guide, here is the rough direction for the reduction factors for the corresponding loads:

- *static permanent loads:* 0.03–0.06 (for elastic joints),
- *dynamic loads:* about 0.10–0.30,
- *thermal loads in the range 80–100 °C:* thermosetting and elastic adhesive layers: approximately 0.50, thermoplastic adhesives: approximately 0.20–0.40,
- *moisture loads:* approximately 0.20–0.80 (if no bond-line corrosion occurs), and
- *variation in adhesive quality:* 0.50 (from experience in aerospace).

(Note: The ranges given here for the reduction factors are for rough guidance only; exact values must be obtained from the adhesive manufacturers or determined and verified by laboratory tests).

9.5.4.2 Dimensioning of Elastic Adhesive Bonds

As already described, increasing the overlap length to transfer larger loads is not possible for metal bonds with high-strength adhesives, if the substrates are above the elastic limit. Due to their nature, elastic bonds behave significantly differently. Thus, in contrast to structural bonding, the stiffness of the substrates has no decisive influence on the load-bearing capacity of the bond. This is because the strengths of the adhesives used for this purpose are generally orders of magnitude below the strength of the substrates, so that the deformations caused by mechanical stress occur almost exclusively in the adhesive layer and usually do not have any or a small influence on the deformation of the substrates. Thus, the stresses occurring in the components are relatively low and well distributed compared to the stresses in high-strength bonds, which is also favored by the significantly greater adhesive layer thickness in such elastic bonds. However, although elastic adhesives produce joints with homogeneous stress distributions, the possible formation of stress peaks should be taken into account in the design of the adhesive bond and minimized as far as possible.

The decisive factor for the occurrence of stress peaks at the overlapping ends is the ratio of the shear modulus G of the adhesive layer and the elastic modulus E of the joining part material G/E. The smaller this ratio, the lower the formation of such stress peaks. Compared to low-deformation high-strength adhesive layers, elastic adhesive layers are characterized by a significantly lower shear modulus and, using the same substrates, the G/E ratio decreases accordingly, resulting in a much better stress distribution over the bonded joint length. This represents the main advantage of elastic bonding and is the main reason for using this option in various applications where load transfer is not the highest priority. The adhesive strengths of the elastic systems, at <10 MPa, are lower than those of the thin, low-deformation, high-strength adhesive layers with bond strengths often exceeding 30 MPa. However, due to the lower and often-absent stress peaks, elastic bonding allows the entire bonding surface to be used for load transfer. As a result, the overlap length can be dimensioned linearly to the load to be transferred and the maximum transmissible load does not approach a limit value.

In Table 9.1, a comparison is made between the characteristic features of structural and elastic adhesive bonds.

Table 9.1 Comparison of the characteristic features of structural and elastic adhesive bonds.

Structural adhesive bonding	Elastic adhesive bonding
High strength (10–40 MPa)	Moderate strength (<10 MPa)
Low elongation at break (5–150%)	High elongation at break (300–800%)
Thin adhesive layers (0.1–0.2 mm)	Thick adhesive layers (1–3 mm)
Low compensation of different thermal material extensions	High compensation of different thermal material extensions
Low tolerance compensation	High tolerance compensation
Low vibration damping	High vibration damping
High resistance to environmental loads	Moderate resistances to environmental loads

In the design of the bonded joints, the following significant differences arise between high-strength *structural* and *elastic* bonding:

- *Structural bonding* – For the thin, low-deformation adhesive layers used in structural bonding, the bond strength is strongly dependent on the adhesive layer thickness, and the highest strengths are achieved at low thicknesses in the range of 0.10–0.30 mm. However, deformations to accommodate surface roughness are very limited in this thickness range, and increasing the thickness leads to stress peaks. A very characteristic feature of structural bonding is that increasing the overlap length does not allow higher loads to be transferred when the substrates are above their elastic limit.
- *Elastic bonding* – Compared to structural bonding, there is no maximum bond strength as a function of overlap length in elastic bonding. Due to the uniform stress distribution and the resulting largely linear relationship between overlap length and adhesive strength, it is possible here to compensate for the low adhesive force values by increasing the overlap lengths.

9.5.4.3 Dimensioning of Adhesive Bonds with Plastics

When dimensioning bonded joints with plastics, the same relationships apply as for metal-bonded joints, but because of the viscoelastic behavior of these materials, a number of points must be taken into account. If the components are to be designed, the principles that apply to metal bonding also apply here, such as the presence of sufficient bonding surfaces and the avoidance of stress peaks, especially due to the occurrence of peel stresses. Because of the similar strength of the substrates and adhesive layers, plastic bonds are not suitable for supporting larger loads and can also – as metal bonds – only be dimensioned to a limited extent by increasing the bonding area. In contrast to metal bonding, however, adhesive bonds can also be designed as butt joints.

When designing adhesive bonds with plastics, the following differences from metal bonding must be taken into account:

- the strength of plastics is more than 10 times lower than that of metals,
- show very different deformation behavior from polymer to polymer,
- in particular, the estimation of the time-dependent strength values required for a design requires precise information on the deformation behavior of the plastic in question,
- since the factors influencing the bonding of plastics are more complex than for metals, there is no generally applicable equation for calculating the transmissible loads that can be used,
- corrosion does not have a critical influence on the quality of the bond, as it does for metals, and
- unlike metals, water diffusion is possible through the bonding surfaces, which leads to an increased moisture load on the adhesive layer.

To limit the mechanical stress exclusively to the deformation in the elastic area of the joined parts, the corresponding stress–strain behavior of the substrates must always be taken into account in the design of plastic joints. Based on the fact that the strengths of substrate and adhesive are comparable, the overlap length for plastic components should be 2–5 times the material thickness. When dimensioning the bonded joints, appropriate reduction factors should be taken into account depending on the application, and the adhesives should be matched to the special properties of the plastic material. For example, the level of the curing temperature must be taken into account and care must be taken to ensure that comparable glass transition temperatures and similar elasticity or shear moduli of the bonded part and adhesive layer are set for optimum load transfer.

9.5.4.4 Concluding Remarks on the Structural Design and Dimensioning of Bonded Joints

For a long time, calculations and representations of the stress distribution in bonded single-lap joints were based on the simplified assumption of a quasi-elastic deformation behavior of the adhesive layer in the bond, which had its justification in the relatively brittle materials used for bonded joints in the past. However, the adhesives used today often contain specially developed plasticizing components, such as elastomer particles in the epoxy adhesives, which now ensure that the shear modulus has become smaller and the proportion of plastic deformation larger. The actual stress distribution is thus much more uniform and can be further improved toward a more homogeneous stress distribution by viscoelastic post-deformation under permanent load, elevated temperature, and under the influence of moderate swelling processes due to moisture.

As already mentioned, an adhesive bond must be dimensioned in such a way that the load in practical use is smaller than the determined load-bearing capacity taking into account reduction and safety factors. As a general rule, the optimum overlap length of an adhesive bond is given when the ratio of the yield strength of the substrates $R_{p0.2}$ and the adhesive strength τ_B is equal to the ratio of the overlap length l_{ov} and the bonded part thickness s. In practice, it has been shown that the optimum overlap lengths for metals are 10–20 times the thickness of the bonded part, and for plastics 2–5 times.

9.5.5 Examples for Dimensioning of Adhesive Bonds

9.5.5.1 Example 1 – Joining Two Stainless-Steel Plates (Structural Bonding)

For this joint of stainless-steel plates and a 2K epoxy adhesive used in outdoor structural bonding application (Figure 9.55), the maximum load-bearing capacity under temperature and moisture loads, the required overlap length (l_o), and the adhesive utilization factor δ must be calculated.

To be determined:

- load-bearing capacity under temperature and moisture load,
- reduction factor for the temperature influence $f_{temp,}$
- reduction factor for the moisture influence $f_{hum,}$
- total reduction factor $f_{tot,}$

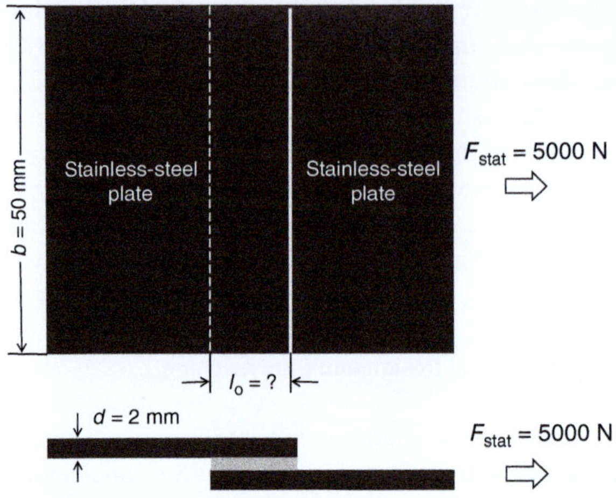

Figure 9.55 Joining two stainless-steel plates (example 1). Question mark means that l_o has to be determined.

- maximal load-bearing capacity $\tau_{\text{Bond real}}$ under temperature and moisture load,
- required overlap length l_{ov}, and
- adhesive bond utilization factor δ.

Known are:

- substrate 1: stainless steel (1.4301/X5 CrNi 18-10)
- substrate 2:stainless steel (1.4301/X5 CrNi 18-10)
- yield strength of substrates ($R_{p0.2}$): 190 MPa
- joining part thickness (s): 2 mm
- overlap width (b): 50 mm
- permanent static load to be transmitted (F_{stat}): 5000 N
- service temperature: 80 °C
- relative humidity in operation:80%
- adhesive type: 2K epoxy adhesive, RT curing
- adhesive strength according to DIN 53283 (τ_{Bond}): 26 MPa.

Calculations:

- *Load-bearing capacity (outdoor use)* – This is the maximal possible adhesive strength of the joint $\tau_{\text{Bond real}}$, which can be expected after reducing the original adhesive strength τ_{Bond} by multiplying it with the appropriate reduction factors and possibly a safety factor. In this example, it is calculated using the reduction factors for temperature and moisture load.
- *Reduction factor for the influence of temperature* – From the data sheet, the adhesive strength of the adhesive is given as 12 MPa at 80 °C. Putting the value in relation to the adhesive strength at 23 °C – which is 26 MPa – gives the reduction factor: $\boldsymbol{f_{\text{temp}}} = \mathbf{12/26\,MPa} = 0.46$.
- *Reduction factor for the influence of moisture* – Moisture aging tests, carried out in the laboratory, show that a strength reduction of 25% is to be expected after application of this load. The reduction factor for moisture influence is therefore: $\boldsymbol{f_{\text{hum}}} = 0.75$.

- *Total reduction factor* – The total reduction factor is obtained by multiplying the individual factors for temperature and moisture loads: $\boldsymbol{f_{tot} = f_{temp} \cdot f_{hum} = 0.46 \cdot 0.75} = \boldsymbol{\mathit{0.35}}$. Due to the influence of temperature and moisture, the load-bearing capacity of this adhesive-bonded joint is therefore only about 35% of the original strength.
- *Maximum load-bearing capacity* – This is the maximum possible adhesive strength of the joint $\tau_{\text{Bond real}}$ that can be expected in real terms by multiplying τ_{Bond} with the appropriate reduction factors. In this example, it is calculated using the total reduction factor f_{tot}: $\tau_{\text{Bond real}} = \tau_{\text{Bond}} \cdot f_{\text{tot}}$. For the example, the result is: $\boldsymbol{\tau_{\text{B real}} = 26\ \text{MPa} \cdot 0.35} = \mathit{9.1\ MPa}$.
- *Required overlap length l_o* – The overlap length l_o is obtained from the relationship: $\tau_{\text{Bond real}} = F_{\text{stat}}/l_o \cdot b$ and $l_o = F_{\text{stat}}/\tau_{\text{Bond real}} \cdot b$. For this example, the overlap length can be calculated with $\boldsymbol{l_o = 5000\ \text{N}/9.1\ \text{MPa} \cdot 50\ \text{mm}} = \mathit{11.0\ mm}$. Thus, the overlap length of this adhesive joint with an overlap width of 50 mm must be dimensioned with approximately 11 mm to withstand a permanent load of $F_{\text{stat}} = 5000$ N under the heat and moisture loads that occur.
- *Degree of adhesive-bond utilization δ* – The degree of adhesive utilization δ is related to the strength of the joining part material and results from the quotient of the tensile stress σ_{pres} present in the substrate at the fracture of the joint and the strength of the substrate, which in the case of metal bonding is given as the yield $R_{\text{p0.2}}$ leading to: $\delta = \sigma_{\text{pres}}/R_{\text{p0.2}}$. From the adhesive strength, the load transmitted by the bond at break can be calculated: $F_{\text{Bond}} = \tau_{\text{Bond real}} \cdot l_o \cdot b = 9.1 \cdot 11 \cdot 50 = 5005$ N. The same load also acts in the joining part material and generates the tensile stress there: $\sigma_{\text{pres}} = F_{\text{Bond}}/s \cdot b = 5.005/2 \cdot 50 = 50.05$ MPa. This results in a degree of adhesive bond use: $\boldsymbol{\delta = \sigma_{\text{pres}}/R_{\text{p0.2}} = 50.05/190\ \text{MPa}} = \mathit{0.26}$.
  For the sheet thickness of 2 mm, the degree of utilization of the bond is calculated as 0.26, which means that only 26% of the substrates are utilized. The bonded parts are therefore clearly in the elastic range, and the component is thus designed to be technically but not economically safe. Reducing the thickness of the substrates from 2 to 1 mm would therefore result in 52% utilization of the substrates and a better economic situation. However, this may not be possible due to the strength requirements for the entire bonded component.

9.5.5.2 Example 2 – Bonding of a GFRP-Element for Interior Design (Elastic Bonding)

An elastic 1K PU adhesive is used for this joint between a GFRP plastic sheet and a stainless-steel frame made for interior design (Figure 9.56). The load-bearing capacity with the reduction factor for the static permanent load and a safety factor of 0.5, the required bonding area (A), the adhesive layer width, and the adhesive thickness (d) of this elastic bonding must be calculated.

To be determined:

- load-bearing capacity $\tau_{\text{Bond real}}$, taking into account the reduction factor for the static mechanical load $f_{\text{stat}} = 0.06$ and a safety factor $s = 0.50$,
- required bond area A,

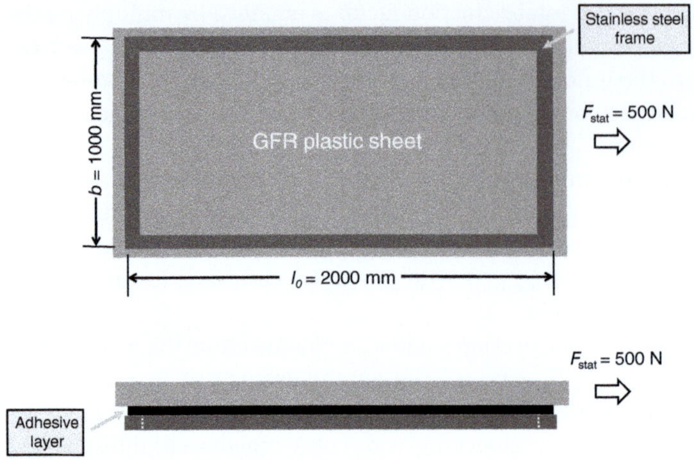

Figure 9.56 Joining a GFR plastic sheet to a stainless-steel frame (example 2).

- required adhesive layer width, and
- required adhesive thickness d.

Known are:

- substrate 1: GFRP plastic sheet
- substrate 2: stainless-steel frame
- permanent load to be transmitted (F_{stat}): 500 N (50 kg weight of the element)
- length of stainless-steel frame (l_0): 2000 mm
- width of stainless-steel frame (b): 1000 mm
- total length of the bond line: 6000 mm
- temperature difference in application (ΔT): 60 K
- coefficient of thermal expansion of steel (α steel): $12 \cdot 10^{-6}$ K^{-1}
- coefficient of thermal expansion of FRP (α FRP): $20 \cdot 10^{-6}$ K^{-1}
- adhesive type: 1K PU, humidity curing
- tan Υ (max) value of adhesive: 0.15
- adhesive strength according to DIN 53283 (τ_B): 3 MPa
- reduction factor permanent static load (f_{stat}): 0.06
- safety factor (s): 0.50

Calculations:

- *Load-bearing capacity (indoor use)* – In this example, the load-bearing capacity $\tau_{Bond\ real}$ is calculated using the reduction factor for the static load $f_{stat} = 0.06$ and a safety factor $s = 0.50$ as follows: $\tau_{Bond\ real} = \tau_{Bond} \cdot f_{stat} \cdot s$ resulting in: $\boldsymbol{\tau_{B\ real} = 3\ MPa \cdot 0.06 \cdot 0.50} = \mathit{0.09\ MPa}$.
- *Required bond area A* – The required bond area results of: $\tau_{Bond\ real} = F_{stat}/A$; $A = F_{stat}/\tau_{Bond\ real}$ and for this example: $\boldsymbol{A = 500\ N/0.09\ MPa} = \mathit{5556\ mm^2}$.
- *Required bond width b* – The bond width is calculated as follows: $b = A/l$; since the total length of the frame is 6000 mm, the example has a minimum bond width

of: $\boldsymbol{b} = \mathbf{5556\,mm^2/6000\,mm} = 0.93\,mm$. Assuming that a much wider bond will result after application of the PU adhesive, the calculated bond width can be approached with a good degree of confidence. It should be noted that only the static load was considered in this calculation. Other reduction factors such as for thermal, dynamic, or moisture loads were not used.

- *Required adhesive thickness d* – Assuming a temperature difference of 60 K during application and considering the longer of the two substrate sides, the calculation gives $\Delta l = l_0 \cdot \Delta\alpha \cdot \Delta T$ and for the example $\boldsymbol{\Delta l} = \mathbf{2000\,mm \cdot 8 \cdot 10^{-6}\,K^{-1} \cdot 60\,K} = 0.96\,mm$. If the element is now free to expand on both sides, there will be a change in length of $1/2 \cdot 0.96\,\text{mm} = 0.48\,mm$ on both sides. This length expansion is used to calculate the joint thickness as follows: $d = \Delta l \cdot 0.5/\tan \Upsilon\,(\text{max})$ and for the example $\boldsymbol{d} = \mathbf{0.96\,mm \cdot 0.50/0.15} = 3.2\,mm$.

  Thus, for a tan Υ (max) value of 0.15, which is different for each adhesive and can be taken from the data sheets or found out from the adhesive manufacturer, an adhesive thickness of *3.2 mm* is thus required to compensate for the differences of the thermal expansion of the two materials over a temperature range of 60 K.

9.5.5.3 Example 3 – Bonding of a Glass Pane with an Aluminum Frame (Elastic Bonding)

A high-performance acrylic foam tape is used for joining a glass pane with an aluminum frame (Figure 9.57). The load-bearing capacity with the reduction factor for the static permanent load and a safety factor of 0.5, the required bonding area (*A*), the adhesive layer width, and the adhesive thickness (*d*) of this elastic bonding must calculated.

To be determined:

- load-bearing capacity $\tau_{\text{Bond real}}$ taking into account the reduction factor for the static mechanical load f_{stat} and a safety factor $s = 0.5$,

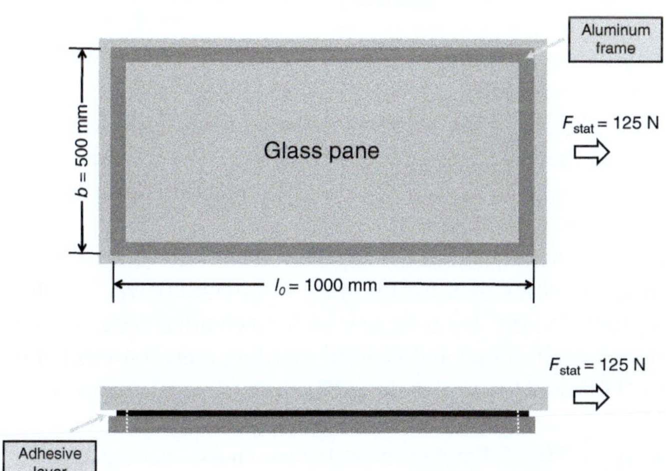

Figure 9.57 Joining of a glass pane to an aluminum frame (example 3).

- required bond area A,
- required adhesive layer width, and
- required adhesive thickness d.

Known are:

- substrate 1: glass pane
- substrate 2: aluminum frame
- permanent load to be transmitted (F_{stat}): 125 N
- length of glass pane (l_0): 1000 mm
- width of glass pane (b): 500 mm
- thickness of glass pane: 10 mm
- total length of bond line: 3000 mm
- temperature difference in the application (ΔT): 40 K
- coefficient of thermal expansion of aluminum (α aluminum): $23 \cdot 10^{-6}\ K^{-1}$
- coefficient of thermal expansion of glass (α glass): $8.5 \cdot 10^{-6}\ K^{-1}$
- adhesive type: acrylic foam tape
- tan ϒ (max) value of the adhesive: 3.0
- adhesive strength according to DIN 53283 (τ_{Bond}): 0.25 MPa
- reduction factor permanent static load (f_{stat}): 0.06
- reduction factor temperature 50 °C (f_{temp}): 0.6
- reduction factor humidity (f_{hum}): 0.4
- safety factor (s): 0.5

Calculations:

- *Load-bearing capacity (indoor use)* – In this example for use in indoors, it is calculated using the reduction factor for the static load $f_{stat} = 0.06$, the reduction factor for the temperature $f_{temp} = 0.60$ a safety factor $s = 0.50$: $\tau_{B\,real} = \tau_B \cdot f_{stat} \cdot f_{temp} \cdot s$. For this example, the result is: $\boldsymbol{\tau_{Bond\,real} = 0.25\,MPa \cdot 0.06 \cdot 0.60 \cdot 0.50 =}$ *0.0045 MPa.*
- *Load-bearing capacity (outdoor application)* – If the bonded component were to be used outdoors, the reduction factor $f_{hum} = 0.40$ for the moisture effect would have to be considered in addition to the reduction factors for the static load and the temperature effect, and the maximum possible bond strength for outdoor use would be: $\tau_{Bond\,real} = \tau_{Bond} \cdot f_{stat} \cdot f_{temp} \cdot f_{hum} \cdot s$. For this example, the result is: $\boldsymbol{\tau_{B\,real} = 0.25\,MPa \cdot 0.06 \cdot 0.60 \cdot 0.40 \cdot 0.50 =}$ *0.0018 MPa.*
  These values calculated for indoor and outdoor application are in good agreement with the following manufacturer's data for the maximum possible $\tau_{Bond\,real}$ for the acrylic foam tape used: 25 mm^2 bonding area at a mechanical load of 10 N (= 0.0040 MPa) for indoor applications and 60 mm^2 bonding area at a mechanical load of 10 N (0.0017) for outdoor applications. The following calculations are made for indoor use.
- *Required bond area A* – The following calculations must be performed: $\tau_{Bond\,real} = F_{stat}/A$ and $A = F_{stat}/\tau_{Bond\,real}$; for this example: $\boldsymbol{A = 125\,N/0.0045\,MPa}$ $= 27\,778\ mm^2$.

- *Required bond width b* – This is calculated by $b = A/l$; since the total length of the bond line is 3000 mm, the required adhesive tape width is: $\boldsymbol{b} = \mathbf{27\,778\,mm^2/3000\,mm}$ = *9.26 mm*. If a 10 mm wide adhesive tape offered by the manufacturers were used for this bonding, the application could be realized with good reliability based on the calculations performed and the somewhat oversized adhesive tape width of 10 mm.
- *Required adhesive thickness d* – Assuming that a temperature difference of 50 K is to be assumed in the application and that the longer of the two sides must be taken into account, the calculation of the adhesive tape thickness is: $\Delta l = l_0 \cdot \Delta\alpha \cdot \Delta T$ and for the example $\Delta l = 1000\,\text{mm} \cdot 14.5 \cdot 10^{-6}\ \text{K}^{-1} \cdot 50\,\text{K} = \mathit{0.725\,mm}$. If the element is now free to expand on both sides, the change on both long sides is: $\frac{1}{2} \cdot 0.725\,\text{mm} = 0.363\,\text{mm}$. The linear expansion is used to calculate the adhesive thickness as follows: $d = \Delta l \cdot 0.5/\tan \Upsilon$ (max) and for the example $d = 0.363\,\text{mm} \cdot 0.50/3.00 = \mathit{0.06\,mm}$. Thus, for the acrylate foam adhesive tape with an excellent tan Υ (max) value of 3.00 (300% linear deformation relative to the tape thickness), an adhesive thickness of only *0.06 mm* is necessary to compensate for the differences in the coefficients of thermal expansion of the two substrates, glass pane and aluminum frame, over a temperature range of 50 K. However, to compensate for the tolerances of the materials involved and to adapt to roughness of the substrate surface, an adhesive tape with a significant greater thickness should be selected.

References

1 ASTM D 3983 (2019). *Standard test method for measuring strength and shear modulus of non-rigid adhesives by the thick-adherent tensile-lap specimen*. American Society for Testing and Materials.

2 Pocius, A.V. (2002). *Adhesion and Adhesive Technology*, 15–19. Munich: Carl Hanser Verlag.

3 ISO 10354 (1992). Adhesives – Characterization of durability of structural-adhesive-bonded assemblies – Wedge rupture test.

4 DIN 65448 (1988). Aerospace; structural adhesives; wedge test.

5 Späth, W. (1956). *Beiträge zur Technologie der Hochpolymeren*. Gerner Stuttgart: Gummi und Kunststoffe.

6 Brockmann, W. and Kollek, H. (1980). *Ermittlung der Langzeitbeanspruchbarkeit von Metallklebungen mit im Maschinenbau gebräuchlichen und zukünftigen Bindemitteln*. Forsch. Heft. Forsch. Kurat. Masch.-Bau H.81.

7 Lees, W.A. (1984). *Adhesives in Engineering Design*, 48–55. Berlin: Springer-Verlag.

8 Fauner, G. and Brockmann, W. (1986). Fügen durch Kleben. In: *Handbuch der Fertigungstechnik*, Band 5, Fügen, Handhaben und Montieren (eds. G. Spur and R. Stöferle), 467–498. München, Wien: Carl Hanser.

10

Development (Step 4) – Establishing a Robust Manufacturing Process Ready for Production

10.1 Introduction

10.1.1 Status of the Project

A key question in the development of an adhesive bond is what the adhesive layer must look like and how it can be reproducibly produced. This question can only be answered on an application-specific basis after the required material and process parameters have been developed. The result comes about empirically and is usually based on a large number of internal and external experiences as well as laboratory and component tests. For this purpose, bonded test joints and later practical prototypes have already been produced, tested, and evaluated in the recent step (Step 3 – Feasibility). Depending on the results, the adhesives used or their processing parameters are adjusted, and finally the most suitable candidate is produced as a prototype and extensively tested. This fine-tuning of the product design can be repeated several times until a satisfactory result is achieved, and in the end leads to a steady increase in the maturity of the adhesive bond and the associated process parameters.

Now, at the beginning of Step 4 – Development, the maturity of the concept design is at its highest, it is validated for practical use and ready to be developed for manufacturing. In Figure 10.1, the path taken so far to increase the degree of maturity from the first sketch of the adhesive bond to the validation for practice – which is the basis for the development of the production process – is demonstrated.

In Figure 10.2, the major tasks with the corresponding results of the development step that must be available to introduce the process into production are shown.

10.1.1.1 Objectives of the Development Step

We have just completed the feasibility step, which involved demonstrating the feasibility of the preferred adhesive design and the validation of the bonded component in the field. Now, the objective of the next step is to develop a robust process for the safe production of the proposed bonded joint. Based on the experience gained so far and related to the production scale, the sub-processes such as application of the adhesive, joining of the substrates, curing of the adhesive, and fixing and removal of the component must be worked out and described with technical precision. Once

Adhesive Bonding in Five Steps: Achieving Safe and High-Quality Bonds, First Edition. Jürgen Klingen.

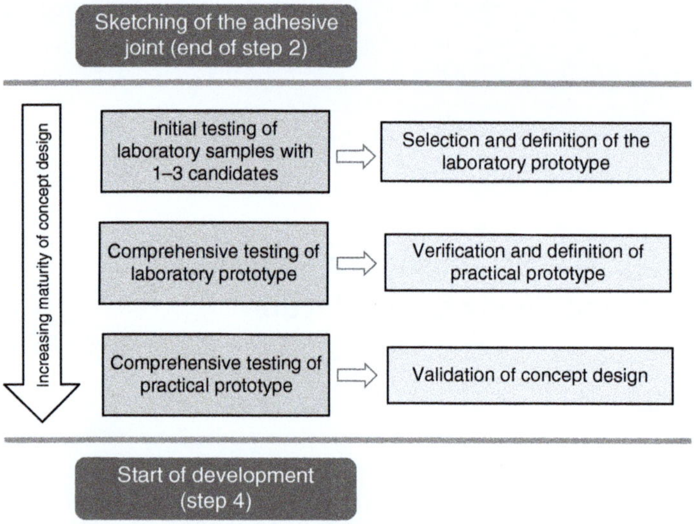

Figure 10.1 Increase of maturity of the concept design from first sketching to validation of the concept design.

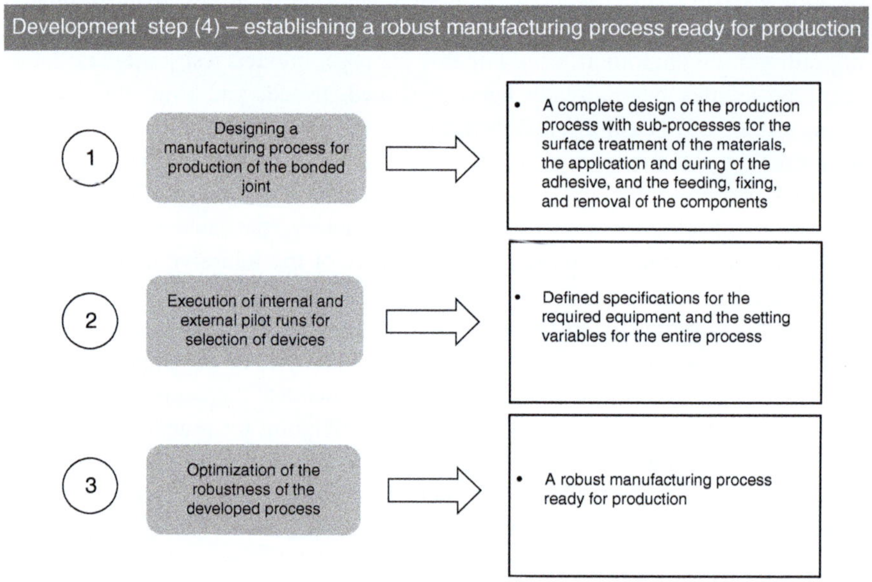

Figure 10.2 Major tasks and accomplishments of Step 4 (development).

this has been done and a detailed description of the planned production process is available, it is first necessary to investigate and evaluate the internal possibilities for setting up the required manufacturing process. If these are not sufficient, investments must be made to create or supplement the necessary facilities. To identify the optimum equipment for the relevant sub-processes, intensive external contacts are necessary. Expert discussions and, in most cases, extensive pilot runs at the potential suppliers and in-house enable the optimum bonding process to be set up. Once the optimum manufacturing process with the necessary sub-processes for production scale has been determined, its suitability in principle must first be demonstrated by appropriate pilot runs. Then, after successful completion, the process must be optimized using statistical and experimental methods, which ultimately concludes the development of the robust manufacturing process for producing the bonded joint.

10.1.1.2 Requirements for the Successful Completion of the Development Step

Due to the interdisciplinary nature of adhesive-bonding technology between engineering and natural sciences, intensive coordination between application, material, process developers, and production specialists on the one hand and suppliers of the required adhesives and equipment on the other must be carried out during the development step to ensure successful development of the process. In addition, theoretical and practical training of the relevant employees is a mandatory prerequisite for the successful implementation of the individual process steps. For this purpose, in cooperation between industry and scientific institutions, a comprehensive training concept in three stages has been developed in Europe some years ago, which meets the specific requirements of adhesive-bonding technology. As a result, corresponding courses are offered by various scientific institutions today. Finally, a comment on the requirements regarding the premises where the bonding processes are to take place. Since *adhesive-friendly* material surfaces are a mandatory requirement for the successful production of an adhesive bond, great importance should be attached to clean and well-cleaned premises, which must of course strictly comply with the requirements of occupational safety and environmental protection.

10.2 Adhesive-Bonding Manufacturing Process (Overview)

Like mechanical joining, adhesive bonding is a manufacturing technology for connecting materials. However, there are some special features that must be taken into account when designing the bonding process with regard to the property profile of the adhesive layers produced. First of all, it should be understood that the adhesive layer contains a polymer that differs in many properties from the metallic materials commonly used in classical bonding technologies. The main difference from welding, riveting, or clinching is that an *unfinished* often reactive chemical material is used to produce high-quality adhesive bonds. In mechanical joining technologies, the molten metal largely regains its original properties after cooling, as in welding, or the metal material remains unchanged, as in riveting and clinching. With adhesive bonding, things are different. Here, the final properties of bonded joints are decisively determined by the processing of the adhesive in the manufacturing process. Thus, the manufacturing process is not only responsible for the pure joining of the individual components, but also for their final properties.

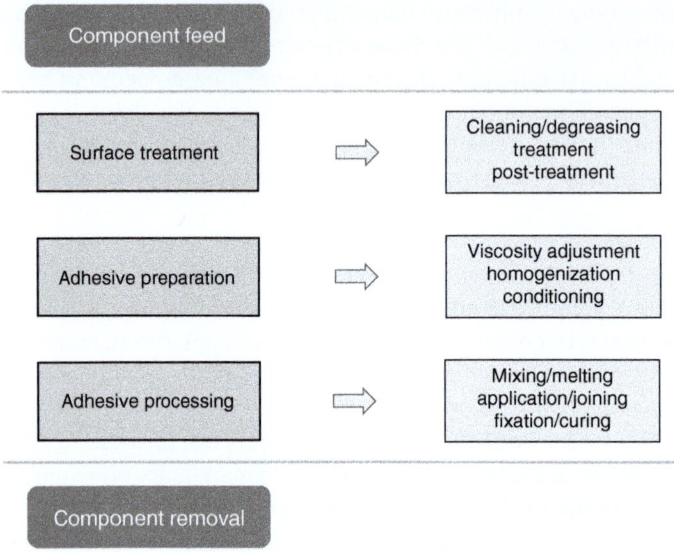

Figure 10.3 Manufacturing process chain from component feeding to removal of the joined part.

Based on decades of experience, this process is in principle suitable for a wide range of applications in industry and trade, provided that the following criteria are observed during planning and implementation:

- use of simple and continuously controllable production technologies,
- use of reproducible quality standards for process reliability,
- preparation of mature dimensioning and standards for the components to be manufactured and for the processes used,
- in the case of large-scale production, consideration of automation and integration into downstream manufacturing processes,
- use of empirical values to assess the general suitability of the process,
- development of an overall concept relevant to personal and environmental protection, and
- fulfillment of economic requirements.

The manufacturing process for producing an adhesive bond consists of a chain of successive individual sub-processes (Figure 10.3). After the components are fed in, surface treatment usually takes place, followed by adhesive processing, which reaches component removal at the end of the process chain.

The surface treatment of materials has already been discussed in Section 7.3.

In this section, great attention is now paid to the activities related to adhesive preparation and adhesive processing, since they contribute more or less to the final properties of the bonded joint:

Adhesive preparation:

- storage of the adhesives, especially temperature and time,
- tempering of the adhesives prior to bonding to ensure suitable and reproducible curing,
- homogenization of adhesives containing fillers to avoid concentration gradients,

- adjustment of viscosity for optimum processing and build-up of the bond on the substrates,
- heating conditions for hotmelt adhesives, e.g. to prevent degradation of the adhesive, and
- concentration of humidity in the environment during wet curing of 1K adhesives, e.g. based on polyurethanes (PUs), cyanoacrylates, and silicones.

Adhesive processing:

- mixing of the adhesive components to ensure homogeneity and mixing ratio of the components without bubble formation,
- applying the exact amount of adhesive,
- fixing the parts to be joined to avoid disturbing the structure of the adhesive layer by displacement,
- curing of reactive multi-component systems under control of temperature, pressure, and time, and
- drying of solvent-borne and water-borne systems under control of temperature, pressure, and time, as well as prevention of bubble formation.

10.2.1 Process Objectives for the Production of a Bonded Joint

Almost all types of adhesives undergo a transition from the liquid to the solid state during the manufacture of an adhesive bond. The exception is pressure PSAs, which remain as they were delivered and exhibit permanent tack. The viscous properties of the adhesives are necessary for the formation of molecular interactions on the substrate surface, which are essential for adhesion build-up. The material properties after solidification, on the other hand, are responsible for the cohesive strength of the adhesive bond.

The successful production of an adhesive bond is thus based on two process objectives with their specific steps, which are essential components of the overall process for producing a safe and high-quality bonded component (Figure 10.4):

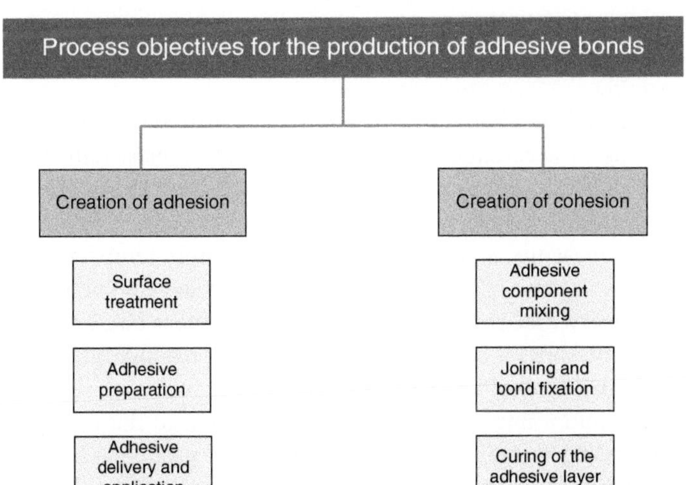

Figure 10.4 Process objectives and process steps for the production of adhesive bonds.

- *achieving adhesion build-up on the substrate surfaces,* and
- *achieving cohesion build-up within the adhesive layer..*

In addition to the surface treatment of the materials and the adhesive selection already described, the build-up of adhesion depends strongly on the preparation and application of the adhesive. On the other hand, the cohesion strength is largely determined by the quality of the mixture of the adhesive components and the control of the process parameters during curing, such as temperature, time, and pressure.

10.3 Processes for Adhesion Build-up

For good adhesion build-up to the material surfaces, special attention must be paid to the following process steps:

- surface treatment of the substrates,
- adhesive preparation, and
- adhesive delivery and application.

The surface treatment of the substrates has already been dealt with in detail before, and the knowledge gained there can be used in the development phase and for production. Therefore, in this chapter, the process steps for generating adhesion will start with the preparation of the adhesive.

10.3.1 Adhesive Preparation

For optimum development of the adhesive forces on the material surface, the adhesive should be homogenized and adjusted to the appropriate viscosity range before it comes into contact with the surface. To fulfill this, a number of points must be taken into account already in the early phase of *adhesive handling*. To this end, the appropriate process steps must be planned depending on the type of adhesive base material and its processing-specific parameters.

10.3.1.1 Storage/Tempering

To ensure constant processing temperatures, the adhesive must be adequately tempered before processing in accordance with the requirements, especially if it is stored in non-air-conditioned rooms during cold or warm seasons. This requirement for sufficient temperature control before bonding naturally also applies to the substrates used. This measure is necessary because different temperatures would lead to viscosity fluctuations in the adhesive and thus to deviations in the bonding properties; a viscosity that is too high has a negative effect on bonding, and in extreme cases even prevents it. Particularly in the case of reactive adhesive systems, the temperature and time of storage play a major role in their shelf life and processability. It is therefore important to pay close attention to the relevant manufacturer's instructions, e.g. found in technical data sheets or product information. Another important point must be strictly observed when storing waterborne adhesive dispersions. Storage temperatures that are too low could lead to freezing of these systems, which would result in irreversible destruction of the dispersion and the adhesives could thus no longer be used.

10.3.1.2 Homogenization

When using adhesives containing fillers, it must be ensured that the solids are homogeneously distributed in the adhesive before use. This is usually done once by stirring after annealing and before further processing of the adhesive. If there is a risk of further sedimentation, homogenization must be continued during processing. Fillers are solid and non-volatile substances that do not interfere with the reactions of the monomer and polymer molecules present in the adhesive. When adding fillers, it is important to achieve the most homogeneous distribution possible and good wetting of the filler particles by the liquid components of the adhesive. Much attention must be paid to avoid cohesion-reducing voids in the cured adhesive layer and adhesion-inhibiting filler accumulations on the material surface. Careful homogenization of the adhesive thus compensates for concentration differences and, if necessary, breaks up agglomerates to ensure both good adhesion to the material surface and the required cohesion in the finished adhesive layer.

10.3.1.3 Setting the Viscosity

The viscosity of the adhesive system is crucial for the formation of molecular interactions on the surface of the substrate and the displacement of air from the capillaries or surface cavities. While viscosity in solvent-free systems is largely determined by chain length, molecular structure, and the presence of polar groups in the adhesive polymer, in solvent-containing adhesives it depends on the concentration of the adhesive polymer in the solvent and the type of solvent. The name viscosity goes back to the typical viscous sap of the berries of the medlar plant (Latin: viscum), from which bird glue was obtained. In Roman times, this viscous and sticky mass was used to make glue rods, which were used to catch birds. Viscosity is commonly perceived as *thickness,* or resistance to pouring. It describes the internal resistance to flow of a fluid and can be considered a measure of fluid friction. For example, water is *thin* and has a low viscosity, while vegetable oil is *thick* and has a high viscosity. The reason for an increase in viscosity is internal friction, which can be simply thought of as the movement of two interlocking layers of molecules. During flow, the molecular chains pull past each other and a certain force is required to overcome the interlocking. The force F required to move the layers is proportional to their area A and their velocity v, and inversely proportional to the distance x between the layers:

$$F = \eta \cdot A \cdot v/x$$

The constant of proportionality η is called dynamic viscosity – usually simply viscosity – and is a property representing the internal friction of the fluid. The unit of viscosity is pascal-second (Pa·s) or, for low-viscosity systems milli-pascal-seconds (mPa·s).

Table 10.1 Typical viscosity values for commonly used materials and adhesives.

Material	Viscosity (mPa·s)	Adhesive	Viscosity (mPa·s)
Water	1	Cyanoacrylate adhesive	3–21 000
Mercury	1.5	UV acrylate adhesive	$10–10^6$
Blood at 37 °C	4–25	UV epoxy adhesive	>100
Glycol	10^3	1K PU adhesives	20 000–500 000
Liquid honey	10^4	2K epoxy adhesives	6 000–640 000
Polymer melts	10^6	1K epoxy adhesives	15 000–200 000
Bitumen	10^9		
Glass	10^{23}		

According to the SI system, viscosity is defined as follows:

- a fluid located between two plates has a viscosity of 1 N·s/m^2 if, for a size of 1 m^2 and a plate spacing of 1 m, a force of 1 N is required to move the plates against each other at a speed of 1 m/s (1 N·s/m^2 = 1 Pa·s).

Typical viscosities of a range of commonly used materials and adhesives are listed in Table 10.1.

10.3.1.4 Newtonian and Non-Newtonian Fluids

According to the criterion of viscosity, fluids are divided into Newtonian and non-Newtonian. The description of viscosity and the relationship with shear stress τ and shear rate D was established by Isaac Newton as early as 1687:

$$\tau = \eta \cdot D$$

If the relationship is linear according to this equation, Newtonian fluids are present. If the viscosity depends on the shear rate D, the fluids are called non-Newtonian.

Non-Newtonian fluids are divided into two groups:

- *Shear-thinning (structural viscous) fluids* – For these substances, viscosity decreases with increasing shear rate. The application of shear forces causes the structure of the substance to break down, making it more fluid. Most fluids, such as polymer melts and emulsions, belong to this group.
- *Shear-thickening (dilatant) fluids* – For these substances, such as water, honey, and plastisols, viscosity increases with increasing shear rate.

Viscosity curves show the relationship between shear rate and viscosity and demonstrate very nicely the differences between Newtonian and non-Newtonian fluids. For Newtonian fluids, which satisfy Newton's dynamic viscosity equation, the viscosity stays constant. Shear-thinning (structurally viscous) and shear-thickening (dilatant) substances are represented in decreasing or increasing curves (Figure 10.5).

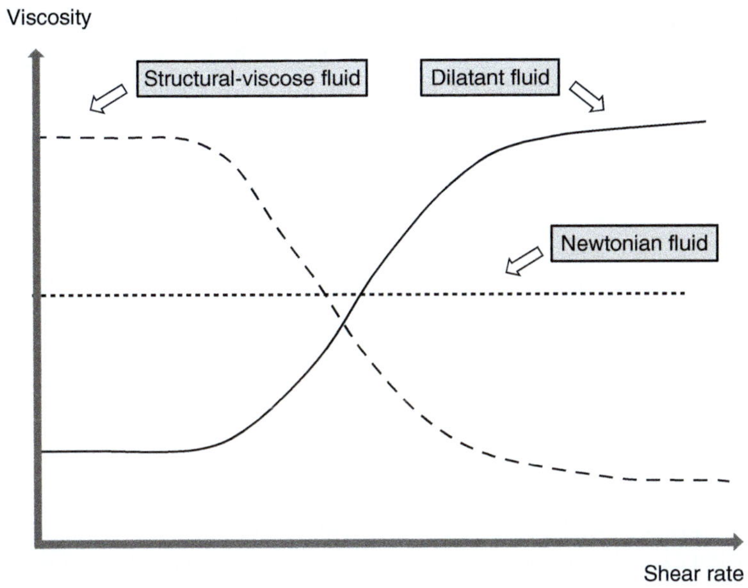

Figure 10.5 Viscosity curves of Newtonian, structural-viscose, and dilatant fluids.

The following causes may be responsible for the structural viscosity:

- fillers with an anisotropic shape are oriented during processing,
- liquid droplets in the material matrix are deformed,
- polymer molecules are stretched when shear forces occur, and
- aggregates formed from particles are destroyed by a dispersing or stirring process.

10.3.1.5 Viscosity of Adhesives

Adhesives are generally non-Newtonian fluids in which the shear stresses occurring between two adjacent layers are not proportional to the shear rates. Only low-molecular-weight adhesives show nearly Newtonian behavior in individual cases.

The following parameters influence the viscosity of adhesives:

- polar and non-polar interactions of the mixed components,
- molecular weight and molecular weight distribution of the mixed components,
- filler content in the adhesive,
- properties of the fillers such as particle size, size distribution, shape, and polarity,
- solvent or plasticizer content in the adhesive,
- shear rate and shear time during processing of the adhesive, as well as
- temperature and pressure.

The requirement for suitable adhesive viscosity to ensure the formation of molecular interactions at the material surface and the displacement of air from the capillaries or surface voids necessary for good wetting has already been addressed. In addition, there are other process-related requirements for the adhesive viscosity that must be taken into account during the development phase. For example, spray application requires a low-, while application by screen printing requires a high-viscosity setting of the adhesive. If the viscosity is too low for the process used, an adjustment can be made by adding thickeners, such as fumed silica products. In the case of solvent-borne systems with too high viscosity, dilution of the adhesive with appropriate solvent additives ensures the right viscosity. However, if the viscosity of solvent-free reactive adhesives is to be lowered, the user's hands are usually tied. In exceptional cases, viscosity can be reduced by adding reactive diluents contained in the formulation, however, a usually resulting change in the properties of the adhesive must be taken into account. To prevent absorption of the adhesive by the surface when bonding porous substrates, higher-viscosity adhesives are advantageous. These also prevent excessive leakage at the edges of the bonded joint and also improve their gap-bridging properties.

10.3.1.6 Thixotropy of Adhesives

To achieve advantages in the production of an adhesive bond, adhesives are often specifically adjusted to be thixotropic. Thixotropy is a very common property of liquid substances that occurs when shear stress is applied (e.g. during stirring, shaking, brushing, or coating). In this case, they enter a state of lower viscosity and return again to their original higher viscosity when the action is stopped and the stress is removed. In simplified terms, this means that a thixotropic adhesive becomes thinner with the duration of the (constant) stress. In viscosity measurements, this property is noticeable by different results in successive measurements over time. Here, the corresponding viscosity curves at constant shear rate show decreasing viscosities with time. The thixotropic behavior should not be confused with the structural viscosity of shear-thinning substances, where the viscosity decreases with increasing shear rates, but remains constant over time. In contrast, for thixotropy, viscosity is time-dependent at constant shear rates. Silica products are often used to adjust the thixotropic behavior of adhesives [1], and the so-called pyrogenic silicon dioxides produced by flame pyrolysis are very effective. This synthetically produced colloidal material with a particle size ranging from 5 to 50 nm consists entirely of amorphous, often partially hydrophobic particles. The effectiveness of these materials is due to the extremely large specific surface area, which is in the range of 50–600 m^2/g.

The following advantages are achieved by thixotropic adjustment of adhesives:

- no run-off of the adhesive from a vertical substrate surface,
- no or only slight suction of the adhesive of a porous joining material,
- improved application of the adhesive,
- enabling higher adhesive layer thicknesses, and
- reduced flow of the cured adhesive layer.

10.3.1.7 Setting the Room Humidity

When using moisture-curing or moisture-catalyzed 1K adhesive systems such as PU, silicone, or cyanoacrylate adhesives, it is important to ensure constant and

Table 10.2 Dependence of water vapor contents (in g/m^3) on temperature and relative humidity in the air.

Temperature in °C	Relative humidity in %									
	10	20	30	40	50	60	70	80	90	100
10	0.9	1.9	2.8	3.8	4.7	5.6	6.6	7.5	8.5	9.4
20	1.7	3.5	5.2	6.9	8.7	10.4	12.1	13.8	15.6	17.3
30	3.0	6.1	9.1	12.1	15.2	18.2	21.3	24.3	27.3	30.4
40	5.1	10.2	15.2	20.3	25.4	30.5	35.5	40.6	45.7	50.8

suitable humidity in the process rooms. For example, low air humidity, which occurs particularly in the winter months, prolongs the usually very short curing time of cyanoacrylate adhesives or delays the solidification of moisture-curing PU or silicone adhesives. Table 10.2 shows the water vapor content in the air as a function of temperature and relative humidity.

These data can be used, for example, to estimate the air volume required for the complete curing of a 1K PU adhesive:

- At least 1 mol (18 g) of water is required for the reaction of 1 mol of a reactive isocyanate group in the adhesive with a molecular weight of 42 g/mol. To ensure complete curing of the adhesive system in winter at a relative humidity of, for example, 20% and a temperature of 10 °C in the production facility, the total moisture content of approximately 10 m^3 of air (19 g of water) is required. In the summer months, on the other hand, at a relative humidity of, e.g. 70% and a temperature of 20 °C, only approximately 1.50 m^3 of air (18 g water) will lead to complete curing.

10.3.2 Adhesive Delivery and Application

10.3.2.1 Adhesive Delivery to the Surface

To deliver the liquid adhesive to the material surface, conveying, metering, and mixing processes must be used. Metering refers to the measuring or addition of a specific amount of material, which in adhesive-bonding technology is done from:

- tubes,
- cartridges,
- dual cartridges,
- hobbocks (20 l buckets),
- 200 l drums, and
- 1000 l reusable containers.

In industry and craft, tubes and cartridges are mostly used for manual production of joining parts with small quantities, while pails, drums, and containers are used for automatic metering for the production of large quantities, such as in the automotive industry. Today, modern metering and mixing systems precisely and reproducibly process a wide variety of adhesive systems with a wide range of viscosities, mixing ratios, and pot lives. Different degrees of difficulty occur, depending to a not inconsiderable extent on the magnitude of the physical properties of the adhesive or its mixture.

Here is a rough estimate of the degree of difficulty in adhesive processing depending on the physical properties of the adhesive:

- *Viscosity of the adhesive* – Processing of the adhesive is unproblematic at viscosities between 100 and 50 000 mPa·s, difficult below 100 mPa·s or between 50 000 and 200 000 mPa·s, and very problematic at high viscosities greater than 200 000 mPa·s.
- *Viscosity differences of the adhesive components* – Adhesive mixtures with a relatively small viscosity ratio of components A:B from 1 : 1 to 20 : 1 can be processed without problems. Processing at viscosity ratios of the mixtures of 20 : 1 to 100 : 1 is difficult, and very problematic at very large viscosity differences of over 100 : 1.
- *Volumetric component mixing ratios* – Processing mixtures from 100 : 100 to 100 : 20 is unproblematic, from 100 : 20 to 100 : 5 is difficult, and ratios greater than 100 : 5 are very problematic.
- *Pot life of reactive mixtures* – Unproblematic are adhesives with a pot life greater than 10 minutes, difficult between 10 seconds and 10 minutes, and very problematic for very low pot lives between 5 and 10 seconds.

In the case of 2K systems, metering of the adhesive components enables:

- compliance with the specified mixing ratio through the precise measurement of the individual adhesive components in relation to onc another, and
- the exact application of the total quantity of adhesive to maintain the required quantity per bonded component.

Reactive adhesives with a very short pot life are applied directly to the bonded parts with the aid of metering units in which the reactive components are fed from separate containers and then mixed. The metering units used for this purpose take over the control of all physical and/or mechanical parameters such as pressure, temperature, time, and volume. They ensure that the adhesive applied to the workpiece is precisely metered either directly volumetrically or indirectly via a time–pressure control system. The metering accuracy depends on the viscosity of the adhesive to be processed, which in turn is strongly influenced by the temperature and shear behavior.

The essential functional units of a metering system include storage tanks, metering pumps with the corresponding drives, metering valves, and – in the case of 2K systems – the mixing system (Figures 10.6 and 10.7). The stocking of the adhesive before dosing depends on the application and the planned quantities and is in different orders of magnitude.

Figure 10.8 shows an automated adhesive application and curing system for electronic applications on a larger scale, where a photoinitiated 1K adhesive is applied to the electronic component and then cured by UV radiation.

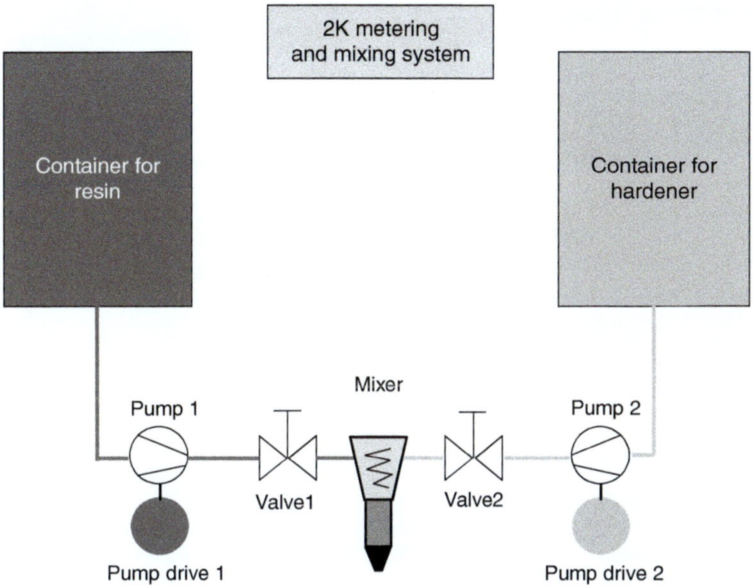

Figure 10.6 Principle of a metering and mixing system for 2K adhesive systems.

Figure 10.7 Large-scale 2K application system according to the principle illustrated in Figure 10.6. Source: Dr. Jürgen Klingen, Aboso-Consulting.

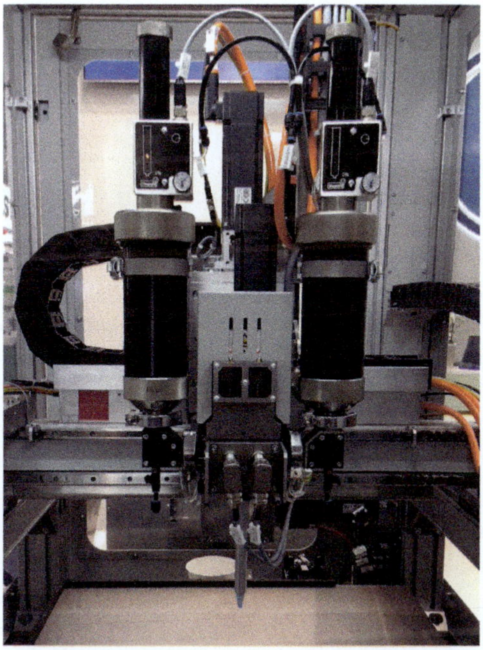

(a) (b)

Figure 10.8 Automatic application and curing system for electronic components, where the UV photoinitiated adhesive is applied first (a) and then cured (b). Source: Dr. Jürgen Klingen, Aboso-Consulting.

10.3.2.1.1 Pumps

The pumps used for adhesive delivery and metering are based on the rotating or oscillating principle:

- *Gear pumps (rotating)* – The permanently metering gear pumps are not self-priming, operate according to the positive displacement principle, and require a constantly applied maximum pressure of up to approximately 100 bar. Their metering capacity is proportional to the drive speed. The characteristic value of a pump is the specific displacement (the displaced volume per revolution), which results from the geometric data of the gear teeth. When metering adhesives, which can range from 0.01 (low viscosity) to 100 Pa·s (high viscosity), the gear pump enables material to be conveyed with virtually no pulsation and ensures a constant throughput. It is also capable of conveying filled systems with a Mohs hardness <5. The slip of the gear pump increases with service life and is not visible from the outside. Gear pumps are also available in heated versions.
- *Piston pumps (oscillating)* – This self-priming piston pump is offered both in the single-acting version as a shot pump and in the double-acting version as a differential piston pump. The pump works up to an operating pressure of 400 bar, and its performance is proportional to the speed of the piston rod. The metering is carried out intermittently (interrupting), but permanent metering is ensured when using a double-acting differential piston pump. The piston pump, which enables pulsation-free material delivery of adhesives in the range from 0.01 (low viscosity)

to 1000 Pa·s (high viscosity), is also available in a heated version and able to deliver filled adhesive systems with abrasive particles of Mohs' hardness >5. Unlike the gear pump, the wear of the piston pump is visible from the outside.

- *Diaphragm pumps (oscillating)* – The self-priming diaphragm pump can be hydraulically, mechanically, or electrically driven and operates up to a maximum working pressure of 3500 bar, with 50–100 bar being the usual range. Its performance is proportional to the number of strokes and the material discharge is pulsating, but this can be adjusted by appropriate dampers. A system upgraded in this way is capable of applying adhesives in the range from 0.01 (low viscosity) to 100 pa·s (high viscosity) without pulsation. It is also capable of conveying systems containing fillers with a Mohs hardness <5. Possible diaphragm defects, which can occur after a long service life, are detected by a sensor. A heated version of the diaphragm pump is usually not available.
- *Peristaltic pumps (rotating)* – The permanently metering peristaltic pump operates according to the positive displacement principle and continuously delivers the mass located in the hose via an external rotor. The pump is self-priming, but can only deliver up to an operating pressure of 10 bar. In this case, its performance is proportional to the drive speed and the metering accuracy is rather low. Such pumps, which are not offered in a heated version, allow metering of adhesives between 1 (very low viscosity) and 1000 mPa·s (medium viscosity), whereby a low pulsating material discharge must be accepted. They are also capable of pumping systems containing fillers with a Mohs hardness <5.
- *Eccentric screw pumps (rotating)* – The eccentric screw pump belongs to the group of rotary positive displacement pumps and consists of a rotating part, the rotor, and a stationary part, the stator, in which the former rotates. The geometric shape of the two parts creates chambers with a defined volume, which move the material due to the movement of the rotor. The mode of operation of this pump is comparable to that of gear pumps and offers advantages when processing very highly viscous or viscous paste-like masses. The permanently metering heated eccentric screw pump is self-priming and operates up to a maximum operating pressure of 100 bar, with the metering capacity proportional to the drive speed. When metering adhesives between 0.01 (low viscosity) and 100 Pa·s (high viscosity), the eccentric screw pump results in virtually pulsation-free material discharge and is also capable of pumping filled systems with a Mohs hardness <5. The slip of the eccentric screw pump increases with service life and is not visible from the outside.

10.3.2.1.2 Valves

The valves used in adhesive metering have the task of reproducibly dispensing the intended metering quantities, with adhesive quantities down to <1 μl being handled in special applications.

- *Needle and "Snuff Back" valves* – These heated valves can be used up to an operating pressure of 400 bar and allow very accurate dispensing. They consist of an actuator and a metering system and are also capable of metering adhesives filled with abrasive particles with a Mohs hardness >5 in a viscosity range from 1 to 1000 Pa·s. *Snuff Back* valves are similar to needle valves, but by creating a slight vacuum in the outlet needle, they allow improved breakaway of the adhesive strand when dispensing is interrupted. Such valves operate at pressures up to 50 bar and are capable of handling adhesives in a viscosity range of 1–10 000 mPa·s.

- *Pinch valves* – Since the closing mechanism does not come into contact with the adhesive, these valves are preferred for dispensing cyanoacrylate and anaerobic curing adhesives. The pinch valves can only handle unfilled adhesives and operate up to an operating pressure of 10 bar in a viscosity range of 1–1000 mPa·s.
- *Proportional valves* – These valves are designed so that the variable valve cross-section allows the variation of the output. Proportional valves work according to the tap principle, the volume flow is controlled by varying the stroke, which enables them to respond very quickly to changes in setpoint requirements, for example, by a robotic system. Proportional valves operate at up to 300 bar and are also used for filled adhesives with a Mohs hardness <5 in a viscosity range of 10–1000 Pa·s. However, abrasive-filled systems change the characteristic curve of the valve.

10.3.2.2 Adhesive Application to the Substrate

When applying the adhesive to the substrate, which can be done manually or automatically and with movable or fixed nozzles, the shape of the adhesive mass on the surface to be coated is determined by controlling the associated physical and/or mechanical parameters. Manual application requires little or no investment, but the quality of the process execution depends on the person performing it and their daily form. This method still requires a very high level of attention to occupational health and safety, both in planning and execution. Automatic application methods require considerable investment costs, but due to the smaller spread compared to the manual method, they lead to a better quality of the bonded component and also shorten the cycle times.

10.3.2.2.1 Moving-Nozzle System

Automatic application systems mainly use moving nozzles. The major advantage of this method is that surface treatment can be carried out at the same time as adhesive application.

Advantages of the moving-nozzle system:

- less space required for large components,
- lower safety risk, because there is no rapid movement of heavy components,
- trouble-free application even on smooth parts,
- simultaneous treatment of the substrate and adhesive application possible, and
- no weight restriction for the components.

10.3.2.2.2 Fixed-Nozzle System

The alternative application method with a fixed nozzle enables dispensing systems with short piping systems, which is particularly advantageous for reactive adhesives.

However, the need to integrate special grippers into the process for component movement increases investment costs. In the fixed-nozzle system, the robot can perform additional joining operations as well as assembly work, and there is also the option of integrating surface treatment, usually without significantly extending the cycle time.

Advantages of the fixed-nozzle system:

- more compact dispensing systems with short hoses, thus lower investment costs,
- adhesive application over head by turning the component is possible, and
- several additional production steps are possible with a robot system, such as adhesive treatment of the material surface, assembly work, and component removal.

The type of adhesive and its physical state or behavior play an important role in the choice of application method. For example, the adhesive temperature, viscosity, fillers used, and, in the case of reactive systems, the pot life, open time, and mixing ratio of the components have a significant influence on the selection of the appropriate application method. Other parameters that influence the selection process are the adhesive thickness or adhesive quantity, the accuracy and reproducibility of the quantity to be dispensed, the geometry of the joining surface, and the desired degree of automation.

Adhesive application processes can be divided into:

- *partial-surface* application methods and
- *full-surface* application methods.

Bead and dot applications are methods in which the adhesive is partially applied to the substrate. In contrast, the objective of roller, spray, and laminating adhesive application methods is to cover the entire surface of the substrate with adhesive for the subsequent bonding process (Figure 10.9).

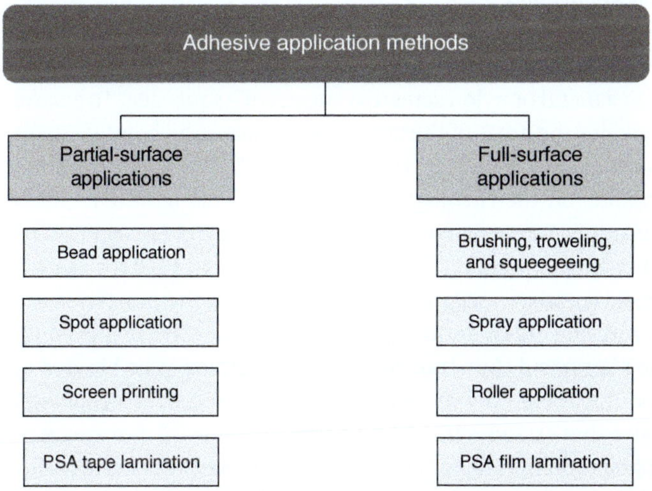

Figure 10.9 Methods for partial- and full-surface adhesive application.

The partial-surface application methods for adhesives include:

- bead application,
- spot application,
- screen-printing application, and
- PSA tape lamination.

10.3.2.3 Bead Application

For higher-viscosity adhesive systems, bead application is the most commonly used application method. Here, the adhesive is applied from nozzles, which are often heated. The type of nozzle depends on the type of adhesive and the desired application quantity. The resulting application pattern is determined by the design of the nozzle, the shape of its tip, the nozzle length, and the application method (manual, automatic), as well as by the process parameters, such as distance to the surface and application speed. If the applicator is not moved or the adhesive supply is interrupted at regular intervals, the result is spot or intermittent adhesive application.

10.3.2.4 Spot Application

Here, the adhesive application is usually accomplished by applying a droplet to the material, assisted by gravity, with high-viscosity adhesives requiring contact with the surface to apply the droplet. The method is applied when using adhesives with short open time, such as cyanoacrylate adhesives, where the droplet is usually applied from small bottles by gravity and the bonding area is enlarged by pressure after joining. Another reason for using this method is the need to bond small areas with, for example, photoinitiated, solvent-based, or water-based adhesives in many areas such as the electronics industry, household, or do-it-yourself.

10.3.2.5 Screen-Printing Application

For flat joining surfaces such as printed circuit boards, and when using adhesives with a long pot life, the dot application can be done by screen printing, in which the adhesive is forced through a metal or nylon screen by means of a squeegee. To ensure that the adhesive only reaches the areas of the substrate surface to be bonded, these screens are sometimes provided with an impermeable coating. The screen-printing process places high demands on the rheological properties of the adhesive and is used, for example, for the application of anaerobic adhesives to cylinder sealing surfaces or the partial application of liquid UV-initiated adhesives to PP door sealing films for passenger cars. In another variant of spot application, the adhesives are first applied to a stamp with the aid of a squeegee. Here, the adhesive is picked up at the raised areas of the stamp and then transferred to the surface to be bonded at about 25% of the original thickness. The prerequisite for this process is a rheological behavior of the adhesive that excludes threading during the upward movement of the stamp. The stamp application is used, for example, in the electronics industry for bonding electronic components, but has the disadvantage of limited reproducibility and low flexibility.

10.3.2.6 PSA Tape Lamination

Probably the most diverse applications with partially applied adhesive products are carried out with PSA tapes. These products are very popular in industry and trade due to their economical, fast, and clean handling compared to other fastening systems. The backing of the adhesive tapes consists of plastic, foam, paper, textile, or metallic materials. The adhesive tape can have an adhesive layer on one or both sides of the carrier. To avoid entrapment of air bubbles, adhesive tapes should be applied using an elastic squeegee or a roller, starting from one side of the adhesive surface and using the *laminating* method. The application temperature should be between 18 and 35 °C, as initial strength is reduced at lower temperatures. To improve the initial adhesion and remove the moisture absorbed on the surface, the parts to be bonded can be preheated to approximately 35–40 °C. Since the strength of the bond depends directly on the degree of surface contact with the substrate, a short, high pressure should be applied with a squeegee, pressure roller, or pressure device immediately after application of the adhesive tape. The type and strength of the contact pressure depends on the surface structure of the part to be joined, the mechanical properties of the material, and the geometry of the part. As a rule of thumb, a pressure of 20 N/cm^2 is very suitable for forming a good bond.

Full-surface applications of adhesives include:

- brushing, troweling, or squeegeeing,
- spray application,
- roller application, and
- PSA film lamination.

10.3.2.7 Brushing, Troweling, and Squeegeeing

Very often, adhesives are applied by the manual method of working by brushing, troweling, or squeegeeing. These are simple methods that can be done without much preparation or equipment setup. The disadvantage of these methods is that the layer thicknesses are difficult to control, which to some extent, can be accomplished by using notched trowels with triangular or square edge cutouts. Applying adhesives with brushes has the great advantage that curved surfaces can also be coated and corners or cavities are easily reached.

10.3.2.8 Spray Application

In spray application, the adhesive is finely atomized by aerosol formation, which requires air extraction at the relevant workplaces. This process is preferred when processing hotmelt adhesives, including PSA, based for a wide variety of laminating applications. Here, the spray air is often preheated to achieve the longest possible open times for the adhesives. One variant of spray application is so-called vortex spraying, in which the adhesive is formed into a cone by a stream of air exiting the nozzle at the same time as the adhesive. This enables sharp and clear delineation of the adhesive surface. Another variant is spin-spray application, which is used, for example, in the manufacture of sandwich joints. In this process, thin adhesive layers are formed over a large area by adhesive threads without atomization of the adhesive and formation of aerosols.

10.3.2.9 Roller Application

The roller application method is only suitable for web-shaped substrates and enables very thin coat thicknesses to be applied evenly over the entire surface. This makes it the basic method for laminating a wide variety of flexible materials onto substrates. In the roller process for laminating films, for example, a rubber-coated roller is first immersed in a reservoir and, after the excess adhesive has been removed by a doctor blade, transfers the remaining quantity to a structured roller (anilox roller). From there, the adhesive is then transferred to the substrate in a well-dosed manner.

10.3.2.10 PSA Film Lamination

For adhesive technology applications in surface design, surface protection and retroreflective traffic signs, self-adhesive films, often based on PVC, are applied to the substrates over their entire surface. In most cases, such self-adhesive products are bonded to aluminum alloys or plastic materials. The process is carried out using rubber rollers, pressure rollers, or squeegees and requires absolutely clean grease-free substrates and processing temperatures of at least 18 °C. To avoid the inclusion of air bubbles and to ensure the mobility of the film product until it reaches its final position on the substrate, self-adhesive films are often applied wet. In this process, the material surface is sprayed with water prior to application and the self-adhesive film is then applied. After exact positioning of the film, the water is then pressed out starting in the center of the adhesive surface with the aid of elastic rollers or squeegees, resulting in a bubble-free adhesive bonded component. Even without the use of water, modern self-adhesive films can be bonded effortlessly and in a short time without the inclusion of air bubbles or wrinkles. This is made possible by microfine air channels in the adhesive of such products, which allow the air under the film product to escape during bonding. At the same time, tiny glass beads in the adhesive layer ensure problem-free positioning of even large film areas. In this way, even inexperienced users succeed in making trouble-free and fast adhesive bonds with such products. This method thus enables a significant reduction in scrap when carrying out corresponding applications. Hotmelt adhesive films are frequently used for a wide variety of two-dimensional applications with low heat resistance requirements. In this process, the film is inserted between the parts to be joined and applied with pressure and heat to approximately 20–30 °C above the softening range of the hotmelt adhesive, which enables the surfaces of the parts to be joined to be wetted by the adhesive. After cooling, the bond has sufficient strength for appropriate further processing.

10.4 Processes for Cohesion Build-up

Having discussed the process steps relevant to building adhesion, we now turn to the process steps responsible for ensuring cohesion:

- adhesive component mixing,
- joining and fixation of the bond, and
- curing of the adhesive layer.

10.4.1 Adhesive Component Mixing

When using 2K adhesives, the adhesive components must be mixed to perform the chemical reaction required for curing. This process step takes place after the adhesive components have been precisely metered and before the adhesive is applied to the material surface. 2K adhesives have the advantage that, after mixing, the curing reaction usually takes place at room temperature and no curing ovens need to be provided, which has a very positive effect on both the economy and the complexity of the bonding process. During curing, the parts to be joined must be fixed against movement at least until they reach a handling strength, which can be done either by suitable devices or additionally applied adhesive strips. During mixing, different requirements are placed on the mixing quality during processing, depending on the adhesive chemistry used. For example, in the case of free-radical curing systems, such as the 2K acrylate adhesives, relatively small amounts of hardener and accelerator are added to the base monomers to catalyze and accelerate the reaction. Due to the curing mechanism, these adhesive systems cure in a relatively wide range of mixing ratios without serious loss of strength, with the curing time decreasing with increasing addition of hardener.

However, unlike the radical-curing systems, polyaddition-curing adhesives, such as 2K epoxy adhesives, must adhere to the ratio of the two components specified by the adhesive formulation to achieve optimum bond strength. In case of deviation, residual amounts the components in the adhesive layer would act as plasticizers and thus damage the adhesive bond. It is not possible to shorten the curing time with these systems by increasing the addition of the second component, often also referred to as the hardener. When mixing the components, care must always be taken to ensure that they are evenly and homogeneously distributed in the finished adhesive batch and that any deviations in the mixing ratio do not exceed or fall below a value of 5%. If a small amount of dye is applied to a component for visual control of the mixing quality, the mixing process must be optimized until a uniform color of the mixed adhesive is achieved. When mixing adhesive components manually, the required mixing ratios should be specified in units of weight, since in this case greater accuracy can be achieved than with units of volume. In automated mixing and metering systems, on the other hand, mixing is usually volume-based due to the available control options, and both the mixing ratio and the homogeneity of the finished adhesive mixture are continuously ensured by appropriate control mechanisms.

For homogenization of 2K adhesive systems, the following mixing methods are used:

- batch process with agitators,
- static mixers, and
- dynamic mixing heads.

10.4.1.1 Mixing in Batch

In batch processing, the components are metered and mixed in an appropriate container exactly according to the manufacturer's specifications. With the first contact of the components in the mixing container, the pot life begins, which is the time limit of the workability of the adhesive. If the prepared mixture is not to be further processed, it is possible to stop the chemical curing reaction in the adhesive mixture by freezing the batch. This process is used, for example, in the optical industry to perform extensive quality assurance measures on the adhesive mixture before it is released for production.

10.4.1.2 Static Mixing

Static mixers are simple mixing devices that are effective in a wide range of applications. They consist of a mixing tube with mechanical mixing elements and are very often used in the continuous processing of 2K systems such as epoxy or PU adhesives. Since these mixers do not use moving parts, they operate very quietly. To achieve the desired mixing results, the static mixer uses only the energy of the flow, which results in a pressure drop across the mixing section. Here, the type of static mixer, its size, the number of elements, and the Reynolds number of the fluid are the main factors that affect the pressure drop and the performance of the static mixer. Once the optimum type and size of static mixer is selected, the number of elements required for the mixing process is chosen, which is usually a function of the Reynolds number of the fluid. Better coefficients of variation are achieved by adding additional elements.

In static mixing, the adhesive components to be mixed are forced into the mixer in the desired ratio and then the mixing elements divide the original material stream, turn the streams, and recombine them. If the first element divides the two-layer material flow of the adhesive components, there will be at least 4 layers after merging. After the next, there are 8 layers, then 16, and so on. In this way, better mixing is achieved at each stage. Thus, static mixers are energy-saving and efficient mixing devices suitable for a wide range of applications. There are many different types of static mixers on the market, such as the commonly used Kenics and Sulzer mixers.

The mixing tubes demonstrated in Figure 10.10 are characterized by the following features:

- *(a)*: 16 elements, 65 536 layers, about 2.6 g of adhesive in the mixer, 63 mm of length.
- *(b)*: 16 elements, 65 536 layers, approximately 2.9 g of adhesive in the mixer, 78 mm in length.
- *(c)*: 20 elements, 1 048 576 layers, approximately 5.1 g of adhesive in the mixer, 132 mm in length.

Kenics mixers use round-shaped mixing spirals, each offset by 90°, which divide the components to be mixed at the edges of the mixing spirals (Figure 10.11a). An alternative is static mixers based on the Sulzer principle with rectangular cross-sections, which allow a more compact design of the mixer (Figure 10.11b). The inlet positions for parts A and B of the adhesive in such a mixer are shown in Figure 10.12.

10.4.1.3 Quality of the Mixture

In a static mixer, the desired mixture homogeneity can be determined in advance by the number of mixing coils. Depending on the number of coils, the number of

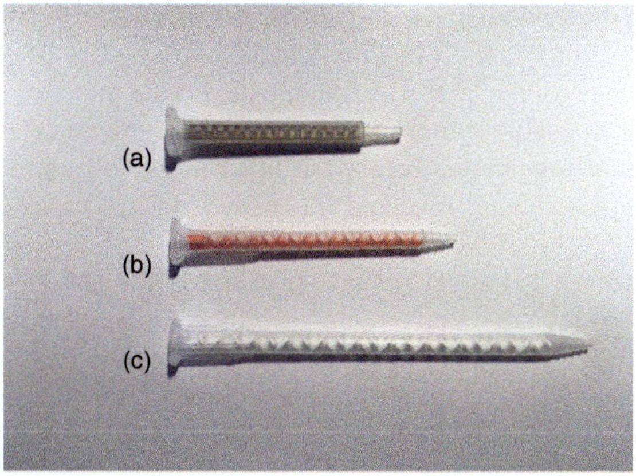

Figure 10.10 Static mixers for 50 ml cartridges according to the Sulzer (a) and the Kenics principles (b and c). Source: Dr. Jürgen Klingen, Aboso-Consulting.

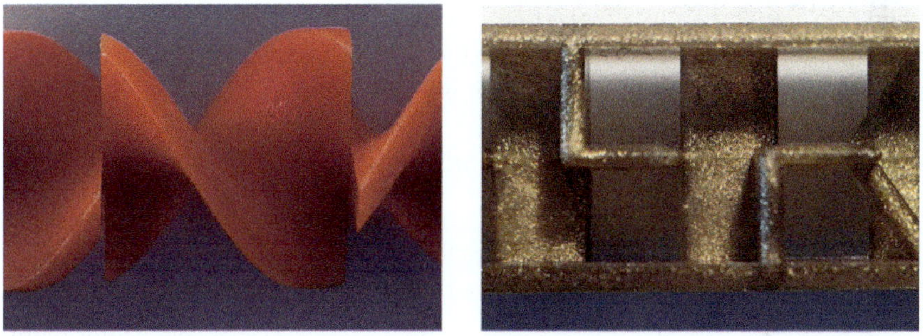

Figure 10.11 Mixing coils of static mixers according to the Kenics (a) and Sulzer (b) principles. Source: Dr. Jürgen Klingen, Aboso-Consulting.

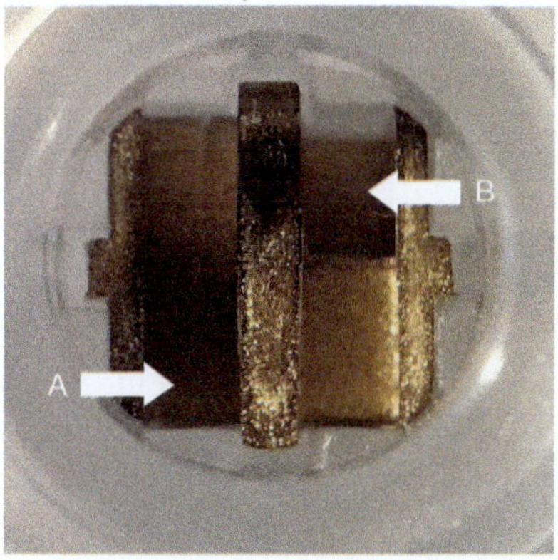

Figure 10.12 Adhesive inlet of components A and B into a mixer according to the Sulzer principle. Source: Dr. Jürgen Klingen, Aboso-Consulting.

resulting layers at the end of the mixing process is calculated according to the law of geometric series: $s = 2^n$ (where s = number of layers and n = number of coils). Thus, in a mixer with 16 mixing coils, the result is a radial mixture of the adhesive consisting of 65 536 layers.

For the mixing quality of adhesives processed with such systems, the number of mixing elements lined up and also the viscous properties of the adhesive mass are responsible. To ensure that the mixers are matched to the adhesive system used and thus enable optimum mixing qualities, only mixers approved by the adhesive manufacturer should be used. If this cannot be guaranteed, it is essential to check the mixing quality on cured adhesive samples, as a visual assessment is not sufficient.

10.4.1.4 Process Technology of Static Mixing

Due to their simple design and low unit costs, static mixers are mainly used as disposable mixers and disposed of after one time use. If the pot life at the nozzle outlet is exceeded due to interruption of the process chain, or if the viscosity of the adhesive mass in the mixing tube has increased to such an extent that the pressure exceeds the set limits, the mixer must be replaced. For extending the pot life of the adhesive during longer production interruptions, it is possible to store the mixing tube and application gun in a freezer at approximately −25 °C, which reduces the reactivity of the mixture. To avoid clogging by cured adhesive and to ensure a quick start-up of production after interruptions, it is also possible to flush the mixing tube with one of the two components.

In general, the following applies to the trouble-free processing of 2K systems with static mixers:

- To avoid curing in the mixer, two to three times the adhesive mass present in the 100% filled mixing tube should have passed through the mixing system within the pot life of the adhesive system used.

Here is an example:

- Using a hand mixer for 50 ml cartridges, a 2K epoxy adhesive with a pot life of 90 minutes, mixer with 16 elements, is to be metered, mixed, and applied to a substrate using a Kenics mixer. The adhesive mass in the 100% filled mixer is 3 g and according to the rule described above, at least 6–9 g must be processed within the pot life of the adhesive, which in this example is 90 minutes. If 2 g of adhesive is applied to a component to be bonded in 5 seconds and then paused for 10 seconds, the total processing time for 2 g of adhesive is 15 seconds. Thus, 720 g are processed through the Kenics mixer within the pot life of the adhesive, which is 120 times more than the minimum allowable amount (6 g) within the pot life of the adhesive used. Thus, for 6 g of adhesive, the time in the mixer is 45 seconds, which is well below the adhesive's pot life of 90 minutes. So, the adhesive used in this example is very well suited to the application.

The question often arises as to how long an adhesive cartridge can work with only one mixing tube.

Here is the next example:

- As described in the example above, when processing an adhesive with a pot life of 90 minutes, at least 6 g must be forced through the mixer in that time. Thus, an entire 50-g cartridge must be used in 750 minutes (approximately 12 1/2 hours) when using a mixer filled with 3 g of adhesive. It is important to note that this only applies if at least about 0.7 g of adhesive per 10 minutes is uniformly processed during this time.

Summary static mixing of adhesives:

- static mixing is performed by fixed mixing elements arranged in a mixing tube,
- the mixing energy is extracted from the flow energy,
- mixing is performed by alternating clockwise and counterclockwise mixing elements, which divide the adhesive flow into layers until homogeneity is achieved,
- the number of layers can be determined by 2^n (n, number of mixing elements),
- requires high attention when pressing the two components of difficult to mix adhesives into the mixer,
- is carried out with very inexpensive mixers, which are maintenance-free, but have no possibility of adjustment,
- to avoid curing in the mixer, requires that two to three times the adhesive mass present in the 100% filled mixing tube should have passed through the mixing system within the pot life of the adhesive system used,
- offers many types and sizes to suit the mixing task,
- can be performed at mixing ratios from about 100 : 1000 to 100 : 3, and
- viscosity differences of the components up to about 10 : 1.

10.4.1.5 Dynamic Mixers

In the case of large viscosity differences or extreme mixing ratios of the components, the more complex dynamic mixing process can be used to mix the adhesive components. Here, they are mixed in the chamber of the application head by rotating mixing elements and, in contrast to static mixing systems, the mixing energy is thus supplied from outside. To prevent undesirable curing of the adhesive in the mixer after longer production breaks, cooling of the system is often necessary. Clogging of the rotor is avoided by flushing the mixer with suitable flushing media after processing. When bridging shorter pauses, the mixer can also be flushed with fresh adhesive.

Summary dynamic mixing of adhesives:

- dynamic mixing is performed by moving mixing elements in a closed mixing chamber,
- the mixing energy is supplied from outside,
- the mixing speed is a decisive parameter for the quality of the mixture,
- the size of the mixing chamber should be selected to ensure a residence time of the adhesive of at least one second,
- intermediate rinsing is necessary in case of interruptions,
- is a technically complex process with a correspondingly high price,
- allows many variants and sizes for adaptation to the mixing task,
- can be used with ratios of about 100 : 100 to 100 : 1, and
- viscosity differences up to about 100 : 1.

10.4.2 Joining and Fixation of the Bond

10.4.2.1 Joining

The joining of components can be performed manually or automatically, whereby the choice of method can only be made with regard to the overall concept for the desired joining process. Automatic joining, for example, is often an integral part of the overall process through the use of robots. There are many examples of this in the automotive industry, where very large quantities of joined components are almost always produced. However, when planning the automation of the joining process, the economic efficiency related to the individual case must always be carefully examined.

10.4.2.2 Fixation

Fixation of the component after joining, which is necessary for proper and reproducible curing of the adhesive, is carried out by various measures or by exploiting different effects. The simplest method of fixing is to utilize the effect of gravity, with the advantage that no additional devices are required. However, the fixation of freshly bonded components can also be achieved relatively easily by constructive precautions, such as the installation of additional snap connections or groove/spring systems. So-called hybrid processes are also frequently used to fix bonded joints by combining adhesive bonding with mechanical joining processes such as spot welding, riveting, flanging (hemming), or bolting. This is done in automotive engineering, where such bonding processes have been very well established for many years. Here, adhesive bonding is combined with spot welding or flanging to produce body-in-white and add-on parts such as doors, hoods, trunk lids, and flaps. Very popular and also possible without great expense is the fixing of bonded components before curing using double-sided adhesive tapes. The use of separate equipment specially designed for this requirement is the most complex and expensive method of fixing and should, for economic reasons, only be used where other methods are not possible.

10.4.3 Curing of the Adhesive Layer

After application of the adhesive and joining of the materials, the adhesives must be cured. Contact adhesives, where joining takes place after the adhesive has cured by evaporation of the solvent, and PSAs, which are applied and joined as supplied, are the exceptions here. The conditions under which curing takes place are decisive for the chemical and physical states of the adhesive layer and thus for its properties such as cohesion and adhesive strength. Temperature, time, and pressure are often the determining process parameters. Depending on the adhesive chemistry selected, however, UV radiation and moisture can also play a decisive role in the curing of the adhesive layers. In curing, an extremely important process step for the quality of the adhesive layer, a distinction is made between physically and chemically curing systems.

10.4.3.1 Physically Curing Systems

In physically curing systems, the macromolecules are already fully reacted and enclosed in the adhesive system. To cure the adhesives, only volatile components (such as organic solvents or water) must be evaporated by the influence of temperature and time in the case of solvent/water-based adhesives or, as in the case of hotmelt adhesives, solidification must be carried out by lowering the temperature.

So, the final strength is achieved by the following process steps:

- evaporation of water or organic solvents at room or elevated temperature after joining as in the case of wet bonding,
- evaporation of water or organic solvents at room or elevated temperature before joining and pressing together two substrates containing already dried adhesive layers as in the case of contact bonding,
- solidification of a polymer melt by lowering the temperature as with hotmelt adhesives,
- adhesion building of thermoplastic adhesive films by elevated temperature and simultaneous pressure application as with hot melt adhesive films, and
- bonding of dry self-adhesive products by applying pressure at room temperature exclusively, as in the case of PSA tapes and self-adhesive films.

10.4.3.2 Chemically Curing Systems

In contrast to physically curing systems, chemically curing systems do not contain reacted polymers. To introduce the required adhesion and cohesion properties into the adhesive layer, chemical reactions must take place under the influence of suitable process parameters to form the polymers and ensure crosslinking. In this context, the development of adhesion and cohesion is a time- and often temperature-dependent function that is specific to the curing mechanism in question. Of crucial importance is the turnover of the components involved in the reaction and thus the degree of curing. For this reason, the process parameters must be optimized to achieve almost 100% conversion.

The following methods are used to cure these chemically curing adhesives:

- reaction with moisture, as with cyanoacrylate, 1K PU, and 1K silicone adhesives,
- reaction under exclusion of air and simultaneous metal contact for anaerobic adhesives,
- reaction due to elevated temperature as in the case of 1K epoxy adhesives,
- reaction due to the influence of visual or UV radiation as in the case of photoinitiated acrylate and epoxy adhesives,
- reaction after mixing of two or more components as in the 2K epoxy, 2K acrylic, and 2K PU adhesives,
- solidification of a melt as in the case of non-reactive hot melt adhesives,
- solidification of a melt and subsequent reaction with atmospheric moisture or under heat, as with reactive PU and epoxy hot melt adhesives, and
- reaction of adhesive films by elevated temperature and simultaneous pressurization, as in the case of reactive adhesive films.

For a rough estimate of the influence of temperature on reaction time, the Arrhenius equation can be used for temperature-dependent curing reactions, which states that a temperature increase of 10 °C leads to approximately a doubling of the reaction rate.

For example, when using 2K adhesive that requires a curing time of 240 minutes (6 hours) at 20 °C to achieve hand strength, it would be possible to reduce the required curing time by making the following temperature adjustments:

- at 30 °C to 120 minutes,
- at 40 °C to 60 minutes,
- at 50 °C to 30 minutes,
- at 60 °C to 15 minutes,
- at 70 °C to 7.5 minutes, and
- at 80 °C to 4 minutes.

Thus, when performing a bond with such a reactive 2K adhesive system, the curing time required for hand strength can be reduced from 6 hours to 4 minutes, for example, by increasing the temperature from 20 to 80 °C. This effect of shortening the reaction time by increasing the temperature is used, for example, in the automotive industry to quickly achieve the so-called *hand strength* with 2K epoxy adhesives, allowing bonded parts to be transported, processed, and stored within a production process immediately after joining, without changing the positions of the substrates used relative to each other. In the case of automotive hem flange bonding, this method is applied for the production of car doors, trunk lids, engine hoods, and flaps using special induction-curing 2K epoxy adhesives. Immediately after adhesive application, the prepared fresh bonds experience a temperature rise to around 100 °C, which leads to partial curing of the adhesive layer and thus to rapid build-up of the cohesion required for *handling strength*. This temperature rise is caused by the action of induction heating for a few seconds, which ensures rapid further processing of the bonded components and thus enables short cycle times as well as low waste rates.

Reference

1 Klingen, J. and Patrick G. Zimmerman (1987). Pressure-Sensitive Adhesive Tape Containing Hydrophobic Silica. US Patent 4 710 536.

11

Start of Production (Step 5)

11.1 Work Done So Far

Now that we have completed the development step, an optimized and validated adhesive-bonding process is ready for production of the adhesive bond with sub-processes for surface treatment of the materials, application and curing of the adhesive, and feeding, fixing, and removal of the components.

To get to this point, the following stages had to be passed through, steadily increasing the maturity of the project:

- preparing the plan and project contract (Step 1),
- study substrates/adhesives and selection/definition of concepts (Step 2),
- demonstrating the feasibility and validation of the preferred concept (Step 3), and
- establishing a robust manufacturing process ready for production (Step 4).

In the final stage, we procured the optimum production equipment for all relevant sub-processes in preparation for the start of production, after extensive test runs had been carried out at the suppliers' and our own facilities to prove their suitability. Subsequently, the selected process was optimized using statistical and experimental methods, which ultimately led to a robust production process for manufacturing the bonded joint.

This is now available for production, which only needs to be planned and started. Figure 11.1 shows the essential tasks with the corresponding results that must be available to introduce the process into production.

Adhesive Bonding in Five Steps: Achieving Safe and High-Quality Bonds, First Edition. Jürgen Klingen.

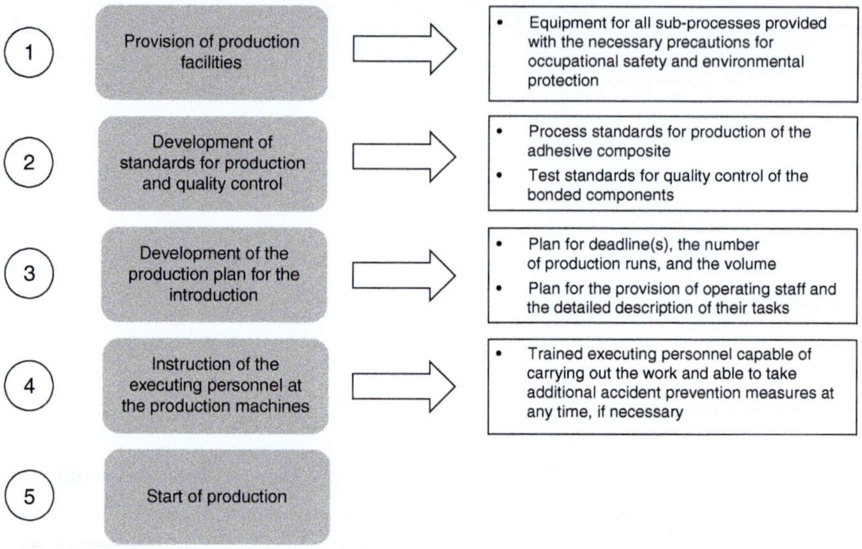

Figure 11.1 Essential tasks to be accomplished for the start of production.

12

Contemporary Adhesive-Bonding Applications

12.1 Introduction

This chapter deals with some selected contemporary adhesive applications in industry and craft, focusing on the use of adhesive-bonding technology in lightweight and facade construction as well as bonding applications with new structural 2K adhesives and high-performance self-adhesive tape systems. Today, adhesive manufacturers offer customized adhesive systems for a wide range of applications. Here, in modern and contemporary applications, user-friendly processing conditions of the adhesives during production play an important role in addition to the achievable strengths and material properties.

Process-oriented adhesive classifications are therefore based not on the chemical basis but above all on their processing possibilities and distinguish, for example, between:

- processing as solids, solutions, dispersions, or melts,
- physically or chemically curing systems,
- curing by drying, mixing, irradiation, influence of moisture, exclusion of oxygen, or addition of heat,
- short or long pot lives, and
- low or high viscosity.

12.1.1 Adhesive Market Development

Under the influence of increasing environmental awareness and environment-friendly emissions legislation, adhesive systems that meet the relevant criteria and largely dispense with the use of solvents are increasingly being used today. The highest requirements for strength and temperature resistance in structural adhesive bonds can still only be achieved with reactive adhesives. Since high-quality raw materials are generally required for their manufacture, the market share of this group in terms of value is higher than its market share in terms of volume compared with dispersion, solvent-based, or hotmelt adhesive systems. The construction,

Adhesive Bonding in Five Steps: Achieving Safe and High-Quality Bonds, First Edition. Jürgen Klingen.

craft, paper, and packaging industries are now among the largest users of adhesives in terms of volume. While largely stable sales volumes have been observed in these sectors and also in wood processing in recent years, the use of adhesives in vehicle construction, mechanical and apparatus engineering, electronics, and medical technology has increased significantly over the same period. The reason for this is that, due to modern manufacturing processes and designs, the use of adhesives instead of conventional mechanical fastening systems has resulted in significant advantages in the manufacture of the corresponding components.

12.1.2 Customer Requirements for Adhesive Bonding

All bonding applications have specific handling, material, and strength requirements that determine the particular type of construction. A cruise ship, for example, is particularly stable and also very heavy due to the materials used. A mobile home, on the other hand, is very light, functional, and inexpensive, but also much less stable. In these two examples, the requirements for handling, strength, and the material to be used for the bonded component are very different, which has a significant effect on the user's choice of adhesive. In any case, contemporary industrial and craft manufacturing requires the reduction of production cycle times, production costs, and – very important – weight of the components. For all these requirements, adhesive bonding has decisive advantages, which should ensure the use of adhesives for the manufacture of industrial products to an even greater extend in the future.

12.2 Adhesive Bonding in Lightweight Construction

12.2.1 Introduction

Today and in the future, energy must be saved in many areas of life due to limited resources. To this end, many options are being discussed in public, but these often come at the expense of quality of life and therefore lack broad acceptance. However, there are also ways without this disadvantage. One of these is the reduction in the weight of technical products, which is far from being exhausted and where adhesive-bonding technology offers many innovative possibilities. For example, electric welding of the light metal aluminum requires about 4 times as much energy as that of steel, which is why it would be better to use adhesives to bond the light metal.

12.2.2 Lightweight Requirements in Transportation

If we look at the transportation industry, such as aircraft and vehicle construction as well as railroad and shipbuilding, the movement of their products consumes energy in addition to the energy used for manufacturing. In relation to their payload, means of transport are very heavy, which must be improved by new design methods, where bonding can play an important role. In this way, fuel consumption can often be significantly reduced through weight reduction, one reason why adhesive-bonding technology is already widely used in the transportation sector. It has been known for many years from aircraft construction that the structural weight of a commercial aircraft can be reduced by at least 20% through the consistent use of adhesive-bonding technology compared with other bonding methods [1]. Automotive manufacturers see a similar savings potential, and in railroad construction this figure is probably even higher. It is therefore worth to reflect on the possibilities of adhesive-bonding technology in these and other areas of industrial production and to include its innovative and contemporary developments into these considerations.

12.2.3 Lightweight Bonding with Modern "Crash-Resistant" 2K Adhesives

Compared to welded seams, adhesive-bonded joints in the automotive industry have numerous advantages, especially in the ever-increasing use of lightweight construction. Thus, the adhesives demonstrate higher resistance to fatigue, are about 30% stiffer than a pure spot weld bond, and increase the so-called torsional stiffness of the body, which in turn leads to an improved driving behavior of the automobile. In addition, there is improved corrosion protection and its sealing function. Adhesive-bonded components created by the use of very different materials, which include for example aluminum or plastics, also lead to weight savings. It has already been discussed, that according to the experts, the weight of a vehicle could be reduced by about 20% in the future, if the adhesive technology would be consistently used. Above all, however, adhesive bonds positively influence the behavior of a vehicle in the event of an accident in the form of an impact. In this case, numerous microscopic cracks form in the adhesive, which have a high damping effect. So, the energy from the impact is virtually absorbed by the adhesive-bonded component, resulting in good *crash behavior*.

Until a few years ago, these properties could only be achieved by heat-curing 1K epoxy adhesives. For some time now, however, there have been new, modern, room-temperature-curing 2K epoxy adhesives for the automotive industry that, in addition to excellent crash resistance (Figure 12.1), offer significant advantages, especially for lightweight construction. The reason for this is the significantly poorer weldability of the new light metals used in lightweight construction, such as aluminum and new very thin steel sheets. To successfully bond such materials with conventional 1K epoxy adhesives in body construction and avoid component distortion, additional spot welding in a hybrid process is necessary, which increases the complexity and cost of the process.

The great advantage of the new modern 2K epoxy adhesives is that they can perform this application without the additional step of spot welding, which avoids warpage of the bonded components and thus simplifies the process and also reduces costs. They thus meet very well the requirements of the automotive industry, which demands a high level of quality and reproducibility for the growing proportion of bonded, safety-relevant components and trouble-free automation in automotive production.

Figure 12.1 Bonded components after a crash test in which the joint bonded with a conventional 2K system is destroyed without significantly absorbing energy (right) and leaving an intact component that absorbed the maximum energy during the impact when the component is bonded with the new crash-resistant 2K adhesive (left). Source: 3M.

Unlike 1K adhesives, whose strength build-up starts at 180 °C, the new 2K products already cure at room temperature, which can be accelerated by a short inductive curing step at 130 °C in the range of seconds to produce handling strength. This is a decisive advantage, because the very early fixation of the bonded component at significantly lower temperatures reduces stresses in the component, resulting in higher process quality and the avoidance of rejects.

In recent years, considerable progress has been made to replace the 1K systems with the new low-temperature curing 2K products. In this context, crash-resistant and oil-absorbing adhesive variants have led to a leap in innovation, so that these products are not only able to simplify manufacturing processes, but also help to minimize manufacturing tolerances and thus avoid production scrap. In addition, these new 2K systems enable completely new design concepts. For example, a car could be quickly assembled from completely prefabricated individual parts with already-finished (painted) surfaces using adhesive-bonding technology. To ensure that the car can then also roll off the assembly line quickly, the adhesive is cured at relatively moderate temperatures in the seconds range to make the bonded component manageable in further processing. This can be done, for example, by brief inductive heating at relatively low temperatures of 130 °C, for example. Complete curing of the adhesive layer is then achieved at room temperature without the need for an additional process step.

12.2.4 Elastic Bonding in Truck Construction

Similar ideas exist in truck construction, where large containers made of different, already-finished materials are often joined using 1K PU adhesives without rivets or screws (Figure 12.2). This application has proven itself in tough use for many years. Modern self-adhesive high-performance foam tapes are also being used more and more frequently for such tasks, serving not only as an elegant joint

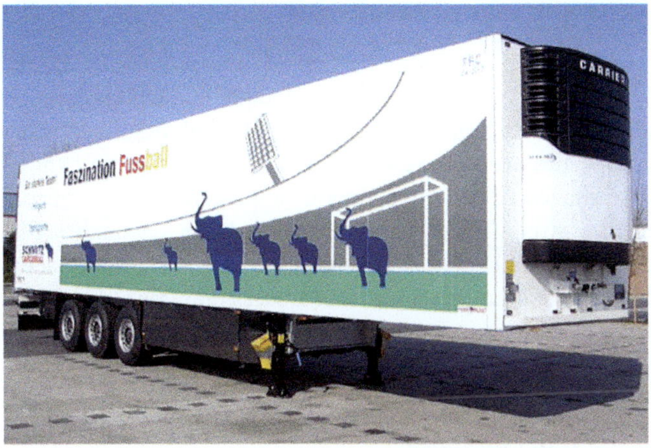

Figure 12.2 Trailer assembled with the help of a high-performance acrylic foam tape without using rivets or screws. Source: 3M.

seal with corrosion protection, but also as a spacer. These products are also used in roof fastening as a clean, easy-to-use, and high-performance substitute for reaction-curing adhesives.

12.2.5 Printable Adhesives

Printable adhesives are frequently used to reduce weight and waste in the automotive and electronics industries [2]. These innovative, solvent-free products can be printed in small quantities and positioned exactly where they are needed, thus helping to reduce weight and waste, for example in the production of cell phones. Once these products have been applied – often by screen printing – the photoinitiated adhesive is cured by visible light or UV radiation.

12.3 The Role of Adhesives in Modern Facade Construction

Bonding in the facade area, for example using concrete, steel, aluminum, and not least glass, is of the greatest interest to adhesive-bonding technology. In this field of application, the use of lightweight systems leads to considerable weight savings, with thin exterior cladding, for example made of resistant natural stone or also glass, also being very visually appealing. Glass, with its largely absent plastic deformability, high strength, and a specific weight of approximately 2.8 g/cm^3 (similar to aluminum), is an excellent, long-term resistant lightweight construction material whose application limits are far from being reached. Due to the sensitivity of glass to local stress concentrations, adhesive bonding is a particularly suitable joining method for this material, since the bonded surfaces used in this joining technique are generally large and thus allow uniform stress distribution in the glass component.

12.3.1 Structural Glazing with Silicone Adhesives

Structural glazing refers to linear glass bonding of facade elements with elastic adhesives, whereby only the adhesive bond supports the weight of the glass component. Special silicone adhesives with high elasticity, relatively high holding power, and very good sealing properties are usually used for this application, but these products have some disadvantages. For example, they are applied in liquid form and cure only slowly to their final strength after application. To obtain a geometrically defined joint with a reproducible thickness, it is usually necessary to use a spacer, which is filled manually with the liquid silicone. This process requires a high level of expertise, particularly due to the horizontal positioning of a facade element, and leads to a very high reject rate.

12.3.2 Structural Glazing with PSA Tapes

Because of this problem, in addition to liquid silicones, self-adhesive high-performance foam tapes are increasingly being used for adhesive applications in exterior facades (Figure 12.3) [3]. These products do not have the disadvantages described for silicone adhesives and thus eliminate the need for a spacer, which is required when using liquid adhesives. However, the greatest advantage of using these adhesive tapes is the possibility of storing the bonded facade elements vertically immediately after bonding or transporting them directly to the construction site. This saves the storage space previously required, since when bonding with silicone adhesives, the elements often had to be stored horizontally for days to cure before further processing. Thus, modern self-adhesive foam products meet the high demands of the construction industry. They have the same functional advantages over mechanical fasteners as silicones but at the same time allow easier

Figure 12.3 Glass facade in Brazil, bonded with a high-performance acrylic foam tape. Source: 3M.

Figure 12.4 Pendulum impact test applied on a glass construction, bonded with high-performance acrylic foam tape. Source: 3M.

and faster processing, which thus leads to greater efficiency and cost reductions in this contemporary application.

In South America, structural glazing has been implemented with self-adhesive foam tapes for many years since the early 1990s, with Brazil leading the way. Another example of the use of these tapes in this application area is the glass roof structure of a highly frequented British railroad station, where extremely high safety requirements had to be met.

In such applications, the performance of the elements under impact loads often has to be demonstrated, and glass structures made with high-performance acrylic foam tapes were also tested. The test is designed to ensure the safety of the glass elements, with the soft pendulum simulating the impact of a human body. Thus, a double-tire pendulum was dropped as a 50 kg elastic impact body from different heights against a flat, taped glass pane of approximately 1.7 m^2 in size (Figure 12.4).

Before applying the adhesive tape, the glass pane and the frame were first cleaned and both parts were bonded under pressure after applying the adhesive tape. Subsequently, the test could be started after a short time, since – compared to silicone applications – adhesive tapes ensure a quick build-up of the adhesive force. Here, the bonded glass construction survived the impact of the pendulum with a mass of 50 kg from a drop height of 900 mm. Even when the mass was doubled to 100 kg and the drop height increased to 1200 mm, the adhesive tape system easily absorbed the enormous forces of the impact and withstood the strong vibrations.

12.4 Adhesive Bonding on Low-Energy Plastics

12.4.1 Introduction

To be able to meet the trend toward lightweight construction and the associated increased use of plastics, an important goal in research and development is to further improve adhesives for bonding to low-energy surfaces. Modern plastics are based on materials that have extremely low surface tension, which is why they are also referred to as low energy. For this reason, it has been very difficult to achieve molecular interactions on the surfaces of these materials, which are necessary for the required adhesion. Another problem is that modern plastics are often additionally structured to achieve a better look or feel. Additives, for example to achieve self-cleaning properties or a high scratch resistance of the surface, also have a negative effect on adhesion, as they usually reduce the surface tension. Finally, bonding processes carried out with plastics often have great difficulty with silicone-containing release agents, which are frequently used, for example, in injection-molding processes to produce plastic articles. Other relatively low-energy surfaces are represented, for example, by powder and so-called self-healing coatings, with the powder-coating technique in particular having gained increasing importance in recent years. Here, in addition to the low environmental impact, the high scratch and impact resistance as well as the very good solvent resistance of these coatings are decisive for the increased use of this low-emission coating process in a wide range of applications, for example in facade, window, and furniture construction as well as in the machinery and electrical sectors.

12.4.2 R&D for Bonding to Low-Energy Surfaces

To meet the challenges for good bonding to low-energy surfaces, complex analyses of these materials had to be performed [4]. These allowed the determination of the chemical nature of the substances on the material surfaces, which determines the low surface tension. For this purpose, modern methods such as electron spectroscopy for chemical analysis (ESCA) were used to precisely determine the individual molecules in the first nanometers of the low-energy surface. In addition, information on the migration rate of these molecules to the surface was obtained, which allowed estimating the long-term evolution of adhesion on the corresponding surfaces based on the time-dependent change in surface chemistry. Finally, the accurate determination of the surface topography of the critical materials was achieved by confocal laser microscopy. This R&D work enabled the development of new adhesive systems that fit the surfaces like a *lock-and-key* system, allowing good molecular interactions at the surfaces, which then lead to good permanent adhesion. In addition, on the basis of these intensive investigations, the rheology and here in particular the flow behavior of these novel adhesives on structured surfaces was precisely adapted to enable complete surface contact and thus significantly improved adhesion. The results of this research and development work are novel adhesives that meet the challenges of low-energy surfaces described above. These new systems exhibit very high adhesion values on these surfaces and prove themselves in applications in which only mechanical bonding was previously possible. They are thus able to replace mechanical joining methods such as riveting, screwing, or welding and enable modern designs using low-energy materials that were not possible in the past.

12.4.3 Self-Adhesive Transfer Tapes

In a wide range of applications, modern self-adhesive transfer tapes enable even slimmer designs for very different products or components, from cell phones to

Figure 12.5 Use of self-adhesive transfer tapes for bonding controls in the automotive industry. Source: 3M.

automotive controls (Figure 12.5). Even on critical substrates such as PE and PP, the products exhibit strong adhesion, which is also the case with glass, nylon, powder coatings, and UV coatings. Important for modern economical production is the very high initial tack of these new tapes, which allows easy and fast further processing after bonding. In automotive engineering, the new products open up completely new possibilities for the elegant design of control elements and can be also used, for example, for the assembly of plastic housings or components, for the attachment of decorative panels in vehicle interiors, or for stamped parts used for noise damping.

These adhesives also offer advantages in the design of slim housings and components in the electronics industry, where die-cut parts for liquid crystal displays (LCDs) or the attachment of lenses are typical applications. With its good adhesion even to powder coatings, the adhesive is also the bonding technique of choice when name and specification plates, decorative trim, or other decorative elements need to be attached in the production of household appliances. As in the production of electronic components and products, its temperature resistance of up to 180 °C is also very advantageous for this application. The long durability of the joints made with these new adhesive tapes also allows them to be used in the production of high-quality components and devices, such as those used in medical technology.

12.4.4 Structural Adhesives for Low-Energy Surfaces

For a long time, structural bonding with PE and PP was not possible without surface treatment; however, after corresponding R&D activities, room-temperature-curing 2K acrylate adhesives for high-strength joining of low-energy materials have now been offered for some time. These adhesives are suitable for structural bonding of many plastics such as ABS, GFRP, PS, PMMA, PE, and PP as well as for high-strength bonding of metals such as aluminum, copper, brass, and steel. The modern adhesives are easy to process from double cartridges in combination with static mixers, with metering, mixing, and application taking place in just one operation. Surface treatment of the plastic surfaces is not required.

These systems have the following properties:

- very short processing time,
- fast strength build-up,
- high adhesive strength,
- good peel strength,
- excellent resistance to water, moisture, and media, as well as
- good impact strength.

12.5 Adhesive Bonding with Structural 2K Adhesives

If small-volume structural bonding applications with manageable batch sizes are to be realized in industry and craft, adhesives that do not require complex metering, application, and mixing equipment are preferred. Thus, mainly easy-to-handle 2K epoxy and 2K acrylic adhesives with relatively high reactivity are used for these applications.

12.5.1 Guinness World Record

The performance of structural 2K adhesives is nicely demonstrated by the setting of a Guinness World Record on 21 June 2012 in Neuss, Germany. An adhesive joint with a diameter of just 7 cm – the diameter of a beverage can – held a 10-ton truck on a crane 1 meter above the ground for one hour. This was achieved thanks to a 2K epoxy adhesive that impressively demonstrated its performance. The world record was recognized in the category *heaviest vehicle lifted with adhesive*. The core of the holding device consisted of two steel bolts, which were bonded in advance with the hard-set 2K epoxy-based construction adhesive under the supervision of the record judge. The finished samples remained under lock and key until the world-record attempt. The adhesive, which is offered on the market, had been developed for bonding metals such as aluminum, steel, and brass, which are exposed to very high temperatures.

12.5.2 Structural Bonding in Sports

Today, more than ever, industry is concerned with making products lighter, smaller, or faster, with aspects such as energy efficiency and price playing an ever-greater role. This requires innovative materials that can often only be joined with the help of adhesive-bonding technology. One example is carbon fiber composites, which are characterized by their unique combination of low weight and extremely high stiffness at the same time. Applications in which this material is joined with structural adhesives based on epoxies have immense potential and are used, for example, in vertical stabilizers for aircraft but also in automotive racing or innovative racing bicycles. For example, a new bicycle with a frame made of carbon fibers allows 15% more stiffness at 600 g less weight compared to the previous model. However, since some highly stressed parts of the bicycle, such as the rim brake, seat post, or bottom bracket ball bearing, must still be made of metal, modern *toughened* epoxy-based 2K structural adhesives enable them to be bonded to the composite frame. These consist of modified epoxy resins into which special elastomers have been incorporated to make them to a toughened adhesive. Here, the chemically incorporated elastomer particles act as *crack stoppers* in the event of impacts, enabling improved adhesives with enhanced dynamic properties to be used while still retaining very good temperature resistance, strength, and aging behavior. Such systems are predestined for very demanding applications such as those described above, as they provide a degree of stiffness in the component structure while absorbing vibrations, shocks, and impacts.

12.5.3 Structural Bonding in Art

Created by the renowned Cologne artist Raymund and supported by adhesive technology, the 3.20 m high and 1.75 m wide art object *Uelfe-Display* stands in the middle of a former natural outdoor swimming pool near Radevormwald/Germany and has defied wind, weather, and heat for many years (Figure 12.6) [5]. With the help of stainless-steel mirrors painted on both sides, the color sculpture reflects landscape, water, walkers, and the forest, making the familiar surroundings appear in a new light.

Since the front of the display had to remain absolutely smooth in this application, mechanical fasteners, such as screws or welding, were out of the question when mounting the artwork. The tight time frame, but especially the size and weight of the artwork, demanded the utmost attention and a high degree of speed and accuracy

Figure 12.6 Art object *Uelfe Display* bonded with 2K structural adhesive. Source: 3M.

from the workers performing the work. In addition, the two painted stainless-steel mirrors are very sensitive components. The task was to fit the two mirror plates, each weighing 250 kg, with three stainless steel rails. These rails had to be hooked into the 6-meter-high steel frame, which was then screwed onto a concrete base. The adhesive of choice was a toughened thixotropic 2K epoxy adhesive that worked well on the metal surfaces and built up both high strength and optimum layer thicknesses. In addition, the construction adhesive used was characterized by high shear strength, good peel forces, and high-impact strength. Most importantly, its weatherability made it a favorite for use in the harsh and humid environment of a lake. The metal rails were joined to the mirror plates and after a curing period of five days at about 18–20 °C, the artwork was spectacularly installed by crane in the middle of the lake. Due to the excellent weather resistance of the adhesive used, it is ensured that visitors to the *Uelfe Display* will enjoy the eye-catching installation for a long time.

12.5.4 Bonding with 2K Acrylates in Advertising

2K acrylate adhesives are frequently used in advertising, where the highest demands on bonding must constantly be combined with the requirements of innovative design. These adhesives are, e.g. successfully used for bonding plastic panels in metal frames, especially when the bonded components are exposed to structural loads and only a small bonding area is available (Figure 12.7).

Without additional treatment of the surfaces, these adhesives are able to quickly build up structural strength on painted metals, plastics such as PMMA, PC, or GFRP, and on wood or ceramics. After about 7–10 minutes, these solvent-free and

Figure 12.7 Typical application of a 2K acrylate adhesive (bonding plastic panel in a metal frame). Source: Dr. Jürgen Klingen, Aboso-Consulting.

viscous adhesives already show hand strength, and full curing is achieved after 24 hours at room temperature. The adhesives are processed manually with very user-friendly application systems consisting of 50-ml cartridges with mixing tubes, whereby the adhesive is metered, mixed, and applied in one step.

12.5.5 Highly Transparent Bonding

Highly transparent variants of the 2K acrylic adhesives enable highly transparent, barely visible, and at the same time structural bonding of metals, plastics, and glass. This opens up completely new design possibilities for manufacturers of high-quality store fittings, transparent sales displays, and modern signage, without having to use visually disruptive and expensive mechanical joining processes. The toughened UV-resistant adhesives cure very quickly at room temperature, and structural strengths are achieved without prior treatment of the surfaces, with plastic materials such as PC, PMMA, and PVC, and with joints involving metals such as aluminum, steel, or other joining partners. These adhesives build up bond strengths of about 10 MPa (e.g. 9.5 MPa on ground steel and 8.0 MPa on glass). After 4–5 minutes, the composites exhibit hand strength and full cure occurs within 6–8 hours at room temperature.

12.6 Inductively Curing Adhesives

Spot welding often combined with 1K epoxy adhesives in a so-called hybrid bonding process is an established, easily automated process for constructively joining two sheets of metal that is heavily used in automotive production. However, due to requirements in lightweight construction and the strong need for the production of optically perfect and accurately fitting components, this classic joining process and its combination with 1K adhesives is reaching its limits. In spot welding, for example, thermal stresses lead to deformation of the components or a negative influence on the appearance of the surfaces, which often requires extremely expensive reworking or leads to a reduction in quality [6].

12.6.1 Advantages of Inductively Curing Adhesives

A major contribution to solving these problems is made by new inductively curing 2K epoxy adhesives that also adhere very well to oily surfaces, as required for many applications in automotive industry. Compared with spot welding, the structural adhesive bonds produced with these adhesives also exhibit higher torsional stiffness, a property that is very interesting for lightweight construction. This means that not only thinner and thus lighter sheets can be used, but also – following the *multi-material design* concept – aluminum, magnesium, and high-strength steel. Here, the new adhesives achieve similar strengths to welding, but the bonded component surfaces are not mechanically affected as in spot welding and are therefore not susceptible to corrosion and cracking. 1K structural epoxy adhesives, which are used as an alternative to spot welding, require high temperatures of around 180 °C and a relatively long time for curing, which is a major disadvantage in view of the ever-shorter cycle times on the production lines. Since such adhesive composites only cure at a very late stage in the production process, the bonded components remain unstable for a long time, can shift, and therefore often have extremely poor fitting accuracy. The result is quality problems or expensive rework. Therefore, the bonding of structural components still has to be combined with the very expensive welding or punching to ensure their stability immediately after the joining process. This is not necessary with the new adhesives. Unlike 1K adhesives, the new 2K products cure at room temperature but more importantly, a handling strength of 1.5–2.0 MPa can be achieved immediately after joining by inductive *spot* curing at 100 °C for 40 seconds. This fixes the components before they pass through subsequent manufacturing processes, which is considered a decisive advantage over 1K adhesives, which go through the manufacturing process *as soft as honey* until they are cured late at 180 °C, causing the problems discussed previously. Thus, the new adhesives are meeting the requirements of modern industrial production, which, for the growing proportion of adhesive-bonded structural components, requires a high degree of quality and reproducibility of the related adhesive-bonding processes.

12.6.2 User Benefits

In view of the stringent requirement for shorter cycle times, such fast-curing adhesives are becoming increasingly important in industrial production, as they can significantly shorten cycle times due to the quickly achieved handling strength of the components. This is achieved through the application of heat over a very short period of time. Conventional heating methods, on the other hand, are critical because the complete components are often heated over a longer period of time, which is less energy efficient and often causes stress or relative movement in the bonded components. This cannot happen with induction heating, as very high heating rates can be achieved here, which can also be precisely limited locally (Figure 12.8).

Depending on the induction temperature, a handling strength of 1.5–2 MPa is ensured after slightly more than 20 seconds (see Figure 12.9). If no induction were applied and the bonded component were stored at room temperature, the handling strength would be reached after about 90 minutes. Through the support of induction, this is already possible, for example, after 24 seconds at 130 °C and after about

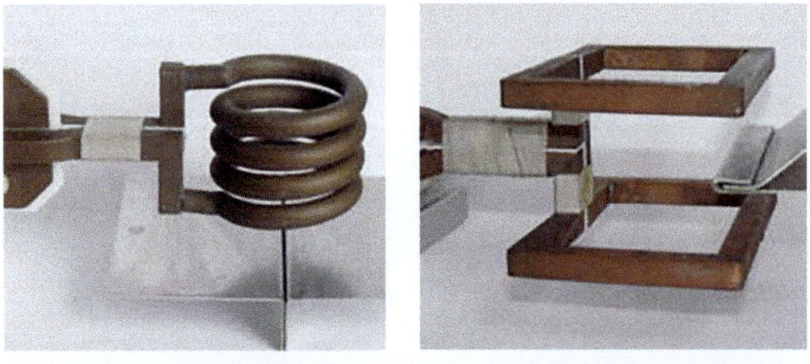

Figure 12.8 Inductive heating of an adhesive composite in the laboratory with a single- (left) and a double-sided induction curing unit (right). Source: 3M.

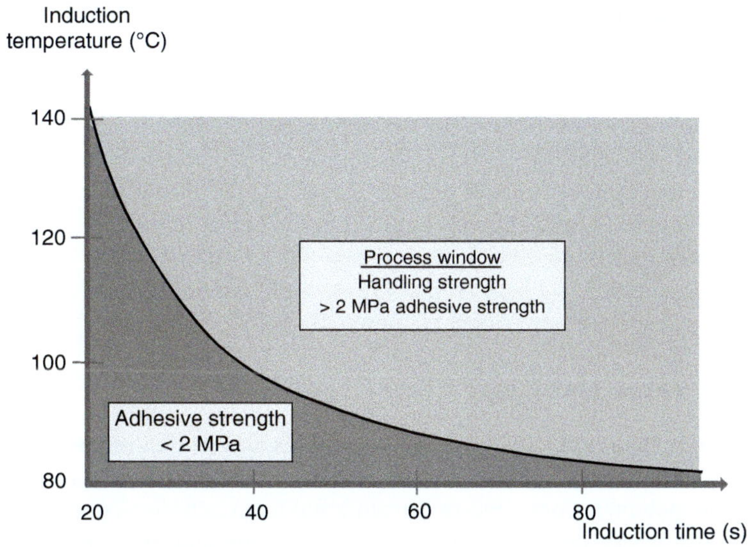

Figure 12.9 Dependence of the handling strength of the new 2K adhesive on the induction temperature and time.

65 seconds at 85 °C. The advantage for the user is that after such a very short heat treatment, the bonded component can then be safely transported through the subsequent manufacturing processes. After the induction step, curing then takes place in about 24 hours exclusively at room temperature, resulting in an adhesive bond with high bond strength of over 20 MPa and very good structural properties.

12.6.3 Potential for Innovative Bonding Solutions

This new adhesive-bonding technology enables the industry to develop innovative manufacturing concepts that offer significant quality and cost advantages over conventional, so-called hot joining processes. The additional *suspenders* used in the past, which ensured handling strength in the form of spot-welded joints, could be eliminated, making it possible for the first time to produce structural components that are joined using adhesives alone, particularly in the automotive industry. In terms of quality and production costs, the new technology could therefore have a positive impact on a large number of applications in which spot-welding, riveting, or *heat-curing* adhesives are still used today. In addition to numerous applications in automotive construction, further examples can be found in apparatus engineering, in the assembly of control cabinets, in terminals and elevators, in the *white goods industry*, and in steel construction. These new 2K adhesives could also further advance modular production using already-finished parts, which in the future would allow, for example, an automobile or other complex component to be quickly assembled from prefabricated individual parts with already finished surfaces at the end of the assembly line. In this process, *spot curing* of such bonded components to achieve handling strength would take place at high-speed using induction, followed by complete curing at room temperature.

12.7 Adhesive Bonding with High-Performance Acrylic Foam Tapes

High-performance acrylic foam tapes were developed in the 1980s to meet the growing demand for self-adhesive products in highly demanding applications. Today, they are used in almost all areas of industry and craft, including automotive, rail, aircraft, and shipbuilding, as well as in the electrical and electronics industries, in metal construction, and in advertising technology. As a result, it is very often found – invisibly, so to speak – where rivets, screws, and welding technology or structural adhesive systems were previously required to join materials.

12.7.1 Characteristics and Properties

Acrylic foam adhesive tapes are usually about 1 mm thick and consist of a white, gray, or black closed-cell acrylic foam coated on both sides with an adhesive made of modified acrylate polymer. Compared to conventional foam adhesive tapes, e.g. based on PE, these adhesive tapes consist of adhesive through and through, thus enabling them to flow into the roughness of surfaces and thus create optimum conditions for the intermolecular interactions necessary for good adhesion. The very easy-to-apply products are particularly suitable for textured materials and are ideally positioned for bonding different materials, such as metals, glass, plastics, or even painted materials, creating a stress-free adhesive bond to the surfaces that can withstand impact loads even at low temperatures. They also exhibit high resistance to UV radiation and moisture, as well as to plasticizers and other chemical influences. The adhesive tapes are permanently resistant to high temperatures, some of them even up to 260 °C for short periods. Compared with mechanical bonding methods, their viscoelastic properties make them much better able to absorb stresses, dynamic forces, and vibrations occurring in the bonded component. An excellent special feature of these products is that they allow compensation for differences in linear expansion of the bonded materials up to 3 times the adhesive tape thickness.

12.7.2 User Benefits

High-performance foam adhesive tapes offer decisive advantages in the design of many components because, unlike riveting and screwing, they are able to create invisible joints. When adhesive tapes are used, there is also no need to drill holes for fastening, which minimizes the risk of corrosion and, thanks to their additional sealing function, provides significantly better protection against the ingress of dirt or water into the joint structure. Application examples for these adhesive tapes – which enable simple and fast joining to speed up production processes and thus reduce labor costs – include roof structures of truck trailers, the manufacture of complex high-rise facades, and solar modules.

12.7.3 Acrylic Foam Tapes for Powder-Coating Processes

Based on the technology described above, special self-adhesive foam tapes have been developed, which are suitable for holding parts together during processing in high-temperature powder-coating processes. Thus, parts bonded with these special tapes can be powder coated and cured up to 230 °C without any problems. Such products exhibit high temperature and good solvent resistance as well as very good adhesion to metals as well as other high-energy surfaces. For the assembly of already-coated components, variants with special properties for powder-coated surfaces are available, which can be used without treatment of the surfaces. Cleaning the surface with isopropanol is usually sufficient, thus eliminating the need for primer or mechanical surface treatment, e.g. sanding, which saves the user time and thus costs. These products can be used wherever powder coatings are used, with applications ranging from the production of small household appliances to large constructions in mechanical engineering.

12.7.4 Flame-Retardant Acrylic Foam Tapes

Flame-retardant variants of foam tapes replace screws, welds, or rivets even in areas where strict fire protection regulations previously made bonding impossible. These temperature-resistant products are used – for short periods up to approximately 150 °C, for long periods up to just over 90 °C – in industries where fire protection is a high priority, e.g. in the automotive or aerospace industries, in rail vehicle and shipbuilding, or in electrical and electronics production. In interior construction, insulation devices, fire protection windows and doors, interior partition walls, or elevators are manufactured with these products.

12.7.5 Visually Demanding Applications

One interesting application that makes optimum use of the advantages of the acrylic foam tapes is the attachment of a retrofit rear-view camera to the inside of the rear window of motor homes and vans, whereby the bodywork must not be damaged or modified during retrofitting (Figure 12.10). In addition, the vehicle's standard rear wiper must be able to clean the camera's field of view at the push of a button in adverse weather conditions. The installation task is very demanding, since on the one hand the adhesive, which is located between the aluminum housing and the slightly curved rear window, must securely fix the device, and on the other hand changes in shape due to different temperature expansions must be compensated for. In addition, the aftermarket camera had to be as easy as possible to install in the

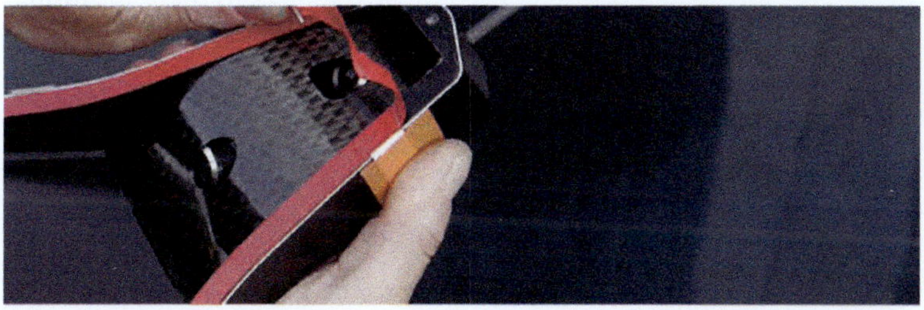

Figure 12.10 Bonding a rear-view camera with a high-performance acrylic foam tape. Source: 3M.

field, and since the camera is attached to the inside of the rear window, the adhesive surface is visible from the outside, which required a *good-looking* adhesive system. The solution is a black high-performance foam tape that performs the bonding task very well while hermetically shielding the custom camera housing from dust and moisture.

12.7.6 Bonding of Touchscreens

A special application of acrylic foam tapes is the attachment of touchscreens to plastic housings of hand-held auto-refractometers (Figure 12.11). These devices are needed to measure the eye function of young children in a simple and reliable way. Here, binocular measurement of refraction is possible from a distance of about 1 m, with the infant sitting relaxed on the lap of the mother or father during the examination. Using a touch screen on the operator's side, the device is aligned with the infant's eyes and measures refraction in less than one second. The results immediately indicate whether ametropia, corneal irregularity, or squinting is present. The goal of this application was to bond a capacitive touchscreen with a glass panel into a plastic injection molded housing. Since the device is exported to many countries, it was necessary that the adhesive bond could permanently withstand the different climatic loads in use. In the end, the optimum solution was to bond the two substrates with die-cuts of a variant of the acrylic foam tape. To ensure that the glass could be inserted with a precise fit, the injection molded part was provided with special guide ridges to facilitate positioning. Initially, the bonding process was carried out by hand. Later, however, an automatic process was established in which all operations could be program-controlled with precisely defined contact pressure and contact time, thus ensuring good and consistent bonding quality.

Figure 12.11 Touch screen of a handheld auto-refractometer bonded with die-cut pieces of high-performance acrylic foam tape. Source: 3M.

12.8 Closing Wounds with Hotmelt Adhesives

Researchers at the Technion – Israel Institute of Technology in Haifa, Israel – recently have developed a medical hot glue system that can be used to bond wounds. A special skin-friendly hotmelt adhesive is applied to the wound to be closed at relative low temperatures using a hotmelt gun. After application, the adhesive hardens quickly and firmly joins the edges of the wound. During the healing process, the adhesive remains flexible and then dissolves after a few weeks without leaving any residue. The major advantage of this biocompatible, biodegradable, and elastic polycaprolactone-based hotmelt adhesive is that it melts at temperatures just above body temperature and is therefore easy to apply without leaving burns. The desired mechanical properties, durability, and body tissue compatibility were achieved through intensive R&D activities without significantly altering the chemical properties of the adhesive.

12.8.1 User Benefits

When applied, the adhesive melts under minimal pressure and can thus be applied directly to the wound using a special medical adhesive gun. It then cures quickly and allows significantly higher adhesive forces than the adhesives previously used for wound closure. Further advantages for the user are that the new adhesive remains flexible throughout the healing process thanks to its specially adjusted rheology and dissolves completely in the tissue after a few weeks. The biocompatibility of the innovative adhesive system has already been confirmed in corresponding laboratory tests. There is a good chance that the new adhesive will significantly help to reduce the use of staples or needles for wound healing in the future, speed up the healing process, and reduce scarring.

References

1 Brockmann, W. (2007). *Leichter Kleben.* In: 3M Blickpunkt Kleben 2007, p. 1.
2 3M (2008). *Punktgenau Kleben ohne Materialverlust.* In: 3M Blickpunkt Kleben 2008, p. 4.
3 Geiß, P.L. (2008). *Geklebte Bauwerke.* In: 3M Blickpunkt Kleben 2008, p. 6.
4 Klingen, J. (2010). *Sie haften auch da, wo bisher nichts hielt.* In: 3M Blickpunkt Kleben 2010, pp. 4–5.
5 3M (2014). *Spiegelobjekt im See.* In: 3M Ideen Verbinden 2014/1, pp. 8–11.
6 Klingen, J. and Hasenberg, D. (2010). Punktuell Kleben mittels Induktion. *Adhäsion Kleben und Dichten* 54: 31–33.

General References

W. Brockmann, P.L. Geiß, J. Klingen, B. Schröder (2005) *Klebtechnik – Klebstoffe, Anwendungen und Verfahren*. Wiley-VCH, Weinheim.

W. Brockmann, P.L. Geiß, J. Klingen, B. Schröder (2009) *Adhesive Bonding – Materials, Applications and Technology*. Wiley-VCH, Weinheim.

C.A. Dahlquist (1966) *Tack, Adhesion, Fundamentals and Practise*. McLaren and Sons Ltd., London.

G. Habenicht (2005) *Kleben - Grundlagen, Technologie, Anwendungen*. 5, Springer-Verlag, Berlin.

G. Habenicht (2016) *Kleben – erfolgreich und fehlerfrei*. 7, Springer Vieweg, Wiesbaden.

B. Müller, W. Rath (2015) *Formulierung von Kleb- und Dichtstoffen*. 3, Vinzent Network, Hannover.

A.V. Pocius (2002) *Adhesion and Adhesive Technology*, 2, Carl Hanser Verlag, Munich.

M. Pröbster (2013) *Elastisch Kleben – aus der Praxis für die Praxis*. Springer Vieweg, Wiesbaden.

D. Satas (ed.) (1989) *Handbook of Pressure Sensitive Adhesive Technology*. Van Nostrand Reinhold, New York.

Adhesive Bonding in Five Steps: Achieving Safe and High-Quality Bonds, First Edition. Jürgen Klingen.

Index

a

A-B Process, 2K acrylic adhesives 260
abrasive fleece hand pads 242
abrasive grit 111
accelerated aging methods 235
acetate systems 144
acidic cleaners 108
acrylic-based PSAs 184, 186
acrylic foam tapes 18, 19, 187
acrylic PSAs 180
acrylonitrile butadiene rubber (ABR) 170
acrylonitrile butadiene styrene (ABS) 36, 90
α-cyano-acrylic acid ester 134
additives 104
adhesion build-up process
  adhesive preparation 326
    delivery and application 331–340
    homogenization 327
    room humidity setting 330–331
    viscosity 327–330
  surface treatment of substrates 326
adhesion destructive 99
adhesion failure (AF) 206
adhesion to surface 26–28
adhesive bonding 1, 2, 21, 31, 104
  advantages and disadvantages of 33–36
  applications 40, 41
    all-purpose adhesives 42–43
    all-rounders for structural 43–44
    easy to use hotmelt adhesives 41–42
    high-speed cyanoacrylate adhesives 41
    PSA tapes for quick connect and disconnect 46
    self-adhesive products 45
    strong PSA tape for all cases 47
    user-friendly comformable high-performance PSA foam tapes 44–45
  bearing capacity 223
  customer requirements for 352
  glass 207
  glass bonds 4–5
  industry and craft 6
  limitations of 32
  loads 214
    environmental loads 219–221
    failure of 215
    mechanical loads 215–219
  long-term behavior of 214–215
  main tasks 129
  metal bonds 2–4
  metals 196
  in nature 7
  performance of standard adhesives 32–33
  plastic bonds 4
  plastic materials 204–205
  plastics property 206
  roles in production today 31
  structural and elastic bonding 36–40
  technology
    1845–1930 10–14
    1930–1960 14–18
    first adhesives 9–10
    since1960 18–20
  wood 210
  wood bonds 5
adhesive components 55
  batch processing 341–342
  dynamic mixing 345
  static mixing 342
adhesive market development 351–352
adhesive preparation 107, 178, 324, 326–331
adhesive processing 172, 227, 324, 325, 352
adhesives 1, 129
  bonding process 222
  curing mechanism 130–131
  1K adhesive systems 132
    anaerobic adhesives 161–164
    cyanoacrylate adhesives 133–137
    hotmelt adhesives 146–155
    1K epoxy adhesive 164–169
    1K PU adhesives 137–143

Adhesive Bonding in Five Steps: Achieving Safe and High-Quality Bonds, First Edition. Jürgen Klingen.

adhesives (*contd.*)
1K silicone adhesives 143–146
photoinitiated adhesives 155–161
pressure sensitive adhesives 174–189
solvent-borne and dispersion adhesives 169–174
2K adhesive systems 189
2K acrylate adhesives 193–195
2K epoxy adhesives 190–193
2K PU adhesives 195–196
number of components 132
remarkable properties 133
requirements 224–225
shear properties 286
strength 130
thermomechanical properties 281–286
adhesive specification 228
adhesive requirements 224–225
loads on joint 223–224
manufacturing process 226–228
parameters relevant 222
practical guidelines 229
substrate 223
successful preparation 222
surface treatment method 225–226
adhesive strength 224, 265
ageing conditions for adhesive bonds 263
cataplasma test 274
climate change test 272–273
climate test 272
climatic corrosion test 273
fracture surface evaluation 274–275
immersion test 272
outdoor weathering test 274
salt spray test 273
alcohols 107
aliphatic hydrocarbons 107, 178
alkaline cleaners 108
alkoxy groups, coupling agent 209
alkoxy propanols 107
Allcock's Porous Plaster 10
Allcock, Thomas 10
all-purpose adhesive 11, 14, 15, 42–43
alternating copolymer 79
aluminum 74, 102
applications of 75
oxide 111
properties of 74
resistance of 74–75
strength of 74
surface improvement 75–76
surfaces bonding 202
surface treatment 202–203
amine systems 144
anaerobic adhesives 17, 249
applications 163–164
chemistry/ composition 161–162
curing/processing 162–163
overview 161
processing profile 249
profile of 164
properties 163
animal hides (glutin glue) 9
anodic treatment process 75, 203
anodizing 117
application-oriented adhesive classifications 351
aqueous cleaners 104, 108, 109
aqueous systems 108, 109
Araldit 194 17
atmospheric plasma activation 243
atmospheric plasma treatment 122–124

b

Bakelite 15
bearing capacity 230
Beiersdorf, Paul Carl 11, 12
bending loads, on adhesive bonds 306
birch tar 9, 10
bisphenol A 165, 182
blasting of material surfaces 113
cryogenic blasting 114
equipment 113–114
SACO 114
blasting process, for bonded joint production 243
block copolymer 79, 147, 178
bonded joint 1
bonded joints production 236
adhesion build-up process 326–340
adhesive preparation 241
cohesion build-up process 340–348
execution of 241
with 1K adhesives 244–256
surface treatment 241–244
two-component adhesives 256–263
personal protective equipment, provision of 237–240
preliminary requirements 236
process objectives 325–326
workplace preparation 239–240
bond line corrosion 221
bond wound closure, with hot melt adhesives 369
booster systems 140
boron nitride 111
borosilicate glass 94
branched polymers 80, 81
brushing 111
burger model 84
butt bonding 204

c

cast iron 67, 68
cast steel 67
cataplasma test 274
chain polymerization 78
chemical bridges 209
chemically curing systems 347, 348
chemical surface treatment
  anodizing 117
  chromating 117
  conversion layer 116
  gas phase treatment 117
  phosphates 116–117
  pickling of substrates 116
chloroprene rubber (CR) 170
chromating 117
chromium deposition method 73
cleavage loads, on adhesive joints 306–307
cleavage test 270
client 60
climate change test 272–273
climatic corrosion test 273
cohesion 97
cohesion build-up process
  adhesive component mixing 341–345
  curing of adhesive layer 346–348
  joining and fixation of bond 346
cohesion of an adhesive layer 29
cohesive 98
color-coding method 54
contact adhesive bonding 43
contact adhesive processing 172
contact bonding 254
contact corrosion 221
contemporary adhesive bonding applications 351
  high-performance acrylic foam tapes 366–368
  inductively curing adhesives 363–365
  in lightweight construction 352–355
  low-energy plastics 357–360
  modern facade construction 355–357
  skin-friendly hotmelt adhesives 369
  structural 2K adhesives 360–363
Coover, Harry 18
copolymers 79, 81, 146, 170, 171, 175, 178
copper 102
corrosion resistance 34, 70–73, 168, 225
corundum 111
coupling agents 209
crash-resistant 2K adhesives, light weight bonding with 353
creep behavior, of adhesive bond 289–292
creep corrosion 221
creep resistance 225
crepe tape 13
critical surface tension 24
cross-linked PS (PS-X) 88
cryogenic blasting 114
curing
  of adhesive layer 346
  conditions 227
  1K adhesives
    anaerobic adhesives 249
    cyanoacrylate adhesives 245
    dispersion adhesives 255–256
    epoxy adhesives 249–250
    hotmelt adhesives 250–251
    photo-initiated adhesives 251–252
    PU adhesives 247
    silicone adhesives 247–248
    solvent-based adhesives 253–254
  of 2K epoxy adhesives 260
  mechanism
    adhesive classification 131
    chemically curing adhesives 131–132
    physically curing adhesives 130–131
cyanoacrylate adhesives 18, 41, 42
  applications 135–136
  chemistry/composition 133–134
  curing 245–246
  curing/processing 134
  dosage form 245
  ester groups 135
  medical applications 136
  overview 133
  processing profile 245–246
  profile 137
  properties 134

d

Dahlquist criterion 176, 285
Day, Horace 10
design, of adhesive bonds 296
  adhesive bond utilization factor δ 300–301
  adhesive layer thickness, influence of 303–304
  avoiding stress peaks 304
    bending loads 306
    cleavage loads 306–307
    peel loads 305–306
    plastic deformation, of substrates 307
  bonded joints 307–309
  bonding area, influence of 302–303
  butt joints 297–298
  optimal overlap length 299–300
  overlap length, influence of 301
  requirements 296
  single lap joints 298–299
  single lap joints, optimization of 299–300

design, of adhesive bonds (*contd.*)
  substrate thickness, influence of 301–302
  substrate width, influence of 302–303
  sufficient boding area, provision of 297
development step
  completion, requirements for 323
  goal of 235
  major tasks and accomplishments 321, 322
  objectives 321–323
  project status 321
  tasks and accomplishments 350
diamond 111
diaphragm pump 335
dicyandiamide 152
Di-cyclopentadiene dioxide 165
diethylenetriamine (DETA) 166
differential scanning calorimetry (DSC) 276
diffusion bonding 254
dimensioning of bonded joints
  bond strength 309
  elastic adhesive bonds 310–312
  element for interior design 315–317
  factors influencing 308
  glass pane with an aluminum frame 317–319
  with plastics 312–313
  reduction factors 309–310
  shear strength 308
  stainless steel plates example 313–315
DIN 2304 52, 56
  quality standard 55
  safety classes 63
dispersion adhesives 43, 255
dispersion contact adhesives 170
dispersion forces 26, 27, 29, 81
dispersive interactions 24, 177
dosage form
  1K adhesives
    anaerobic adhesives 249
    cyanoacrylate adhesives 244–245
    epoxy adhesives 249–250
    hotmelt adhesives 250–251
    photo-initiated adhesives 251–252
    pressure sensitive adhesives 252
    PU adhesives 246–247
    silicone adhesives 247–248
    solvent-based adhesives 253–254
  2K adhesives
    acrylic adhesives 260–263
    epoxy adhesives 258–260
Drew, Richard 13, 14
dry bonding 43
dry processing 172, 174
dual-curing adhesives 159
Duct Tape 16, 17, 47
dynamic mechanical thermal analysis (DMTA) 276–279
dynamic overlap shear test 265
dynamic viscosity 82, 327, 328

e

Eastman 910 18, 133
eccentric screw pump 335
elastic adhesive bonds 281, 311
elastic bonding 39
  for heavier loads 39–40
  main tasks in 39
elastic deformation 292
elastic polycaprolactone-based hotmelt adhesive 369
elastomers polymer 81
environmental loads
  bond line corrosion 220
  contact corrosion 221
  corrosive media impact 220–221
  creep corrosion 221
  stress crack corrosion 221
  temperature changes 219–220
  water diffusion 220
environmental protection and occupational safety 227
EPDM seal 136
epoxy adhesives
  1K adhesives 249–250
  2k adhesives 258–260
esters 107
ethylene-propylene-diene rubber (EPDM) 41

f

face surface 96
fast initial strengths 229
feasibility step 235, 236
ferrous materials 67
  classification 68
  mechanical properties 68–69
  steel classification 69–71
  steel surface improvement 71–74
fiber-reinforced birch tar glue 9
final strength 260
fingerprint spectra 280
finite element analysis (FEA) 307
Fischer, August 14
fixed nozzle system 337
flame-retardant acrylic foam tapes 367
fluor rubber (FPM) 41
foamed PS (PS-E) 87
foreign substances 105
Fourier transform infrared spectroscopy 280
fracture surface evaluation 274–275

full-surface applications of adhesives 339
  brushing, troweling and squeegeeing 339
  PSA film lamination 340
  roller application method 340
  spray application 339

g

galvanic zinc coating 72
galvanized steel bonding 199–200
galvanized steels 101
gatekeeper 54
gate reviews 53–55
gear pumps 334
gel point 259
general purpose PS (GPPS) 87
glasses 66, 105
  adhesive bonding 207
  adhesives bonding 210
  bonds 4–5
  classification of 94
  corrosion 5, 208
  long-term resistance bonding 208
  manufacture of 94
  quartz crystal 93
  surface treatment 208
  transition temperature 282
glass-fiber-reinforced plastic (GFRP) 36
glue cooking 9
good adhesion 2, 3, 21, 26, 73, 97, 102, 118, 171, 174
good cohesion 2, 21, 164
gradient copolymer 79
graft copolymer 79
grinding 111–113

h

halogenated hydrocarbons 107
hand strength 260
Hansaplast 13
high-alloy steel 69
high-density (HD-PE) 85
high-energy plastics 103
high-grade steel 70
high-performance acrylic foam tapes 366–368
  bonding of touchscreens 368
  characteristics and properties of 366
  for powder coating process 366–367
  user benefits 366
  visually demanding application 367–368
high-performance plastics 76
high-quality adhesive bonding 96
high-quality bonded joint production 295
high strength fasteners 46
hook and loop fasteners 46
hot dip galvanized steel railing 199
hot dip galvanizing process 73
hotmelt adhesives 146–148, 212, 250–251
  advantages 151
  applications 153–155
  bottles and containers 153–154
  chemistry/ composition 146–148
  curing hot melt adhesives 151–153
  curing/processing 148–149
  disadvantages 151
  heat shrinkable tubing 154
  overview 146
  profile 154–155
  properties 151
  wood bonding applications 154
hybrid bonding processes 20
hydrocarbon tackifiers 181
hydrogen bonds 26

i

immersion degreasing 110
immersion test 272
impact-resistant PS (PS-I) 88
induced dipoles 26
inductively curing adhesives
  advantages of 363–364
  innovative manufacturing concepts 365
  user benefits 364–365
intermolecular interactions 26
internal strength 29
ioining part fixation 227
ISO 9001 quality standard 50

k

Kelvin-Voigt model 83
Kenics and Sulzer mixers 342
ketones 107

l

laminating method 339
lead glass 94
Leukoplast 12, 13
Lifschütz, Isaac 12
lightweight construction, bonding applications in 352
Linear-elastic behavior 82
linear-low density (LLD-PE) 85
linear polymers 80
liquid adhesives 161
load bearing capacity 230, 310
load-carrying capacity 230
loads, adhesive bonding
  environmental loads
    bond line corrosion 220
    contact corrosion 221
    corrosive media impact 220–221

loads, adhesive bonding (*contd.*)
creep corrosion 221
stress crack corrosion 221
temperature changes 219–220
water diffusion 220
mechanical loads
external loads 215
internal loads 215–219
London forces 27
long term aging-resistant bonds 203–204
low-alloy steel 69
low-density (LD-PE) 85
low-energy plastics 103
low-energy surfaces, structural adhesives for 359
low-pressure plasma process 122
low steel 101

m

3M 13, 18
macromolecules 78
manual hot melt adhesive gun 148
manual processing, of 2K epoxy adhesives 256
masking tape 13
maxwell model 83
mechanical loads 224
external loads 215
internal loads 215
dissimilar substrates 217–218
possible damage 217
shrinkage of, adhesive layer 218–219
similar substrates 216–217
mechanical surface treatment
blasting 113–115
grinding and brushing 111–113
metals 64
aluminum surfaces bonding 202
aluminum surface treatment 202–203
atomic structure 65
bonds 2–4
galvanized steel bonding 199–200
long term aging-resistant bonds 203–204
multi-layered structure 196–197
oily steel sheet bonding 198
stainless steel bonding 200–201
steel sheet bonding 197–198
steel surfaces, adhesive bonding 197
surface structure 100–102
metering system, functional units of 332
metering units 332
methyl ethyl ketone (MEK) 107
micro-Brownian motions 283
modern plastics 358
moving nozzle system 336
Mussels adhesive bond 6

n

natural rubber PSAs 176
needle valves 336
neutral cleaners 108
neutral systems 144
Newtonian fluids 328
no-mix process 261
non-alloy steel 69
non-Newtonian fluids 328

o

OH-terminated poly-siloxanes 143
oily steel sheet bonding 198
oligomers 78
1K adhesives, bonded joints with
activation 244
anaerobic adhesives 249
cyanoacrylate adhesives 244–246
dispersion adhesives 255–256
epoxy adhesives 249–250
hotmelt adhesives 250–251
photo-initiated adhesives 251–252
pressure-sensitive adhesives 252–253
PU adhesives 246–247
shelf life 244
silicone adhesives 247–248
solvent-based adhesives 253–254
1K adhesive systems
anaerobic adhesives 161–164
cyanoacrylate adhesives 133–137
hotmelt adhesives 146–155
1K epoxy adhesive 164–169
1K PU adhesives 137–143
1K silicone adhesives 143–146
photoinitiated adhesives 155–161
pressure sensitive adhesives 174–189
solvent-borne and dispersion adhesives 169–174
1K epoxy adhesives 20
addition polymerization 166–167
advantages and disadvantages 168
applications 168–169
chemistry/composition 164–165
curing/processing 167
epoxy resins 167
overview 164
profile of 169
properties 167–168
1K PU adhesives
applications 140–142
chemistry/ composition 137–139
curing/processing 139–140
overview 137
profile of 142–143
properties 140

1K silicone adhesives (RTV-1)
  applications 145
  chemistry 143–144
  curing/processing 144
  overview 143
  profile of 146
  properties 145
operating staff 62
organic solvents 105–110
outdoor weathering test 274
oxime-releasing adhesives 144

p

partial-surface application methods for adhesives 338
  bead application 337–338
  PSA tape lamination 339
  screen printing application 338
  spot application 337–338
peel loads, on adhesive bonds 305–306
peel tests 268–269
peristaltic pump 335
permanent dipoles 26
phosphate deposition process 73
photoinitiated acrylic adhesives 158
photoinitiated adhesives
  applications 159–160
  cationic curing systems 156–157
  chemistry/composition 155–167
  curing/processing 157–159
  overview 155
  profile of 160–161
  properties 159
  radical curing systems 155–156
photo-initiated adhesives 251–252
photoinitiated epoxy adhesives 158
physically curing systems 347
physical surface treatment
  corona treatment 119–120
  flame treatment 118–125
  laser treatment 125
  overview of 126
  plasma process 120–121
  Pyrosil® flame coating 119
physical test methods, for adhesive bonds 263, 275
  differential scanning calorimetry 276
  dynamic mechanical thermal analysis 276–279
  Fourier transform infrared spectroscopy 280
  thermo gravimetric analysis 279–280
pickling process surface treatment 203
pinch valves 336
piston pump 334
plasma process
  atmospheric plasma treatment 122–124
  low pressure plasma treatment 121–122
  plasma cleaning 124
  plasma polymerization 124
  plastic surfaces 121
plastic bonds 4
plastic deformation 292
plastic deformation of substrates 307
plasticized PVC (P-PVC) 86
plastic materials 65
  butt bonding 204–205
  chemistry of 78–80
  diffusion and solubility characteristics 205
  industry and craft 85–93
  long-term resistance, plastic bonding 206
  properties 76–77
  recycled plastic bonding 205–206
  thermomechanical properties 80–82
  viscoelastic properties of 82–84
plastics 102
polar interactions 24
polyaddition 79
polyaddition-curing adhesives 341
polyamide (PA) 91
polycarbonate (PC) 36, 89
polycondensation 79
polyethylene (PE) 36, 85
polyethylene terephthalate (PET) 81, 88
polymerization 78
polymers
  molecular interactions of 26–27
  structure of 79–80
  synthesis 78
polymethyl methacrylate (PMMA) 36, 90
polypropylene (PP) 36, 85
polystyrene (PS) 86
polytetrafluoroethylene (PTFE) 25, 205
polyurethanes (PU) 91, 246
  curing 247
  foams 92
  processing of 246–247
  processing profile 247
polyvinyl acetate (PVAC) 14, 89
polyvinyl butyral (PVB) 95
polyvinyl chloride (PVC) 36, 86
pot life 259
powder coating technique 358
Preiswerk, Eduard 17
premium steel 70
pressure sensitive adhesives (PSA) 252–253
  applications 185–188
  backings 182
  chemistry/composition 175–182
  film lamination 340
  new developments 188–189
  overview 174–175

pressure sensitive adhesives (PSA) (*contd.*)
processing 182–183
profile 188
properties 183–184
tackifiers, thermomechanical effects of 286
tackifying resins 181
tape lamination 339
thermomechanical properties of 284–285
time at room temperature 177
printable adhesives 355
processing profile
1K adhesives
cyanoacrylate adhesives 245–246
dispersion adhesives 255–256
epoxy adhesives 249–250
hotmelt adhesives 250–251
photo-initiated adhesives 251–252
pressure sensitive adhesives 253
PU adhesives 247
silicone adhesives 247–248
solvent-based adhesives 253–254
2K adhesives
acrylate adhesives 262
epoxy adhesives 260
production process 349
project champion 61
project contract
plan elements 59
planning phase 60
project-relevant people 60
project management system
development and launch 53
DIN 2304 55
gate reviews 53–55
planning, concept and feasibility 52–53
5-step 51
project manager 61
project sponsor 60
project team DIN 2304 safety classes 63
proportional valves 336
pumps, for adhesive delivery 334
pure iron 67
Pyrosil® flame coating 119
Pyrosil® process 243

q

quality assurance 228
quality steel 70
quartz glass 93

r

recycled plastic bonding 205–206
Redux 16
reversible creep 290
roller application method 340

s

safe and high-quality bonds 59
ISO 9001 quality standard 50
project management system 51–53
special process 49–50
salt spray test 273
sandblasting and coating (SACO) 114
Scotch tape 14
screen-printing process 338
segin 9
self-adhesive plaster 11
self-adhesive transfer tapes 358, 359
semi-crystalline polymers 80
semi-structural PSAs 182
shear modulus 308
of adhesives 287
shear properties, of adhesives 287
creep behavior 289–292
flow property 289
shear modulus 287
shear stress-strain behavior 278–288
shear stress-strain behavior 288
shear test 265–267, 308
shear-thickening (dilatant) fluids 328
shear-thinning (structural viscous) fluids 328
Shecut, William 10
Sichel, Ferdinand 12
silane coupling agent 209
silicon adhesives 247
silicon carbide 111
silicone adhesives 248
silicones 99
six sigma methodology 51, 52
skin-friendly hotmelt adhesives 369
Snuff Back valves 335, 336
soda-lime glass 94
solvent-based adhesives 253–254
solvent-borne adhesives 170
solvent-borne and dispersion adhesives
applications 173
chemistry/composition 170–171
curing/processing 171–172
overview 169–170
profile 173–174
properties 172–173
solvent-borne contact adhesives 170
solvent-borne wet adhesives 170
spin spray application 339
spray galvanizing process 73
stainless steel 70, 71, 101
stainless steel bonding 200, 201
standard glass 94
standard2K epoxy adhesive 32
standard steel 70
static mixing 342
statistical copolymer 79

steel
classification 69–71
sheet bonding 197–198
surface improvement 71–74
surfaces, adhesive bonding 197
5-step project management system 52
stress crack corrosion 221
stress cracks 104
stress relaxation 225
behavior, of adhesive layer 294
structural adhesive bonds 280–281
structural bonding 38, 39
Structural Bonding Tape 19
structural 2K adhesives
in art 361–362
2K acrylate adhesives in advertising 362–363
performance, Guinness World Record 360
in sports 360–361
structural steel 70
structural *vs.* elastic bonding 312
styrene-acrylonitrile (SAN) 79, 90
styrene-butadiene rubber (SBR) 41
substance-bound method 33
substrate dependence, of shear test 267
substrates
advantages and disadvantages 129
definition of 223
surface and treatment 64
surface characteristics 100–105
surface cleaning 105–110
surfaces
aluminum 74–76
ferrous materials 67
glass 93
material classes 64
plastic materials 76–77
wood 95
surfaces treatment 96–100
surface treatment 110–126
super glues 18, 133
supervisor in charge (SIC) 61, 63
surface adhesion 26, 27, 29
surface cleaning
aqueous systems 108
degreasing processes 109
metal surfaces 108–109
organic solvents 105–110
plastic surfaces 109
surface contamination 99, 100, 113
surface energy of solids 23–25
surface finishing of glass 95
surfaces wetting 21
surface tension
of liquids 23
Young's equation and 21–23
surface treatment
chemical surface treatment 115–118
goal of 98–100
importance of 97
mechanical surface treatment 111–115
physical surface treatment 118–125
surface treatment, for bonded joint production 241
atmospheric plasma 243
blasting 243
grinding 241–243
Pyrosil® process 243
surface treatment method 225–226
Syndetikon 10, 11
syndiotactic PP 81, 86

t

technological test methods
for adhesive bonds 263, 264
cleavage test 270
peel test 268–270
shear test 265–267
tensile test 267–268
wedge test 270–272
tensile test 267–268, 308
tesafilm® 12, 15, 16
tetra-ethylene glycol di-methyl acrylate (TEGMA) 161
thermal loads 224
thermo gravimetric analysis (TGA) 279–280
thermomechanical plastic materials 80–82
thermomechanical properties, of adhesives
elastic region 283
flow region 283
glassy region 282–283
range of adhesives for industrial application 281–282
thermoplastic elastomers polymer 81
thermoplasts polymers 81
thermoset adhesive layers 159, 160, 164, 196, 281
thermosets polymers 81
thixotropy, of adhesives 330
three-point bending test method 270
time-lapse effect 215, 274
time-temperature superposition, principle of 279
titanium 74, 102, 202
tonnage steel 70
tool steel 68, 70, 71
transportation industry, lightweight requirements in 352
triethylenetetramine (TETA) 166
Troplowitz, Oscar 12
truck construction, elastic bonding in 354
2K acrylate adhesives 262

2K acrylate adhesives (*contd.*)
  in advertising 362–363
  applications 194–195
  chemistry/composition 193
  overview 193
  processing 193–194
  3 process types 194
  profiles 195
  properties 194
2K acrylic adhesives 362
2K adhesive processing
  mixing by hand 256–258
  mixing without the mixing tube 258
  mixing with the mixing tube 258
  use of mixing and dispensing systems 257–258
  using disposable mixing tubes 258
2K adhesive systems
  applications of 190
  curing and processing 190
  2K acrylate adhesives 193–195
  2K epoxy adhesives 190–193
  2K PU adhesives 195–196
  metering and mixing systems 333
  overview 189
2K epoxy adhesives 20
  applications 192–193
  chemistry and composition 190–191
  curing/processing 191
  overview 190
  profiles 192–193
  properties 192
2K PU adhesives
  curing/processing 195
  overview 195
  profiles 196
  properties/applications 196

u
UHU® 15
ultrasonic cleaning 110
unalloyed steel 4, 69, 101, 220
unplasticized PVC (U-PVC) 86

v
valves, in adhesive metering 335
vapor degreasing 110
visco-elastic behavior 83
viscoelastic deformation 292, 293
viscous behavior 82
vortex spraying 339

w
water-based (dispersion) adhesives 170
water-based wet (dispersion) adhesives 170
water resistance, testing 220
water-soluble substances 108
wedge test 270–272
wet bonding 43, 171, 253, 254
wet peel test 269
wet processing 171, 172, 174
wetting of surfaces 21–25
white glue 212
wood 66
  adhesive systems 212
  bonding materials 213
  bonding requirements 211–212
  formaldehyde resin dispersions 213
  Hotmelt adhesives 212
  moisture-curing 1K PU adhesives 213
  post-curing PVA dispersions 213
  properties of 95–96
  PVA dispersions 212
  surfaces 96
wood bonds 5

y
yield strength 68, 74, 267, 297–299, 301, 313, 314
Young's equation 21–23
Young's modulus 35, 68, 216, 308

z
zero-defect production 50
zinc deposition 71–72
Zismann method 24–25